Integrated Waste Management

This book addresses multiple focus areas identified and provides solutions with respect to the circular economy, water pollution, potable water availability, reducing population impact on the environment, and better health by integrated waste management. It explains techniques to handle waste generation, characterization, minimization, collection, separation, treatment, and disposal, and includes chapters that address waste management policy, education, and economic and environmental assessments.

Features:

- Introduces waste management, pollution, and toxicity profile of potentially toxic environmental contaminants and industrial wastes.
- Describes field studies on the application of microbes and plants in bio/phytoremediation of environmental contaminants/industrial wastes.
- Reviews eco-friendly remediation techniques such as phytoremediation, vermi-remediation, nano-remediation, and myco-remediation.
- Presents recent advances and challenges in bioremediation research and applications in environmental management.
- Details underlying tools and techniques for sustainable waste management and (nature-based) solutions in each chapter.

This book is aimed at graduate students and researchers in environmental engineering and waste management.

Environmental Nexus in Waste Management

Series Editors: Vineet Kumar, Department of Microbiology, School of Life Sciences at the Central University of Rajasthan, Rajasthan, India and Sunil Kumar, researcher in Environmental Engineering and Science

The book series on "Environmental Nexus for Waste Management" addresses novel approaches and techniques linked to sustainable development paths with an interdisciplinary focus on waste resources monitoring, assessment, management, water/wastewater reclamation, and recycling as well as discussing solutions for more sustainable use of natural resources. It is a broad compendium of waste management related to research and development and discussions on the most up-to-date tackling strategies. The series of books constitutes an information source and facilitator for the transfer of knowledge, focusing on practical solutions, and better understanding towards achieving sustainable development.

Solid Waste Treatment Technologies

Challenges and Perspectives
Edited by Pratibha Gautam, Vineet Kumar and Sunil Kumar

Microbial Nexus for Sustainable Wastewater Treatment

Resources, Efficiency, and Reuse
Edited by Vineet Kumar, Sunil Kumar, Pradeep Verma and Sartaj Ahmad Bhat

Environmental Nexus for Resource Management

Edited by Hanuman Singh Jatav, Tatiana Minkina, Satish Kumar Singh, Bijay Singh and Vishnu D. Rajput

Environmental Nexus Approach

Management of Water, Waste, and Soil
Edited by Sartaj Ahmad Bhat, Vineet Kumar, Fusheng Li, Fuad Ameen, and Sunil Kumar

Integrated Waste Management

Trends, Policies, and Perspectives
Edited By Ali Mohd Yatoo, Pankaj Kumar Gupta, and Rajeev Pratap Singh

For more information about this series, please visit: www.routledge.com/Environmental-Nexus-in-Waste-Management/book-series/CRCENWM

Integrated Waste Management

Trends, Policies, and Perspectives

Edited by Ali Mohd Yatoo,
Pankaj Kumar Gupta, and Rajeev Pratap Singh

CRC Press
Taylor & Francis Group
Boca Raton London New York

CRC Press is an imprint of the
Taylor & Francis Group, an **informa** business

Designed cover image: Shutterstock

First edition published 2025
by CRC Press
2385 NW Executive Center Drive, Suite 320, Boca Raton FL 33431

and by CRC Press
4 Park Square, Milton Park, Abingdon, Oxon, OX14 4RN

CRC Press is an imprint of Taylor & Francis Group, LLC

© 2025 selection and editorial matter, Ali Mohd Yatoo, Pankaj Kumar Gupta and Rajeev Pratap Singh; individual chapters, the contributors

ISBN: 978-1-032-41700-4 (hbk)
ISBN: 978-1-032-41701-1 (pbk)
ISBN: 978-1-003-35932-6 (ebk)

DOI: 10.1201/9781003359326

Typeset in Times
by Apex CoVantage, LLC

Contents

Razia Gull, Tahir Ahmad Sheikh, M. Anwar Bhat,
Rifat Un Nisa, Aanisa Manzoor, Roheela Ahmad,
Ali Mohd Yatoo, and Raies Ahmad Bhat

Humaira, Saba Wani, Qamer Ridwan, Khair Ul Nisa,
and Asia Mushtaq

Shagufta Iqbal, Khair Ul Nisa, Najeebul Tarfeen, Humaira,
Saba Wani, Qadrul Nisa, and Shafat Ali

Neesa Majeed, Burhan Hamid, Tanveer Ahmad Mir,
Ali Mohd Yatoo, and R. Z. Sayyed

Roheela Ahmad, Tahir Ahmad Sheikh, Aamir Hassan Mir,
Ali Mohd Yatoo, Shakeel Ahmad Mir, Ayman Javed Shah,
Rifat Un Nisa, Javed Ahmad Bhat, Razia Gull, and
Aanisa Manzoor Shah

Malik Ahsaf Aziz, Shayesta Islam, Syed Hujjat ul Baligah,
and Burhan Hamid

Contents vii

Preface

The book *Integrated Waste Management: Trends, Policies, and Perspectives* is a topical and timely contribution that is very helpful in highlighting and addressing the assessment, mitigation, and management of waste and waste disposal/polluted sites in a comprehensive and multifaceted way. This book provides different scientific views and concepts, research, reviews, case studies, and so forth on contemporary waste management issues under distinct sections around the globe. The book covers advances in bioremediation approaches, including phytoremediation, biosurfactants, genetically engineered microbes, microbial enzymes, and phycoremediation of various environmental toxicants like microplastics, heavy metals, and other organic and inorganic pollutants, which are lacking in more comprehensive manner in the previous titles. An in-depth discussion on low-cost and clean technologies is at the centre of all chapters of the book. One more interesting feature of the book is that detailed mechanisms and outcomes during the remediation process of industrial and emerging contaminants are explained. This book also encompasses updated information and future directions for researchers working in the field of management and remediation of waste disposal/polluted sites. Apart from that, it will help to address the queries of environmental scientists, researchers, planners, policymakers, NGOs, environmental consultancies, and general people about the sustainable remediation of toxic environmental pollutants.

About the Editors

Ali Mohd Yatoo (Ph.D.) pursued his M.Sc. (Environmental Science) from H.N.B. Garhwal (Central) University, Uttarakhand, and completed his Ph.D. (Environmental Science) from the University of Kashmir, India. He has more than five years of research experience and wide experience in waste valorization, waste and wastewater management, eco-toxicology, bioremediation, vermicomposting, composting, plant pathology, and organic farming. Dr. Yatoo has published more than 30 research and review papers, including book chapters, in highly reputed peer-reviewed scientific journals and has presented and participated in several workshops, seminars, and conferences. He is also a reviewer of various reputed international scientific journals. Dr. Yatoo has qualified for various state- and national-level competitive examinations like UGC NET (five times) conducted by the University Grants Commission (UGC), the Agricultural Scientists Recruitment Board National Eligibility Test (ASRB NET) in Environmental Sciences, and the Jammu & Kashmir State Eligibility Test (JKSET). He has also qualified CSIR NET with a Junior Research Fellowship (JRF) conducted by the Council of Scientific & Industrial Research, Government of India. Dr. Yatoo has been the recipient of various fellowships and awards from several funding agencies.

Pankaj Kumar Gupta is a Ramanujan fellow, upholding one of the honoured scientific positions at the Indian Institute of Technology (IIT) in Delhi, India. As a Ramanujan fellow, he has undertaken an exceptional project "Engineered Microbiome (E-Biome Project)", a unique combination of microbiology, hydrogeology, and chemical sciences for the restoration of polluted sites, funded by the SERB, Government of India. He is also an adjunct assistant professor at the University of Waterloo, Canada.

Dr. Gupta is a contaminant hydrogeologist; after having completed his Ph.D. at IIT Roorkee, he has been a postdoctoral fellow at the University of Waterloo, Canada. He diversified his expertise in hydrogeology and soil-water quality during his Ph.D./PDF. He is the co-founder of the Society of Young Agriculture and Hydrology Scholars of India (SYAHI) and is also an editorial member of the *Frontiers in Water Journal, Biochar, Carbon Research*, and so on. He has also edited three books entitled *Soil-Water, Agriculture, and Climate Change: Exploring Linkages, Advances in Remediation Techniques for Polluted Soils and Groundwater*, and *Fate and Transport of Subsurface Pollutants*.

Dr. Gupta holds his in-depth experience in incorporating novel technologies (viz. geophysical investigations, groundwater modelling, aquifer mapping, water quality assessment, microbiome, etc.) to map soil-water systems of 30+ sites in India. Dr. Gupta has also worked as a field officer at the Centre for Development Communication, Jaipur, and Salasar temple (Churu, Rajasthan) to improve water and wastewater management practices in the desert villages. Dr. Gupta is passionate about interdisciplinary research and teaching to understand multi-scale interactions between different components of the subsurface environment, especially the soil–groundwater–pollutant–microbes system.

He also teaches and guides Master's and Ph.D. students with diversified knowledge. He has received an AGU Travel Grant (2017), JPGU Travel Grant (2018), and EXCEEDSWINDON and DAAD Germany Grant (2018). He has published more than 20 research papers and approximately 30 book chapters, as well as edited two books and two popular science articles. He has reviewed several research articles for reputed journals. He is a core member of the Society of Young Agriculture and Hydrology Scholars of India (www.syahindia.org), a network of early-career researchers from around the globe.

Rajeev Pratap Singh works in the area of waste management and has worked on various kinds of waste, that is, fly ash, sewage sludge, tannery sludge, palm oil mill waste, contained water irrigation, and so on. He has many publications in reputed journals on waste management and similar topics. Dr. Singh has received several international awards, like the "Green Talent" Award from the Federal Ministry of Education and Research (BMBF), Germany; the Prosper.Net Scopus Young Scientist Award, and the DST Young Scientist Award. He has also co-authored 7 books and more than 50 highly cited research and review articles on different aspects of environmental science. Dr. Gupta has recently availed Water Advanced Research and Innovation (WARI) Fellowship, a fellowship supported by the Department of Science and Technology (DST), Government of India, the Indo-US Science and Technology Forum (IUSSTF), University of Nebraska-Lincoln (UNL), and the Robert Daugherty Water for Food Institute (DWFI). Dr. Singh is also a fellow of the National Academy of Agricultural Science (NAAS), New Delhi. He has also been awarded the Prof. Archana Sharma Memorial Award 2020–2023 by ISCA, Kolkata, at the 108th Indian Science Congress, Nagpur, on 7 January 2023. Recently, Dr. Singh was awarded the Prof. C. N. R. Rao Award for Excellence in Scientific Research by Banaras Hindu University.

Contributors

Munmun Agrawal
Department of Metallurgical Engineering
Indian Institute of Technology
   (Banaras Hindu University)
Varanasi, India

Roheela Ahmad
Division of Soil Science and
   Agricultural Chemistry
Faculty of Agriculture
Wadura, SKUAST Kashmir, India

Mukhtar Ahmed
Meteorological Centre
India Meteorological Department
Rambagh Srinagar, India

Shafat Ali
Centre of Research for Development
University of Kashmir
Srinagar, J&K, India

B.S. Anusha
Division of Environment Science
ICAR—Indian Agricultural Research
   Institute
Pusa Campus, New Delhi, India

Malik Ahsaf Aziz
Division of Basic Sciences and
   Humanities
Faculty of Agriculture
Wadura, SKUAST Kashmir, India

Syed Hujjat ul Baligah
Division of Soil Sciences and
   Agricultural Chemistry
Faculty of Agriculture
Wadura, SKUAST Kashmir, India

Suhaib A. Bandh
Department of Environmental Science
Higher Education Department
Government of Jammu and Kashmir,
   India

Amitava Bandyopadhyay
Department of Chemical Engineering
University of Calcutta
Kolkata, India

Shabir Ahmad Bangroo
Division of Soil Science
SKUAST, Shalimar, Kashmir, India

M. Anwar Bhat
Directorate of Extension
SKUAST Shalimar, Kashmir, India

Javed AhmadBhat
Division of Soil Science and
   Agricultural Chemistry
SKUAST, Kashmir, Wadura, Sopore,
   India

Raies Ahmad Bhat
Division of Agronomy
Faculty of Agriculture
Wadura, SKUAST Kashmir, India

Haobam Bidyapati
Department of Civil Engineering
National Institute of Technology
Manipur, India

Shabeer AhmadDar
Centre of Research for Development
University of Kashmir
Srinagar, J&K, India

Zaffar Mahdi Dar
Division of Basic Sciences and
    Humanities
FOH Shalimar
SKUAST Kashmir, India

Thiyam Tamphasana Devi
Department of Civil Engineering
National Institute of Technology
Manipur, India

Pushpa Gautam
Department of Metallurgical
    Engineering
Indian Institute of Technology (Banaras
    Hindu University)
Varanasi, India

Sandip K. Ghosh
Assistant Secretary, Indian Institute of
    Chemical Engineers
Jadavpur
Kolkata, India

Fahima Gul
Department of Botany
S.P. College
Srinagar, J&K, India

Razia Gull
Division of Agronomy
Faculty of Agriculture
Wadura, SKUAST Kashmir, India

Burhan Hamid
Centre of Research for Development
University of Kashmir
Srinagar, J&K, India

Yasir Hanif
Division of Soil Science and
    Agricultural Chemistry
Sher-e-Kashmir University of
    Agricultural Science and
    Technology
Wadura, Sopore, Kashmir, India

Humaira
Centre of Research for Development
University of Kashmir
Srinagar, J&K, India

Shagufta Iqbal
Department of Zoology
University of Kashmir
Srinagar, J&K, India

Shayesta Islam
Division of Environmental Sciences
FOH
Shalimar, SKUAST Kashmir, India

Rohit Jha
Department of Metallurgical
    Engineering
Indian Institute of Technology (Banaras
    Hindu University)
Varanasi, India

Azra N. Kamili
Department of Botany
Central University of Kashmir
Ganderbal, J& K, India

S. Karan
Department of Environmental Science
G.B. Pant University of Agriculture and
    Technology
Pantnagar, India

Inayat Mustafa Khan
Division of Soil Science and
    Agricultural Chemistry
Sher-e-Kashmir University of
    Agricultural Science and
    Technology
Wadura, Sopore, Kashmir, India

Manoj Kumar
Department of Hydro and Renewable
    Energy
Indian Institute of Technology Roorkee
Roorkee, Uttrakhand, India

Soham Kundu
Department of Chemical Engineering
University of Calcutta
Kolkata, India

Neesa Majeed
Centre of Research for Development
University of Kashmir
Srinagar, J&K, India

Afaan A. Malla
Department of Environmental Science
Government Degree College Baramulla
Indira Gandhi National Open
    University, J&K, India

Fayaz A. Malla
Department of Environmental Science
Government Degree College Tral
J&K, India

Aanisa Manzoor
Division of Soil Science and
    Agricultural Chemistry
Faculty of Agriculture
Wadura, SKUAST Kashmir, India

Aamir Hassan Mir
DARS-Rangreth
Srinagar, India

Shakeel Ahmad Mir
Division of Soil Science
Sher-e-Kashmir University of
    Agricultural Science and Technology
Shalimar, Kashmir, India

Tanveer Ahmad Mir
Centre of Research for Development
University of Kashmir
Srinagar, J&K, India

Modhurima Misra
Department of Biotechnology
Heritage Institute of Technology
Kolkata, India

Asia Mushtaq
Department of Biochemistry
University of Kashmir
Srinagar, J&K, India

Irshad AhmadNawchoo
Department of Botany
University of Kashmir
Srinagar, J&K, India

Khair Ul Nisa
Department of Environmental Sciences
University of Kashmir
Srinagar, J&K, India

Qadrul Nisa
Division of Plant Pathology
SKUAST-K, Shalimar
Srinagar, India

Rifat Un Nisa
Division of Agronomy
Sher-e-Kashmir University of
    Agricultural Science and
    Technology
Wadura, Sopore, Kashmir, India

Mudila Dhanunjaya Rao
CSIR—National Metallurgical
    Laboratory
Jamshedpur, India

Aabiroo Rashid
Division of Agronomy
Sher-e-Kashmir University of
    Agricultural Science and
    Technology
Wadura, Sopore, Kashmir, India

Qamer Ridwan
Department of Botany
School of Biosciences and
    Biotechnology
Baba Ghulam Shah Badshah
    University
Rajouri, J&K, India

RZ Sayyed
Department of Microbiology
PSGVP Mandal's Shri SI Patil Arts, GB
    Patel Science, STKVS Commerce
    College
Shahada, India

Aanisa Manzoor Shah
Division of Soil Science and
    Agricultural Chemistry
Sher-e-Kashmir University of
    Agricultural Science and
    Technology
Wadura, Sopore, Kashmir, India

Ayman Javed Shah
Division of Soil Science and
    Agricultural Chemistry
Sher-e-Kashmir University of
    Agricultural Science and
    Technology
Wadura, Sopore, Kashmir, India

Tahir Ahmad Sheikh
Division of Agronomy
Faculty of Agriculture
Wadura, SKUAST Kashmir, India

Tajamul IslamShah
Division of Soil Science
Sher-e-Kashmir University of
    Agricultural Science and
    Technology, Kashmir
Shalimar-190025, India

Kamalesh K. Singh
Department of Metallurgical Engineering
Indian Institute of Technology (Banaras
    Hindu University)
Varanasi, India

Nazir A. Sofi
Department of Agriculture Research
    Information System
Sher-e-Kashmir University of
    Agricultural Sciences and
    Technology
Shalimar Campus, Srinagar, J&K, India

Najeebul Tarfeen
Centre of Research for Development
University of Kashmir
Srinagar, J&K, India

Sumira Tyub
Centre of Research for Development
University of Kashmir
Srinagar, J&K, India

Saba Wani
Department of Biochemistry and Centre
    of Research for Development
University of Kashmir
Srinagar, J&K, India

Ali Mohd Yatoo
Department of Environmental Science
University of Kashmir
Srinagar, J&K, India

1 Microplastic Pollution
Sources, Effects, Monitoring, and Management

Razia Gull, Tahir Ahmad Sheikh, M. Anwar Bhat, Rifat Un Nisa, Aanisa Manzoor, Roheela Ahmad, Ali Mohd Yatoo, and Raies Ahmad Bhat

1.1 INTRODUCTION

Plastic waste is of great environmental concern due to its global distribution, which affects the terrestrial and marine environment. In the contemporary world, plastic usage is widespread in daily activities and has become an important part of human life (Jiang et al., 2019; Zhang et al., 2020a). According to the report of Plastics Europe (2018), plastic production across the world has increased drastically, reaching nearly 350 million tonnes in 2017. Plastics have gained popularity in the recent past because of their tenacity and frugality. Production of plastics has increased from 15 Mt in 1964 to 311 MT in 2014 and will probably double over the coming 20 years (Perumal and Muthuramalingam, 2021). Recently, amid the current global COVID-19 pandemic, increased quantities of plastic litter, especially clinical waste and disposable plastic, have been produced (Prata et al., 2020). This compounded the estimated 19–23 million tonnes of plastic pollution entering marine and aquatic habitats in 2016 (Borrelle et al., 2020). These projections are ascribed to the continuous overuse and utilization of disposable plastics (including personal protective equipment, like gloves and masks) because of the worldwide COVID-19 pandemic (Silva et al., 2021, 2020). The plastics present in the surroundings can be divided into five types based on their size, such as nano plastics (< 5 mm), meso plastics (≥ 5 mm–5 cm), macro plastics (> 5–50 cm) and mega plastics (> 50 cm) (Lebreton et al., 2018; Gigault et al., 2018). Microplastics (MPs) are the smallest pieces (< 5 mm) of synthetic polymers that are extremely tenacious and pervasive in every ecosystem (Kershaw, 2015; NOAA, 2018; Karim et al., 2019). The initial documentation of MP polymers in the ecosystem was reported in the 1970s (Carpenter and Smith, 1972). Since then, various scientific organizations across the world have found that MPs are ubiquitous in aquatic habitats and have a negative impact on aquatic life (Rands et al., 2010; Sutherland et al., 2010). Over the past decade, MP contamination has been the subject of interest and has captured the attention of scientists and the public because of the drastic negative impact it has on aquatic habitats (Cole et al., 2011). MPs have been a rising problem, particularly since 2014. MP pollution and its negative

DOI: 10.1201/9781003359326-1

environmental impacts are widely recognized as a regional and global concern for aquatic biota, economic activity and human health (Gall and Thompson, 2015; Yu et al., 2020; Alpizar et al., 2020). MPs have been shown to adsorb trace chemical contaminants like pharmaceuticals, organochlorine pesticides, polychlorinated biphenyls (PCBs), heavy metals and polycyclic aromatic hydrocarbons (Brennecke et al., 2016; Camacho et al., 2019; Li et al., 2018b; Rochman et al., 2014), industrial supplements such as polybrominated diphenyl ethers used as brominated antiflame agents, phthalate plasticizers and thermal stabilizers for lead (Halden, 2010; Lithner et al., 2011), yielding a blend of pollutants (Rochman, 2015). MPs are highly bioavailable to various species because of their tiny size and widespread distribution in the marine ecosystem. The MP intake by marine organisms is widely observed in fish (Kumar et al., 2018; Pegado et al., 2018), mammals (Bravo et al., 2013) and bivalves (Li et al., 2018a; Su et al., 2018) collected from the aquatic habitat. During the period 2015–2018, MPs were discovered in commercial salts of roughly 128 brands from 38 distinct sources in various nations across the five continents (Peixoto, 2019). As the MP concentrations in the environment rise, there is an increased chance of ecosystem vulnerability and therefore a higher likelihood of hazardous interactions, uptake and impacts on food webs. A schematic overview of the environmental concerns and processes associated with MPs in soil is depicted in Figure 1.1

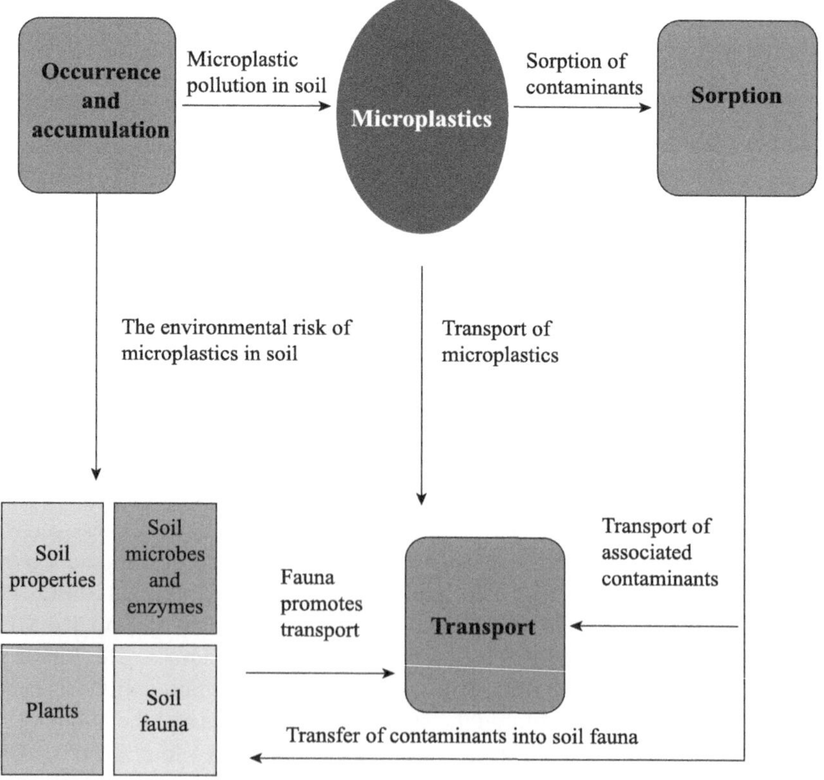

FIGURE 1.1   Environmental concerns and processes of MPs in soil

(Xu et al., 2020). MPs have a variety of impacts on biota. As with macro plastics and macro-fauna, MPs on this relatively smaller scale can cause entanglement and physical discomfort in organisms like zooplankton (Ziajahromi et al., 2017). MPs, if ingested, can cause intestinal blockages and thus hunger or reduced energy budgets (Cole et al., 2015; Wright et al., 2013). It has been assessed that, currently, around 4–12 million tonnes of plastics end up in the rivers and seas per year, with predictions that by 2050 it will exceed the amount of fish (Pico and Barcelo, 2019). It is therefore apprehensible that researchers pay close attention to this topic; especially their ramifications, sampling methods, analytical techniques, global assignments, fate and ambivalent environmental effects are becoming a new era of scientific study (de Souza Machado et al., 2018; Ivar Do Sul and Costa, 2014; Law and oceans, 2014). However, there are some significant research gaps, which include (i) no knowledge of the fate and transportation of MPs; (ii) lack of a standardized method for identifying and characterizing MPs in the different environmental media; (iii) no inputs on atmospheric MPs (outdoor and indoor air); (iv) inadequate research on the soil-borne MPs; and (v) lack of research on the relationship between climate change and MPs.

1.2 TYPES OF MPs

MPs are divided into two categories based on their origin: primary and secondary MPs (Carlos de Sá et al., 2015). Primary MPs are considered tiny plastic particles that enter the environment directly and account for approximately 15–31% of MPs in the oceans (European Commission, 2017). Primary MPs are synthetic polymers of microscopic size and are used as exfoliates for various processes such as sandblasting media, chemical formulations and maintenance of various plastic products and also in the manufacture of synthetic clothing (Chatterjee and Sharma, 2019). Microbeads are another type of primary plastic (size < 2 mm) made of polyethylene (PE), polypropylene (PP) and polystyrene (PS) beads used in cosmetics and healthcare items. Secondary MPs are the fragmented products of meso or macro plastics and are mainly formed under the action of different ecological processes like thermo-oxidative decomposition, photodegradation, thermal decomposition, hydrolysis and biodegradation (Sharma and Chatterjee, 2017). Approximately 69–81% of MPs present in the seas are secondary MPs (European Commission, 2017).

1.3 SOURCES OF MPs

The existence of these dangerous plastic polymers in the habitats (both aquatic and terrestrial) is because of various human activities such as industrial, coastal and domestic activities (Chatterjee and Sharma, 2019). MPs in the ecosystem can also arise from terrestrial sources, including domestic and municipal marine waste along the coasts, untreated sludge discharges, farming operations, coastline tourism and recreational activities (Table 1.1). Terrestrial sources of MPs and microspheres (small plastic particles < 2 mm) constitute about 80% of all plastic wastes in the aquatic ecosystem. MPs in the aquatic habitat can come from primary and

TABLE 1.1
Origins of primary and secondary MPs

Primary microplastics	Secondary microplastics
• Fibre fragments released by synthetic fabrics (Duis and Coors, 2016)	• Fragmentation of massive plastic products (Duis and Coors, 2016)
• Excoriation from car tyres (Duis and Coors, 2016)	• Inadequately managed landfills and unwise landfill management practices (Barnes et al., 2009; Afrin et al., 2020)
• Ingredients used in the production of plastic items (Duis and Coors, 2016)	• Windblown debris or waste further destroyed by recycling plants (Mehlhart and Blepp, 2012; Lambert et al., 2014)
• Remnants from plastics processing companies Granules from the recycling of plastics (Andrady, 2011; Moore, 2008)	• A larger proportion of the plastic output utilized for packing
• Personal care and cleaning items including cosmetics, facial cleansers, body washes, shower gels, toothpastes and so on. (Lassen et al., 2015; Boucher and Friot, 2017)	• Tsunamis, storms, flash floods, high sea waves and other natural calamities (Thompson et al., 2004; Desforges et al., 2014)
• Clinical applications, including dental tooth polishing and other operative pharmaceutical agents (Lassen et al., 2015; Sund et al., 2014)	• Low-density polyethylene (LDP) strips commonly used in plastic mulching to preserve crops, reduce weeds, raise soil warmth and conserve water supply (Sivan, 2011; Rillig, 2012)
• Emission of drilling fluids from gas and oil exploration operations and in industrial abrasives (Lassen et al., 2015; Gregory and Ryan, 1997)	• 5–15 mm polystyrene (PS) fibres used in horticulture as a compost additive (Do and Scherer, 2012)
• Different types of hygiene products, which also release fibre fragments (Duis and Coors, 2016)	

secondary sources (Figure 1.2) (Li et al., 2016). Primary sources include the accidental or intentional spillage of micro-pellets, microfibres, microbeads and other commodities as reported by the United Nations Environment Programme (Gasperi et al., 2018). These products come from industrial trash or are by-products of plastic products and erosion, which include tyres, wheels and boards. Secondary MPs result from degradation caused by biophysiochemical factors such as biodegradation, oxidation, UV light, heat and mechanical forces (Rilling et al., 2017; Zhang et al., 2016).

Over 250,000 tonnes of plastics (approximately 5.25 trillion pieces) float in the world's seas, of which 75–90% are from terrestrial sources and 10–25% from marine sources, which are thought to be a possible source of secondary MPs (Eriksen et al., 2014; Andrady, 2011; Mehlhart and Blepp, 2012). Due to failed and inadequate or defective waste disposal, solid garbage including plastics, metals, glass, textiles, rubber, paper, processed wood, lids, cigarettes, straws stirrers, beverage bottles and caps is continuously dumped into the water bodies by the

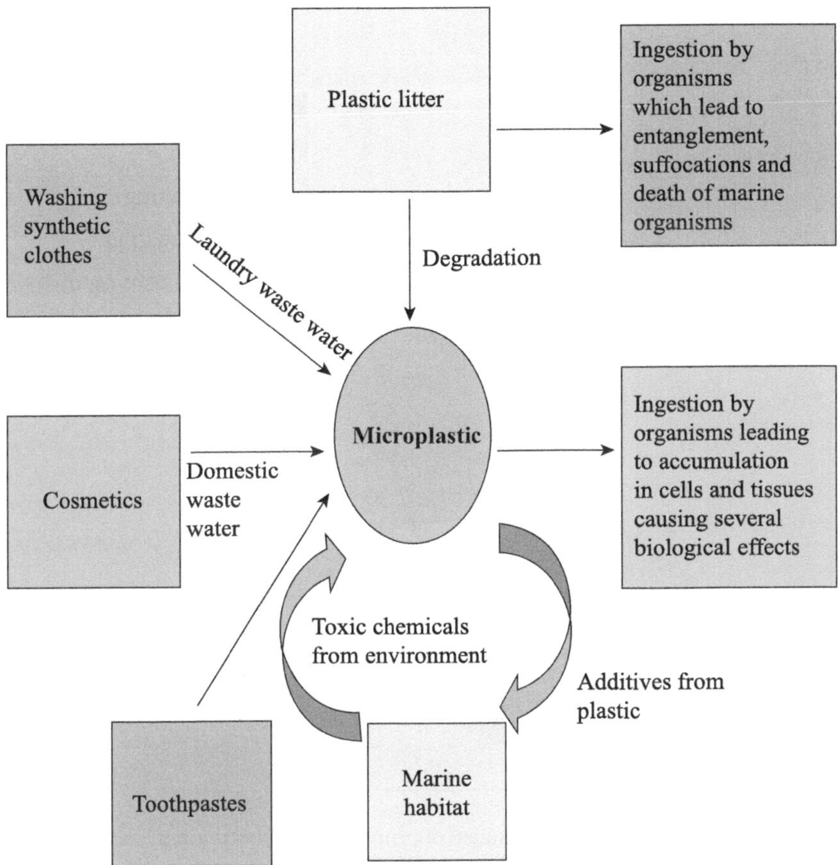

FIGURE 1.2   Sources of microplastics in aquatic environments

growing human population (Alpizar et al., 2020). The buoyancy of most plastic products (e.g., synthetic polymers) often makes it easier for their particles to float and therefore frequently to be carried and washed ashore (Do sul and Costa, 2014). Figure 1.3 reflects the global resource assessment of primary MPs in the oceans.

1.4   EFFECTS OF MPs

1.4.1   EFFECT ON MARINE BIOTA

MPs are prevalent in the aquatic environment and, because of their tiny particulate nature, are mistaken as food by numerous marine organisms, including phytoplankton, zooplankton, corals, lobsters, sea urchins, fish and others. These organisms then transfer the MPs to a higher tropical level (Chatterjee and Sharma, 2019). The impact of MPs on marine biota is of great concern as they lead to entanglement and ingestibility, which may be fatal to marine biota. Table 1.2 shows the observed biological

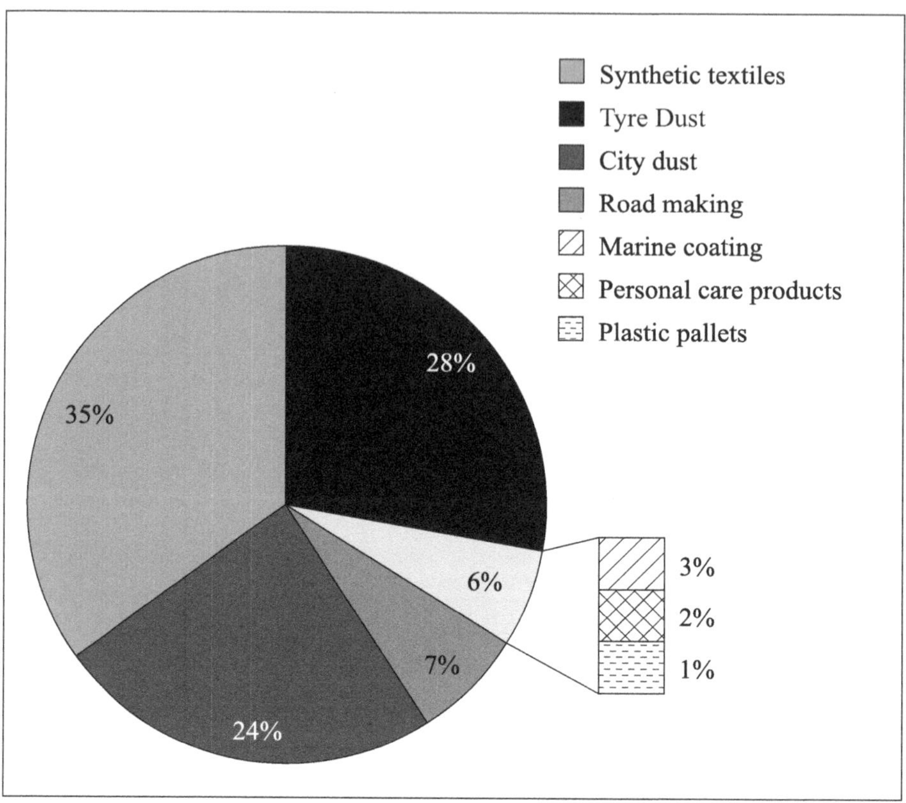

FIGURE 1.3   Global resource assessment of primary microplastics in the oceans

TABLE 1.2
Observed biological effects of microplastics (MPs) on marine biota

Organisms	Effect	Reference
Farrer's scallop (*Chlamys farreri*)	MPs act as both a transporter and a scavenger for bioaccumulation. They increase the detrimental effects of BDE-209 on haemolytic phagocytosis and structural alterations in the digestive gland and gills.	Xia et al. (2020)
Beluga whales (*Delphinapterus leucas*)	MPs were found in the digestive tract. The digestive tract of a single whale contained 18–147 beads of MPs.	Moore et al. (2020)
Zebrafish (*Danio rerio*)	MPs were absorbed, transported through the lumen of the intestine and then excreted.	Batel et al. (2020)
Girella fish (*Girella laevifrons*)	MPs cause inflammation, circulatory disorders and reverse changes in the intestinal lining	Ahrendt et al. (2020) Espinosa et al. (2019)

TABLE 1.2 (*Continued*)
Observed biological effects of microplastics (MPs) on marine biota

Organisms	Effect	Reference
Goldfish (*Carassius auratus*)	MPs lead to fish exposure risks.	Moore et al. (2020) Xiong et al. (2021)
Shrimp (*Paratya australiensis*)	36% of total shrimp contains MPs, with an average of 0.52 ± 0.55 units/individual (24 ± 31 items/g)	Nan et al. (2020)
Sea cucumber (*Apostichopus japonicus*)	MPs had an impact on its growth and physiological state.	Mohsen et al. (2019)
Nile tilapia (*Oreochromis niloticus*)	MPs caused anaemia and perturbations that resulted in the death of tilapia at the early juvenile stage.	Hamed et al. (2019) Xiong et al. (2019)
European sea bass (*Dicentrarchus labrax* L.)	The activity of antioxidant enzymes was reduced by MPs. PVC particles boosted phagocytic and respiratory burst activity. PE particles raised immunoglobulin levels in cutaneous mucus.	Espinosa et al. (2019)
Planktonic crustacean (*Daphnia magna*)	Death rate surged. MPs changed the virulence of pollutants like pesticides.	Zocchi and Sommaruga (2019)
Skeletonema costatum	MPs affect the growth of algae, chlorophyll content and photosynthesis.	Zhang (2017)
Crustacea (*Nephrops norvegicus*)	MPs cause reduction in body weight, metabolic rate, feeding ability and lipid catabolism.	Welden and Cowie (2016)
Crustacea (*Carcinus maenas*)	MPs causes a decrease in the amount of food consumed and the energy required for growth.	Watts et al. (2015)
Microalgae (*Tetraselmis chuii*)	MPs cause reduction in the growth of the microalgae population	Davarpanah and Guilhermino (2015)
Mediterranean mussel (*Mytilus galloprovincialis*)	MPs were present in the stomach, gills and hemolymph tissues. They cause alterations in the gene expression profiles and immune responses.	Avio et al. (2015)
Annelida (*Arenicola marina*)	MPs cause decreased energy efficiency, feeding activity and body weight.	Besseling et al. (2013)
Fish (*Oryzias latipes*)	MPs result in glycogen depletion in the liver, single-cell necrosis in the liver and fatty vacuolation.	Rochman (2013)

effects in marine biota when exposed to MPs. MPs also affect plankton, the most important component of marine habitat. A technical analysis of the effect of marine litter on biodiversity found that more than 80% of documented events between species and marine litter were related to plastics, and 11% of all proclaimed encounters involved MPs (GEF, 2012). MP penetration via phytoplankton cell walls results in reduced chlorophyll absorption (Nerland et al., 2014). Furthermore, when exposed to MPs, heterotrophic planktons phagocytose and store these microscopic plastic bits in their cells. In addition, zooplanktons are known to consume microbeads of size variations from 1.7 to 30.6 μm (Matthew et al., 2013). *Centropages typicus* (a famous copepod) was found to consume MPs (size 7.3 μm) and eventually lose its potential to feed, with a consequent negative impact on its health (Matthew et al., 2013). MPs, macroplastics and microbeads are hazardous to marine species and can result in serious illness when ingested (Fendall and Sewell, 2009) (Figure 1.4). The presence of MPs has been reported in nearly 30% of single fish species (Possatto et al., 2011; Lusher et al., 2013). The prolonged research on *Oryzias latipes* (Japanese medaka) has revealed that even low quantities of MPs (8 ng L^{-1} PE of size 0.5 mm) in the water bodies (San Diego Bay, CA) may have a sublethal effect on the fish (Nerland et al., 2014). MP (<5 mm) accumulation in the fish gut leads to starvation and malnutrition and eventually death (Boerger et al., 2010). Likewise, the intake of various

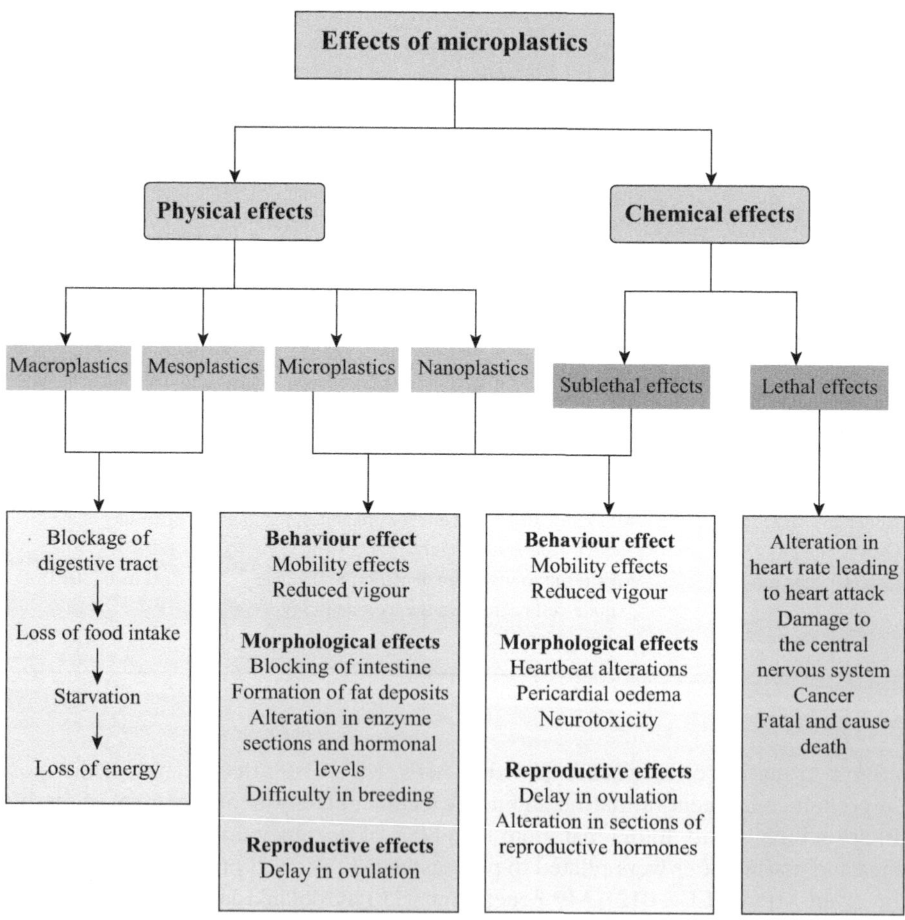

FIGURE 1.4   Effect of microplastics on marine biota

types of MPs was also estimated in Norway lobster, where 83% of lobsters were observed to be contaminated with MPs (Murray and Cowie, 2011).

1.4.2 EFFECT ON TERRESTRIAL ENVIRONMENT

Although researchers have reported the effects of MPs on marine environments, the potential consequences of MPs on terrestrial environments are yet to be investigated. MPs can persist for more than 100 years in the soil even without oxygen and light (Castaneda et al., 2014). Different activities including road run-off, erosion, water supply, littering, plastic mulching and atmospheric deposition are believed to be key sources of MPs in the soil (Blasing and Amelung, 2018; Corradini et al., 2019). Furthermore, MPs interact with different soil organisms, and the negative effects of this interaction can have deleterious effects on their health, which can disrupt various soil functions (Huerta et al., 2017). MPs also allow other terrestrial species to experience changes in their biophysical surroundings (Table 1.3). Various investigations revealed that MPs were found in

TABLE 1.3
Effect on terrestrial biota

Plastic type	Species or biota	Effect	Reference
PS 100 nm and 1300 nm fluorescent fragments	Earthworms (*Eisenia fetida*)	Damages the DNA of earthworms	Jiang et al. (2020)
LDPE	Bacterivorous nematode (*Caenorhabditis elegans*)	Affects reproduction	Schopfer et al. (2020)
Microfibres	Earthworm (*Lumbricus terrestris*)	Modulate the feeding and burrowing habits	Prendergast Miller et al. (2019)
PET microplastic fibres	Terrestrial snails (*Achatina fulica*)	Increases malondialdehyde, whereas decreases glutathione peroxidase and overall antioxidant ability	Song et al. (2019)
LDPE	Soil enzymes (invertase, catalase and urease)	Increases catalase and urease activity considerably, but invertase activities remained unchanged	Huang et al. (2019)
PS microplastics of various sizes (0.47 to 0.53, 27 to 32 and 250 to 300 mm)	Springtails (*Lobella sokamensis*)	Disrupts springtail mobility and behaviour	Kim and An (2019)
PE < 500 mm	Springtail (*Folsomia candida*)	Changes the gut microbial flora and inhibits reproduction	Ju et al. (2019)

(Continued)

TABLE 1.3 (*Continued*)
Effect on terrestrial biota

Plastic type	Species or biota	Effect	Reference
PA (13 to 18 and 90 to 150 mm), PVC particles (106 to 150 mm)	Terrestrial worm (*Enchytraeus crypticus*)	Decreases reproduction	Lahive et al. (2019)
LDPE	Earthworm (*Eisenia fetida*)	Causes oxidative stress	Rodríguez-Seijo et al. (2018)
PE	Redworm (*Eisenia andrei*)	Harms the intestines and the immune system	Rodriguez-Seijo et al. (2017)
Plant species			
PFs, 1.28 ± 0.03 mm long and ~30 mm diameter	Grasses (*Festuca brevipila, Calamagrostis epigejos* and *Holcus lanatus*)	Species dominance observed; *Hieracium* and *Calamagrostis* turned out to be dominant, while reduced biomass is found in *Holcus*	Lozano and Rillig (2020)
PE, PLA	Maize (*Zea mays* L.)	PLA caused reduction in chlorophyll content and biomass in maize; increased pH and concentrations of DTPA-extractable cadmium in soil; changes in structure and diversity of the arbuscular mycorrhizal fungi (AMF) population; and higher phytotoxicity caused by PE and PLA	Wang et al. (2020)
PP, PE, PVC	Garden cress (*Lepidium sativum*)	Oxidative burst occurs, PVC turned out to be most hazardous	Pignattelli et al. (2020)
Biodegradable PLA (density of 1.2–1.3 g cm^{-3})	Perennial ryegrass (*Lolium perenne*)	Reduces plant height and biomass, as well as germination	Boots et al. (2019)
4800 nm MP	Garden cress (*Lepidium sativum*)	Accumulates on root hairs; causes drastic reductions in emergence rate and physical blockage of the seed capsule pores	Bosker et al. (2019)
PS microbeads (0.2 and 1.0 mm)	Lettuce (*Lactuca sativa* L)	Adhesion, absorption, accumulation and translocation of microplastics in vascular tissues	Li et al. (2019)
PA beads 15 to 20 mm, PFs 5000 mm length and 8 mm diameter	Onion (*Allium fistulosum*)	Prominent changes in root characteristics, elemental composition of tissues, vegetative biomass and microbial activity of soil	De souza Machado et al. (2018)

Abbreviations: LDPE, low-density polyethylene; PA, polyamide; PE, polyethylene; PET, polyethylene terephthalate; PLA, polylactide; PP, polypropylene; PS, polystyrene; PVC, polyvinyl chloride.

the digestive systems of 50% of 185 dead sheep (Omidi et al., 2012), 27% of 230 goats and 94% of the 17 dead terrestrial birds in China (Zhu et al., 2018). More importantly, it was also found that in agricultural land low-density microplastics are responsible for decreasing the water-holding capacity of soils (Zhang et al., 2020b).

1.4.3 Effect on Humans

The human population is exposed to MPs from various sources, such as primary MPs in cosmetics, peels, hand cleaners and toothpastes. The Federal Institute for Risk Assessment has assessed the health risks of using facial cleansers, hand cleansers, toothpastes and dental care products containing PE MP particles (BfR, 2015). MP and microbead particles present in toothpaste can be swallowed unintentionally and absorbed through the gastrointestinal tract (Lassen et al., 2015). Besides the toxic effects of MPs, hazardous substances such as phthalates or PCBs contained in MPs or other pollutants adsorbed on the MP surface can contribute to human food exposure (Lassen et al., 2015). It was estimated that the annual MP ingestion through dust in children and adults is around 3000 and 1000 particles, respectively (Dehghani et al., 2017). Alternate ingestion of microparticles can lead to chromosomal changes leading to infertility, obesity and cancer (GESAMP, 2015). In women, oestrogenic mimicking substances can result in breast cancer. High levels of MP contaminants in seafood pose a serious threat to food safety (Van Cauwenberghe and Janssen, 2014). In seafood, the presence of MPs is becoming more prevalent day by day, leading to adverse effects on human health (Yu et al., 2020). MPs have also been found to be consumed by many commercially active aquatic organisms such as mussels (Yu et al., 2020), oysters, crabs (BfR, 2015) and fish (Campanale et al., 2020) and moved across the food chain (Jabeen et al., 2017; BfR, 2015). Additionally, different vegetable and fruit crops can absorb MPs from the soil, and humans can absorb around 80 mg of MPs per day through the food chain (Enyoh et al., 2019; Campanale et al., 2020). Some of the negative effects of MP exposure on human health are presented in Table 1.4.

TABLE 1.4
Effect of microplastics on human health

Plastic type	Effect	References
Functionalized and unaltered poly methyl methacrylate and polystyrene polyvinyl chloride (PVC)	Induce apoptosis of certain human cell types	Mahadevan and Valiyaveettil (2021)
	Decrease viability of cells with a drop in ATP and rise in ROS (reactive oxygen species) concentrations	Liu et al. (2018a)

(Continued)

TABLE 1.4 (*Continued*)
Effect of microplastics on human health

Plastic type	Effect	References
Fluorescent and pristine polystyrene microplastics	Alter the metabolism of amino acids and bile acids	Jin et al. (2019) Luo et al. (2019)
	Create an imbalance of the microbiome in the gut and failure of the intestinal barrier	
Pristine polystyrene microparticle	Reduces hepatic ATP levels and hampers energy metabolism	Stock et al. (2019) Lu et al. (2018)
Microplastics (0.5 to 5 µm)	Raise the likelihood of metabolic ailments in the offspring	Luo et al. (2019)
Polystyrene microplastic particles	Induce liver inflammation and have a negative impact on neurotransmission	Deng et al. (2017)
Amino-modified and Carboxylated polystyrene particles	Enhance energy metabolism (M2) and TGFβ1 (M1)	Fuchs et al. (2016)
Polyethylene particles derived from plastic prosthetic implants	Cause resorption of periprosthetic bone Induce inflammatory response at the implant site	Nich and Goodman (2014)
Polystyrene particles	Upregulate IL-8 expression	Brown et al. (2001)
Unaltered polyethylene particle	Raise IL-1, IL-6 and TNF (tumor necrosis factor) formation in murine macrophages	Green et al. (1998)

1.5 MONITORING AND MANAGEMENT

A significant rise in the total plastic garbage available to the aquatic environment is projected by the end of 2025. This critical concern was also discussed at the "16th Global Meeting of Regional Seas Conventions and Action Plans," which was organized to educate countries about the global menace of dumping plastics in marine habitats, with an estimated 13 billion dollars in financial damage to the marine biota per year (UN Environment, 2014). Considering this present trend of marine plastic contamination, there is a pressing need for targeted research that can aid in the reduction of plastic pollution. The government should adopt some novel steps to raise awareness in the community regarding the ill consequences of plastic garbage on the marine ecosystem. Certain laws and rules that could regulate the widespread usage of plastic products must be put in place immediately; else, the ecosystem's condition will deteriorate in the near future (Sharma and Chatterjee, 2017). There should be a reliable waste management system in place that could control the plastic litter collection. Cosmetics and other products of personal care which include toothpaste, shampoo and face washes containing microbeads should be totally banned. Extended producer responsibility (EPR), a waste management system that uplifts the use of non-plastic packaging materials for beverages and other food items, must be promoted. Different campaigns must be launched by different governmental and non-governmental organizations to sensitize the people about the harmful and deadly impacts of MP pollution.

In developing countries, detailed guidelines are formulated against the use of plastic and related products, including a complete ban on plastic bottles and plastic bags as well as the imposition of fines for their use (Plastic Pollution Coalition, 2016). However, unfortunately, FMCGs continue to sell their products in plastic packaging. In countries like the United States, local governments have framed restrictions and imposed fines for the usage of plastic bags. As a result of this, a reduction of nearly 60–90% in plastic bag usage has been observed (Plastic Pollution Coalition, 2016). Many European countries have completely banned products containing microbeads. The federal administration of various nations like Australia, Belgium, Austria, Canada, Luxembourg, Germany, the Netherlands and Sweden has also imposed a ban on microbead usage in products of personal care (Perschbacher, 2016). The use of this harmful contaminant was completely outlawed under the Canadian Environmental Protection Act. They also enacted restrictions prohibiting the manufacture, export and import of MP-containing personal care items (2016, The Globe and Mail Newspaper). In California, the use of abrasives made up of MP particles, "microbeads", has been completely banned in various household products like creams, facial scrubs, detergents and toothpastes. Furthermore, the California Assembly approved Assembly Bill 888, which imposes strict rules and restrictions prohibiting the use of hazardous MPs and emphasizes strongly the usage of natural and environmentally friendly alternative cosmetic products to safeguard the environment (Plastic Pollution Coalition, 2015). San Francisco turns out to be the first state in the United States to outlaw plastic bottles (Plastic Pollution Coalition, 2016). Several other national and international civil society organizations have asked customers and manufacturers to avoid using microbeads in products of personal care (Plastic Soup Foundation, 2016). Supermarket firms like L'Oréal, Livon and Johnson & Johnson have also pledged to eliminate microbeads and MPs from personal care items (Copeland, 2015). In a nutshell, effective management, recycling and finally an eco-friendly disposal method would all contribute to a plastic-free environment.

1.6 CONCLUSION

Plastic production has been increasing worldwide. Plastic contamination has deteriorated the global environment, due to its long decomposition time. Continuous rise of MP fragments in the environment may cause serious damage to the terrestrial and marine biota. MP contaminants are present in common salt, drinking water and other everyday food items, posing a major hazard to human life. Also, due to their detrimental consequences on marine life, MPs in the aquatic habitat are currently a concern. Thus, there is a pressing need for immediate action against the dispensable usage of plastics and their products. Plastic usage must be rigorously prohibited at both the federal and global levels. New scientific research is needed to elucidate the various elements that determine the prevalence of MPs in the aquatic environment and their biological effect on aquatic biota. There is a need to develop novel research investigations for conservation management and to support various study programmes to protect environment from these harmful fragments. The most important approach to minimize the plastic input entering the environment is collecting and recycling these plastic particles. So, to avert future threats, the prime remedy is to halt its further production and look for alternatives of plastic products.

REFERENCES

Afrin, S., Uddin, M. K., & Rahman, M. M. (2020). Micro-plastics contamination in the soil from Urban Landfill site, Dhaka, Bangladesh. *Heliyon*, *6*(11), e05572. https://doi.org/10.1016/j.heliyon

Ahrendt, C., Perez-Venegas, D. J., Urbina, M., Gonzalez, C., Echeveste, P., Aldana, M., . . . & Galban-Malagon, C. (2020). Microplastic ingestion cause intestinal lesions in the intertidal fish Girella laevifrons. *Marine Pollution Bulletin*, *151*, 110795.

Alpizar, F., Carlsson, F., Lanza, G., Carney, B., Daniels, R. C., Jaime, M., . . . & Tibesigwa, B. (2020). A framework for selecting and designing policies to reduce marine plastic pollution in developing countries. *Environmental Science and Policy*, *109*, 25–35.

Andrady, A. L. (2011). Micro-plastics in the marine environment. *Marine Pollution Bulletin*, *62*, 1596–1605.

Avio, C. G., Gorbi, S., Milan, M., Benedetti, M., Fattorini, D., Errico, G., . . . & Regoli, F. (2015). Pollutants bioavailability and toxicological risk from microplastics to marine mussels. *Environmental Pollution*, *198*, 211–222.

Barnes, D. K., Galgani, F., Thompson, R. C., & Barlaz, M. (2009). Accumulation and fragmentation of plastic debris in global environments. *Philosophical Transactions of the Royal Society B*, *364*, 1985–1998.

Batel, A., Baumann, L., Carteny, C. C., Cormier, B., Keiter, S. H., & Braunbeck, T. (2020). Histological, enzymatic and chemical analyses of the potential effects of differently sized microplastic particles upon long-term ingestion in zebrafish (*Danio rerio*). *Marine Pollution Bulletin*, *153*, 111022.

Besseling, E., Wegner, A., Foekema, E. M., van den Heuvel-Greve, M. J., & Albert, A. K. (2013). Effects of microplastic on fitness and PCB bioaccumulation by the lugworm Arenicola marina (L.). *Environmental Science and Technology*, *47*, 593–600.

BfR. (April 2015). Microplastic Particles in Food. BfR Opinion No. 013/2015 of 30 April 2015. English summary of the report Mikroplastikpartikel in Lebensmitteln, Stellungnahme Nr. 013/2015 des BfR vom 30. German Federal Institute for Risk Assessment.

Blasing, M., & Amelung, W. (2018). Plastics in soil: Analytical methods and possible sources. *Science of the Total Environment*, *612*, 422–435.

Boerger, C. M., Lattin, G. L., Moore, S., & Moore, C. J. (2010). Plastic ingestion by planktivorous fishes in the North Pacific Central Gyre. *Marine Pollution Bulletin*, *60*, 2275–2778.

Boots, B., Russell, C. W., & Green, D. S. (2019). Effects of microplastics in soil ecosystems: Above and below ground. *Environmental Science and Technology*, *53*, 11496–11506.

Borrelle, S. B., Ringma, J., Law, K. L., Monnahan, C. C., Lebreton, L., McGivern, A., . . . & Hilleary, M. A. (2020). Predicted growth in plastic waste exceeds efforts to mitigate plastic pollution. *Science*, *369*, 1515–1518.

Bosker, T., Bouwman, L. J., Brun, N. R., Behrens, P., & Vijver, M. G. (2019). Microplastics accumulate on pores in seed capsule and delay germination and root growth of the terrestrial vascular plant Lepidium sativum. *Chemosphere*, *226*, 774–781.

Boucher, J., & Friot, D. (2017). *In: Primary micro-plastics in the oceans: A global evaluation of sources* (p. 43). IUCN.

Bravo, E. L., Van Franeker, J. A., Jansen, O. E., & Brasseur, S. M. J. M. (2013). Plastic ingestion by harbour seals (*Phoca vitulina*) in the Netherlands. *Marine Pollution Bulletin, 67*, 200–202. https://doi.org/10.1016/j.marpolbul.2012.11.035

Brennecke, D., Duarte, B., Paiva, F., Cacador, I., & Canning-Clode, J. (2016). Microplastics as vectors for heavy metal contamination from the marine environment. *Estuarine, Coastal and Shelf Science, 178*, 189–195. https://doi.org/10.1016/j.ecss

Brown, D. M., Wilson, M. R., MacNee, W., Stone, V., & Donaldson, K. (2001). Size-dependent proinflammatory effects of ultrafine polystyrene particles: A role for surface area and oxidative stress in the enhanced activity of ultrafines. *Toxicology and Applied Pharmacology, 175*, 191–199.

Camacho, M., Herrera, A., Go´mez, M., Acosta-Dacal, A., Martı´nez, I., Henrı´quez-Herna´ndez, L., & Luzardo, O. P. (2019). Organic pollutants in marine plastic debris from Canary Islands beaches. *Science of the Total Environment, 662*, 22–31. https://doi.org/10.1016/j.scitotenv

Campanale, C., Massarelli, C., Savino, I., Locaputo, V., & Uricchio, V. F. (2020). A detailed review study on potential effects of micro-plastics and additives of concern on human health. *International Journal of Environmental Research and Public Health, 17*(4), 1212. https://doi.org/10.3390/ijerph

Carlos de Sá, L., Luís, L. G., & Guilhermino, L. (2015). Effects of microplastics on juveniles of the common goby (*Pomatoschistus microps*): confusion with prey, reduction of the predatory performance and efficiency, and possible influence of developmental conditions. *Environmental Pollution, 196*, 359–362. https://doi.org/10.1016/j.envpol.2014.10.026

Carpenter, E. J., & Smith, K. (1972). Plastics on the Sargasso Sea surface. *Science, 175*, 1240–1241.

Castaneda, R. A., Avlijas, S., Simard, M. A., Ricciardi, A., & Smith, R. (2014). Micro-plastic pollution in St. Lawrence River sediments. *Canadian Journal of Fisheries and Aquatic Sciences, 71*(12), 1767–1771.

Chatterjee, S., & Sharma, S. (2019). Microplastics in our oceans and marine health. Field Actions Science Reports [Online]. http://journals.Openeditionorg/factsreports/5257

Cole, M., Lindeque, P., Fileman, E., Halsband, C., & Galloway, T. S. (2015). The impact of polystyrene microplastics on feeding, function and fecundity in the marine copepod Calanus helgolandicus. *Environmental Science and Technology, 49*, 1130–1137.

Cole, M., Lindeque, P., Halsband, C., & Galloway, T. S. (2011). Microplastics as contaminants in the marine environment: A review. *Marine Pollution Bulletin, 62*, 2588–2597. https://doi.org/10.1016/j.marpolbul

Copeland, C. (2015). *Microbeads: An emerging water quality issue.* www.fas.org/sgp/crs/misc/IN10319

Corradini, F., Meza, P., Eguiluz, R., Casado, F., Huerta-Lwanga, E., & Geissen, V. (2019). Evidence of micro-plastic accumulation in agricultural soils from sewage sludge disposal. *Science of the Total Environment, 671*, 411–420. https://doi.org/10.1016/j

Davarpanah, E., & Guilhermino, L. (2015). Single and combined effects of microplastics and copper on the population growth of the marine microalgae Tetraselmis chuii. *Estuarine, Coastal and Shelf Science, 167*, 269–275.

Dehghani, S., Moore, F., & Akhbarizadeh, R. (2017). Microplastic pollution in deposited urban dust, Tehran Metropolis, Iran. *Environmental Science and Pollution Research, 24*, 20360–20371.

Deng, Y., Zhang, Y., Lemos, B., & Ren, H. (2017). Tissue accumulation of microplastics in mice and biomarker responses suggest widespread health risks of exposure. *Scientific Reports, 7*, 46687.

Desforges, J. P., Galbraith, M., Dangerfield, N., & Ross, P. S. (2014). Widespread distribution of micro-plastics in subsurface seawater in the NE Pacific Ocean. *Marine Pollution Bulletin*, *79*, 94–99.

de Souza Machado, A. A., Kloas, W., & Zarfl, C. (2018). Micro-plastics as an emerging threat to terrestrial ecosystems. *Global Change Biology*, *24*, 1405–1416.

Do, T. C. V., & Scherer, H. W. (2012). Compost and biogas residues as basic materials for potting substrates. *Plant Soil and Environment*, *58*, 459–464.

do Sul, J. A. I., & Costa, M. F. (2014). The present and future of microplastic pollution in the marine environment. *Environmental Pollution*, *185*, 352–364.

Duis, K., & Coors, A. (2016). Micro-plastics in the aquatic and terrestrial environment: Sources (with a specific focus on personal care products), fate and effects. *Environmental Sciences Europe*, *28*, 2. https://doi.org/10.1186/s12302-015-0069-y

Enyoh, C. E., Verla, A. W., & Verla, E. N. (2019). Uptake of micro-plastics by plant: A reason to worry or to be happy? *World Scientific News*, *131*, 256–267.

Eriksen, M., Lebreton, L. C. M., & Carson, H. S. (2014). Plastic pollution in the world's oceans: More than 5 trillion plastic pieces weighing over 250,000 tons Afloat at Sea. *Public Library of Science*, *9*, e111913.

Espinosa, C., Esteban, M. A., & Cuesta, A. (2019). Dietary administration of PVC and PE microplastics produces histological damage, oxidative stress and immunoregulation in European sea bass (*Dicentrarchus labrax* L.). *Fish Shellfish Immunology*, *95*, 574–583.

European Commission. (2017). *Micro-plastics focus on food and health 14 December*. https://publications.jrc.ec.europa.eu/repository/bitstream/JRC110629/jrc110629

Fendall, L. S., & Sewell, M. A. (2009). Contributing to marine pollution by washing your face: Microplastics in facial cleansers. *Marine Pollution Bulletin*, *58*, 1225–1228.

Fuchs, A. K., Syrovets, T., Haas, K. A., Loos, C., Musyanovych, A., Mailänder, V., . . . & Simmet, T. (2016). Carboxyl- and amino-functionalized polystyrene nanoparticles differentially affect the polarization profile of M1 and M2 macrophage subsets. *Biomaterials*, *85*, 78–87.

Gall, S. C., & Thompson, R. C. (2015). The impact of debris on marine life. *Marine Pollution Bulletin*, *92*, 170–179.

Gasperi, J., Wright, S. L., Dris, R., Collard, F., Mandin, C., Guerrouache, M., . . . & Tassin, B. (2018). Microplastics in air: Are we breathing it in? *Current Opinion in Environmental Science and Health*, *1*, 1–5.

GEF. (2012). Secretariat of the convention on biological diversity and scientific and technical advisory panel GEF. *Impacts of Marine Debris on Biodiversity: Current Status and Potential Solutions*, *67*, 9.

Gigault, J., Halle, A. T., & Baudrimont, M. (2018). Current opinion: What is a nanoplastic? *Environmental Pollution*, *235*, 1030–1034.

The Globe and Mail Newspaper. (2016). *Government of Canada labels microbeads 'toxic substance'*. www.theglobeandmail.com/news/national/feds-label-microbeads-as-toxic-substance/article30698903/. Accessed 20 February 2017.

Green, T. R., Fisher, J., Stone, M., Wroblewski, B. M., & Ingham, E. (1998). Polyethylene particles of a 'critical size' are necessary for the induction of cytokines by macrophages in vitro. *Biomaterials*, *19*, 2297–2302.

Gregory, M. R., & Ryan, P. G. (1997). Pelagic plastics and other seaborne persistent synthetic debris: A review of southern hemisphere perspectives. In: Coe, J. M., Rogers, D. B. (Eds.), *Marine Debris – Sources, Impacts and Solutions* (pp. 49–66). Springer-Verlag.

Halden, R. U. (2010). Plastics and health risks. *Annual Review of Public Health*, *31*, 179–294. https://doi.org/10.1146/annurev.publhealth.012809.103714

Hamed, M., Soliman, H. A. M., Osman, A. G. M., & Sayed, A. E. H. (2019). Assessment the effect of exposure to microplastics in Nile Tilapia (Oreochromis niloticus) early juvenile: I: Blood biomarkers. *Chemosphere, 228*, 345–350.

Huang, Y., Zhao, Y., Wang, J., Zhang, M., Jia, W., & Qin, X. (2019). LDPE microplastic films alter microbial community composition and enzymatic activities in soil. *Environmental Pollution, 254*, 112983.

Huerta, L. E., Gertsen, H., Gooren, H., Peters, P., Sal´anki, T., van der, P. M., . . . & Geissen, V. (2017). Incorporation of microplastics from litter into burrows of Lumbricus terrestris. *Environmental Pollution, 220*, 523–531.

Ivar Do Sul, J. A., & Costa, M. F. (2014). The present and future of microplastic pollution in the marine environment. *Environmental Pollution, 185*, 352–364.

Jabeen, K., Su, L., Li, J., Yang, D., Tong, C., Mu, J., & Shi, H. (2017). Micro-plastics and mesoplastics in fish from coastal and fresh waters of China. *Environmental Pollution, 221*, 141–149. https://doi.org/10.1016/j.envpol.2016.11.055

Jiang, C., Yin, L., Li, Z., Wen, X., Luo, X., Hu, S., . . . & Liu, Y. (2019). Microplastic pollution in the rivers of the Tibet Plateau. *Environmental Pollution, 249*, 91–98. https://doi.org/10.1016/j.envpol

Jiang, X., Chang, Y., Zhang, T., Qiao, Y., Klobucar, G., & Li, M. (2020). Toxicological effects of polystyrene microplastics on earthworm (*Eisenia fetida*). *Environmental Pollution, 259*, 113896.

Jin, Y., Lu, L., Tu, W., Luo, T., & Fu, Z. (2019). Impacts of polystyrene microplastic on the gut barrier, microbiota and metabolism of mice. *Science of the Total Environment, 649*, 308–317.

Ju, H., Zhu, D., & Qiao, M. (2019). Effects of polyethylene microplastics on the gut microbial community, reproduction and avoidance behaviors of the soil springtail, Folsomia candida. *Environmental Pollution, 247*, 890–897.

Karim, E. M., Sanjee, A. S., Mahmud, S., Shaha, M., Moniruzzaman, M., & Das, C. K. (2019). Micro-plastics pollution in Bangladesh: Current scenario and future research perspective. *Chem. Ecol.* https://doi.org/10.1080/02757540

Kershaw, P. J., & Rochman, C. M. (2015). Sources, fate and effects of microplastics in the marine environment: part 2 of a global assessment. Reports and studies-IMO/FAO/Unesco-IOC/WMO/IAEA/UN/UNEP joint group of experts on the scientific aspects of marine environmental protection (GESAMP) Eng No. 93.

Kim, S. W., & An, Y. J. (2019). Soil microplastics inhibit the movement of springtail species. *Environment International, 126*, 699–706.

Kumar, V. E., Ravikumar, G., & Jeyasanta, K. I. (2018). Occurrence of microplastics in fishes from two landing sites in Tuticorin, South east coast of India. *Marine Pollution Bulletin, 135*, 889–894. https://doi.org/10.1016/j

Lahive, E., Walton, A., Horton, A. A., Spurgeon, D. J., & Svendsen, C. (2019). Microplastic particles reduce reproduction in the terrestrial worm *Enchytraeus crypticus* in soil exposure. *Environmental Pollution, 255*, 113174

Lambert, S., Sinclair, C. J., & Boxall, A. B. (2014). Occurrence, degradation and effect of polymer-based materials in the environment. *Reviews of Environmental Contamination and Toxicology, 227*, 1–53.

Lassen, C., Hansen, S. F., Magnusson, K., Noren, F., Hartmann, N. I. B., Jensen, P. R., . . . & Brinch, A. (2015). Microplastics: Occurrence, effects and sources of releases to the environment in Denmark. *The Danish Environmental Protection Agency.* www.eng.mst.dk/

Law, K. L., & Oceans, T. R. (2014). Micro-plastics in the seas. *Science, 345*, 144–145.

Lebreton, L., Slat, B., & Ferrari, F. (2018). Evidence that the great pacific garbage patch is rapidly accumulating plastic. *Scientific Reports, 8,* 4666. https://doi.org/10.1038/s41598-018-22939-w

Li, H., Ma, L., Lin, L., Ni, Z., Xu, X., Shi, H., & Rittschof, D. (2018a). Microplastics in oysters Saccostrea cucullata along the Pearl River Estuary, China. *Environmental Pollution, 236,* 619–625. https://doi.org/10.1016/j

Li, J., Zhang, K., & Zhang, H. (2018b). Adsorption of antibiotics on microplastics. *Environmental Pollution, 237,* 460–467. https://doi.org/10.1016/j

Li, L., Zhou, Q., Yin, N., Tu, C., & Luo, Y. (2019). Uptake and accumulation of microplastics in an edible plant. *Chinese Science Bulletin, 64,* 928–934.

Li, W. C., Tse, H. F., & Fok, L. (2016). Plastic waste in the marine environment: A review of sources, occurrence and effects. *Science of the Total Environment, 566–567,* 333–349.

Lithner, D., Larsson, A., & Dave, G. (2011). Environmental and health hazard ranking and assessment of plastic polymers based on chemical composition. *Science of the Total Environment, 409,* 3309–3324. https://doi.org/10.1016/j

Liu, X., Tian, X., Xu, X., & Lu, J. (2018). Design of a phosphinate-based bioluminescent probe for superoxide radical anion imaging in living cells. *Luminescence, 33,* 1101–1106.

Lozano, Y. M., & Rillig, M. C. (2020). Effects of microplastic fibers and drought on plant communities. *Environmental Science and Technology, 54*(10), 6166–6173. https://doi.org/10.1021/acs.est

Lu, L., Wan, Z., Luo, T., Fu, Z., & Jin, Y. (2018). Polystyrene microplastics induce gut microbiota dysbiosis and hepatic lipid metabolism disorder in mice. *Science of the Total Environment, 631,* 449–458.

Luo, T., Wang, C., Pan, Z., Jin, C., Fu, Z., & Jin, Y. (2019). Maternal polystyrene microplastic exposure during gestation and lactation altered metabolic homeostasis in the dams and their F1 and F2 offspring. *Environmental Science and Technology, 53,* 10978–10992.

Lusher, A. L., McHugh, M., & Thompson, R. C. (2013). Occurrence of microplastics in the gastrointestinal tract of pelagic and demersal fish from the English channel. *Marine Pollution Bulletin, 67,* 94–99.

Mahadevan, G., & Valiyaveettil, S. (2021). Understanding the interactions of poly (methyl methacrylate) and poly (vinyl chloride) nanoparticles with BHK-21 cell line. *Scientific Reports, 11,* 1–15.

Matthew, C., Pennie, L., Elaine, F., Claudia, H., Rhys, G., Julian, M., & Tamara, S. G. (2013). Microplastic ingestion by zooplankton. *Environmental Science and Technology.* https://doi.org/10.1021/es400663f

Mehlhart, G., & Blepp, M. (2012). *Study on Land-Sourced Litter (LSL) in the marine environment: Review of sources and literature in the context of the initiative of the declaration of the global plastics associations for solutions on marine litter.* Oko-Institut e.V.

Mohsen, M., Wang, Q., Zhang, L., Sun, L., Lin, C., & Yang, H. (2019). Heavy metals in sediment, microplastic and sea cucumber Apostichopus japonicus from farms in China. *Marine Pollution Bulletin, 143,* 42–49.

Moore, C. J. (2008). Synthetic polymers in the marine environment: A rapidly increasing, long-term threat. *Environmental Research, 108,* 131–139.

Moore, R. C., Loseto, L., Noel, M., Etemadifar, A., Brewster, J. D., MacPhee, S., . . . & Ross, P. S. (2020). Microplastics in beluga whales (Delphinapterus leucas) from the eastern Beaufort Sea. *Marine Pollution Bulletin, 150,* 110723.

Murray, F., & Cowie, P. R. (2011). Plastic contamination in the decapod crustacean Nephrops norvegicus. *Marine Pollution Bulletin, 62,* 1207–1217.

Nan, B., Su, L., Kellar, C., Craig, N. J., Keough, M. J., & Pettigrove, V. (2020). Identification of microplastics in surface water and Australian freshwater shrimp Paratya australiensis in Victoria, Australia. *Environmental Pollution, 259*, 113865.

Nerland, I. L., Halsband, C., Allan, I., & Thomas, K. V. (2014). Microplastics in marine environments: Occurrence, distribution and effects project no. 14338 report no. 6754–2014 Oslo.

News European Parliament. (22 November 2018). Micro-plastics: Sources, effects and solutions. www.europarl.europa.eu/news/en/headlines/society/20181116STO19217

Nich, C., & Goodman, S. B. (2014). Role of macrophages in the biological reaction to wear debris from joint replacements. *Information Sciences, 24*, 259–265.

NOAA (National Oceanic and Atmospheric Administration). (2018). *What Are Microplastics?* https://oceanservice.noaa.gov/facts/microplastics.html

Omidi, A., Naeemipoor, H., & Hosseini, M. (2012). Plastic debris in the digestive tract of sheep and goats: An increasing environmental contamination in Birjand, Iran. *Bulletin of Environmental Contamination and Toxicology, 88*, 691–694. https://doi.org/10.1007/s00128-012-0587-x

Pegado, T. D. S., Schmid, K., Wilnemiller, K. O., Chelazzi, D., Cincinelli, A., Dei, L., & Giarrizzo, T. (2018). First evidence of microplastic ingestion by fishes from the Amazon River estuary. *Marine Pollution Bulletin, 133*, 814–821. https://doi.org/10.1016/j.marpolbul

Peixoto, D., Pinheiro, C., Amorim, J., Oliva-Teles, L., Guilhermino, L., & Vieira, M. N. (2019). Microplastic pollution in commercial salt for human consumption: A review. *Estuarine, Coastal and Shelf Science, 219*, 161–168.

Perschbacher, E. (2016). *Microbeads—Legislative update.* www.ijc.org/en_/blog/2016/03/01/microbeads_legislative_update_story4

Perumal, K., & Muthuramalingam, S. (2021). *Microplastics pollution studies in India: A recent review of sources, abundances and research perspectives.* https://doi.org/10.21203/rs.3.rs-535083/v1

Pico, Y., & Barcelo, D. (2019). Analysis and prevention of microplastics pollution in water: Current perspectives and future directions. *American Chemical Society.* https://doi.org/10.1021/acsomega.9b00222

Pignattelli, S., Broccoli, A., & Renzi, M. (2020). Physiological responses of garden cress (*Lepidium sativum*) to different types of microplastics. *Science of the Total Environment*, 138609.

Plastics Europe. (2018). Plastics—The facts, 2018. *An Analysis of European Plastic Production, Demand and Waste Data.* www.plasticseurope.org/en/resources/market-data. Accessed 21 March 2020.

Plastic Pollution Coalition. (2015). *The golden state gilds its microbead ban.* www.plasticpollutioncoalition.org/pft/2015/10/8/the-goldenstate-gilds-its-microbead-ban

Plastic Pollution Coalition. (2016). *California introduces national trash reduction act.* www.plasticpollutioncoalition.Org/pft/2015/11/6/California-introducesnationaltrash-reduction-act. Accessed on 02 September 2018.

Plastic Soup Foundation. (2016). www.beatthemicrobead.org/en/industry. Accessed on 27 January 2017.

Possatto, F. E., Barletta, M., Costa, M. F., Ivar do Sul, J. A., & Dantas, D. V. (2011). Plastic debris ingestion by marine catfish: An unexpected fisheries impact. *Marine Pollution Bulletin, 62*, 1098–1102.

Prata, J. C., Silva, A. L., Walker, T. R., Duarte, A. C., & Rocha-Santos, T. (2020). COVID-19 pandemic repercussions on the use and management of plastics. *Environmental Science and Technology, 54*, 7760–7765.

Prendergast-Miller, M. T., Katsiamides, A., Abbass, M., Sturzenbaum, S. R., Thorpe, K. L., & Hodson, M. E. (2019). Polyester-derived microfibre impacts on the soil-dwelling earthworm *Lumbricus terrestris*. *Environmental Pollution*, *251*, 453–459.

Rands, M. R. W., Adams, W. M., Bennun, L., Butchart, S. H. M., Clements, A., Coomes, D., . . . & Vira, B. (2010). Biodiversity conservation: Challenges beyond. *Science*, *329*, 1298–1303.

Rillig, M. C. (2012). Micro-plastic in terrestrial ecosystems and the soil? *Environmental Science and Technology*, *46*, 6453–6454.

Rilling, M. C., Ingraffia, R., & De Souza Machado, A. A. (2017). Microplastic incorporation into soil in agroecosystems. *Frontiers of Plant Science*, *8*, 1805.

Rochman, C. M. (2013). Ingested plastic transfers hazardous chemicals to fish and induces hepatic stress. *Scientific Reports*, *3*, 3263.

Rochman, C. M. (2015). The complex mixture, fate and toxicity of chemicals associated with plastic debris in the marine environment. In: Bergmann, M., Gutow, L., & Klages, M. (Eds.), *Marine anthropogenic litter* (1st ed., pp. 117–140). Springer.

Rochman, C. M., Kurobe, T., Flores, I., & The, S. J. (2014). Early warning signs of endocrine disruption in adult fish from the ingestion of polyethylene with and without sorbed chemical pollutants from the marine environment. *Science of the Total Environment*, *493*, 656–661. https://doi.org/10.1016/j

Rodríguez-Seijo, A., da Costa, J. P., Rocha-Santos, T., Duarte, A. C., & Pereira, R. (2018). Oxidative stress, energy metabolism and molecular responses of earthworms (*Eisenia fetida*) exposed to low-density polyethylene microplastics. *Environmental Science and Pollution Research*, *25*, 33599–33610.

Rodriguez-Seijo, A., Lourenço, J., Rocha-Santos, T., Da Costa, J., Duarte, A., & Vala, H. (2017). Histopathological and molecular effects of microplastics in Eisenia andrei Bouche. *Environmental Pollution*, *220*, 495–503.

Schopfer, L., Menzel, R., Schnepf, U., Ruess, L., Marhan, S., & Brümmer, F. (2020). Microplastics effects on reproduction and body length of the soil-dwelling nematode Caenorhabditis elegans. *Frontiers of Environmental Science*, *8*, 41.

Sharma, S., & Chatterjee, S. (2017). Micro-plastic pollution, a threat to marine ecosystem and human health: A short review. *Environmental Science and Pollution Research*, *24*, 21530–21547.

Silva, A. L. P., Prata, J. C., Walker, T. R., Campos, D., Duarte, A. C., Soares, A. M., . . . & Rocha-Santos, T. (2020). Rethinking and optimising plastic waste management under COVID-19 pandemic: Policy solutions based on redesign and reduction of single-use plastics and personal protective equipment. *Science of the Total Environment*, *742*, 140565.

Silva, A. L. P., Prata, J. C., Walker, T. R., Duarte, A. C., Ouyang, W., Barceló, D., & Rocha-Santos, T. (2021). Increased plastic pollution due to COVID-19 pandemic: Challenges and recommendations. *Chemical Engineering Journal*, *405*, 126683.

Sivan, A. (2011). New perspectives in plastic biodegradation. *Current Opinion in Biotechnology*, *22*(3), 422–426.

Song, Y., Cao, C., Qiu, R., Hu, J., Liu, M., & Lu, S. (2019). Uptake and adverse effects of polyethylene terephthalate microplastics fibers on terrestrial snails (*Achatina fulica*) after soil exposure. *Environmental Pollution*, *250*, 447–455.

Stock, V., Böhmert, L., Lisicki, E., Block, R., Cara-Carmona, J., Pack, L. K., . . . & Henderson, C. J. (2019). Uptake and effects of orally ingested polystyrene microplastic particles in vitro and in vivo. *Archives of Toxicology*, *93*, 1817–1833.

Su, L., Cai, H., Kolandhasamy, P., Wu, C., Rochman, C. M., & Shi, H. (2018). Using the Asian clam as an indicator of microplastic pollution in freshwater ecosystems. *Environmental Pollution*, *234*, 347–355. https://doi.org/10.1016/j

Sund, P., Schulze, P. E., & Syversen, F. (2014). Sources of micro-plastic pollution to the marine environment. Report no M-321/2015. Asker: Mepex Consult.

Sutherland, W. J., Clout, M., Cote, I. M., Daszak, P., Depledge, M. H., Fellman, L., . . . & Watkinson, A. R. (2010). A horizon scan of global conservation issues for 2010. *Trends in Ecology and Evolution*, *25*, 1–7.

Thompson, R. C., Olsen, Y., & Mitchell, R. P. (2004). Where is all the plastic? *Science*, *304*, 838.

UN Environment. (2014). *Ocean experts call for greater local government role in fight against marine waste.* https://europaeu/capacity4dev/unep/blog/oceanexperts-call-greater-local-governmentrole-fight-against-marine-waste. Accessed on 13 September 2018.

van Cauwenberghe, L., & Janssen, C. R. (2014). Microplastics in bivalves cultured for human consumption. *Environmental Pollution*, *193*, 65–70.

Wang, F., Zhang, X., Zhang, S., Zhang, S., & Sun, Y. (2020). Interactions of microplastics and cadmium on plant growth and arbuscular mycorrhizal fungal communities in an agricultural soil. *Chemosphere*, 126791.

Watts, A. J., Urbina, M. A., Corr, S., Lewis, C., & Galloway, T. S. (2015). Ingestion of plastic microfibers by the crab carcinus maenas and its effect on food consumption and energy balance. *Environmental Science and Technology*, *49*, 14597–14604.

Welden, N. A. C., & Cowie, P. R. (2016). Long-term microplastic retention causes reduced body condition in the langoustine, Nephrops norvegicus. *Environmental Pollution*, *218*, 895–900.

Wright, S. L., Rowe, D., Thompson, R. C., & Galloway, T. S. (2013). Microplastic ingestion decreases energy reserves in marine worms. *Current Biology*, *23*, 1031–1033.

Xia, B., Zhang, J., Zhao, X., Feng, J., Teng, Y., Chen, B., . . . & Qu, K. (2020). Polystyrene microplastics increase uptake, elimination and cytotoxicity of decabromodiphenyl ether (BDE-209) in the marine scallop Chlamys farreri. *Environmental Pollution*, *258*, 113657.

Xiong, X., Bond, T., Siddique, M. S., & Yu, W. (2021). The stimulation of microbial activity by microplastic contributes to membrane fouling in ultrafiltration. *Journal of Membrane Science*, *635*, 119477.

Xiong, X., Tu, Y., Chen, X., Jiang, X., Shi, H., Wu, C., & Elser, J. J. (2019). Ingestion and egestion of polyethylene microplastics by goldfish (*Carassius auratus*): Influence of color and morphological features. *Heliyon*, *5*, 03063.

Xu, B., Liu, F., Cryder, Z., Huang, D., Lu, Z., He, Y., . . . & Xu, J. (2020). Microplastics in the soil environment: Occurrence, risks, interactions and fate—A review. *Critical Reviews in Environmental Science and Technology*, *50*(21), 2175–2222.

Yu, Q., Hu, X., Yang, B., Zhang, G., Wang, J., & Ling, W. (2020). Distribution, abundance and risks of microplastics in the environment. *Chemosphere*, *249*, 126059. https://doi.org/10.1016/j

Zhang, C. (2017). Toxic effects of microplastic on marine microalgae Skeletonema costatum: Interactions between microplastic and algae. *Environmental Pollution*, *220*, 1282–1288.

Zhang, D., Liu, X., Huang, W., Li, J., Wang, C., Zhang, D., & Zhang, C. (2020a). Microplastic pollution in deep-sea sediments and organisms of the Western Pacific Ocean. *Environmental Pollution*, *259*, 113948. https://doi.org/10.1016/j.envpol

Zhang, K., Su, J., Xiong, X., Wu, X., Wu, C., & Liu, J. (2016). Microplastic pollution of lake-shore sediments from remote lakes in Tibet plateau, China. *Environmental Pollution*, *219*, 450–455.

Zhang, S., Liu, X., Hao, X., Wang, J., & Zhang, Y. (2020b). Distribution of low-density microplastics in the mollisol farmlands of northeast China. *Science of the Total Environment*, *708*, 135091 https://doi.org/10.1016/j.scitotenv

Zhu, L., Bai, H., & Chen, B. (2018). Micro-plastic pollution in North Yellow Sea, China: Observations on occurrence, distribution and identification. *Science of the Total Environment*, *636*, 20–29.

Ziajahromi, S., Kumar, A., Neale, P. A., & Leusch, F. D. L. (2017). Impact of microplastic beads and fibers on waterflea (*Ceriodaphnia dubia*) survival, growth, and reproduction: Implications of single and mixture exposures. *Environmental Science and Technology*, *51*, 13397–13406.

Zocchi, M., & Sommaruga, R. (2019). Microplastics modify the toxicity of glyphosate on Daphnia magna. *Science of the Total Environment*, *697*, 134194.

2 Use of Genetic Engineering in Reclamation of Waste Disposal Sites

Humaira, Saba Wani, Qamer Ridwan,
Khair Ul Nisa, and Asia Mushtaq

2.1 INTRODUCTION

Our use of diverse resources has increased due to the worrisome rise in the world population, which has also raised the demand for energy (Awasthi et al., 2022). By 2050, there will be around 3.5 billion tonnes of municipal solid waste generated, up from 2.2 billion tonnes in 2025 (Zhou et al., 2022; Yatoo et al., 2024; Jadaun et al., 2022). Throw-away creation and therapy are influenced by many socio-economic conditions, with 34% of waste being produced for a population of 16% in high-income countries. Low-income nations produce 5% of the world's waste, yet only 39% of that rubbish is collected; the remainder is disposed of in the open. The environment is impacted and polluted by these open dumping practices. Uncollected garbage costs four to seven times more to treat than it does to collect. Additionally, inappropriate waste disposal practices led to emissions of 1.6 billion tonnes of CO_2 equivalent in 2016 (Ahmed et al., 2018; Awasthi et al., 2022; Zhang et al., 2021).

Efficacious throw-away management practices improve the economy, the habitat, and public health. Nonetheless, landfills and dump yards get more than two-thirds of the waste produced. This is because processing waste can cost anywhere between $35 and $100/t. Although waste treatment has a good impact on the environment, hazardous chemicals may leak out throughout the process. For instance, burning emits poisonous chemicals that harm the environment (Singh et al., 2021; Khan et al., 2022). The need of the hour is to look at waste treatment from a different perspective than the traditional one, where the goal is to take the edge off emissions or move closer to an environmentally friendly plan. Many nations are aiming for climate-positive emissions by 2050, which must only be accomplished if the throw-away administration also implements an environmentally friendly approach. Worthwhile throw-off management can reduce global greenhouse gas emissions by up to 21% (Thakur et al., 2022). The majority of throw-away therapies by no means are budget-arily viable for several reasons, such as the cheap cost of trash disposal, the lack of

DOI: 10.1201/9781003359326-2

garbage collection and processing in many countries, the expense, and the absence or reduction of carbon taxes. Many municipalities have indicated that they may recoup up to 50% of costs, which calls into question the viability and maintainability of the suggested mechanics (Tan et al., 2021; Zhang et al., 2022).

Heavy metal emissions, as well as environmental build-up, are brought on by both natural phenomena, such as volcanic eruptions and mainland dust, and human schemes, such as quarrying, the ignition of crude oils for phosphate fertilizers, combat activities, and metal working businesses. The origins of the metals present in our environment include spontaneous degradation of the earth's crust, surface runoff, quarrying, chemical spills, inner-city overspill, untreated discharge, air debasement consequences, and bug or ailment power chemicals. The pollutants' concentrations can range from distinctly dangerous levels from an unintentional overturn to hardly perceptible levels that, following the prolonged unveiling, can harm people's well-being (Alexander, 1999; Doty, 2008). Dense metals in particular are dangerous in view of the fact that they alter DNA, and this is likely what leads to their cancerous effects in both people and animals (Knasmüller et al., 1998; Baudouin et al., 2002; Hooda, 2007). Because dense metals are not biodegradable and remain in the soil for generations without intervention, there is a potential concern. Most contaminated areas must be cleaned up to restore the region and reduce the number of hazardous substances that enter the food chain. Currently, several engineering-based remediation procedures, for example, soil flushing, incineration, or pump and treat systems, can be utilized to clean up metal-adultered soils (Hooda, 2007).

Excavation, transportation, soil cleaning, eradication, pumping, cleaning dirty water, adding reactants like H_2O_2 or $KMnO_4$, and burning are some of the engineering techniques used to clean up contaminated areas. Commercial premises that are contaminated are frequently left unclean because of the steep price of restoration technology. As opposed to organic pollutants, which can be broken down into innocuous small molecules, harmful elements like Hg, As, Cd, Pb, Cu, and Zn cannot be changed by any bioprocesses and thus persist in the natural environment, making them particularly challenging to remove from the soil, water, and air (Krämer and Chardonnens, 2001). Metal residues in various environments may contaminate groundwater and exterior water or especially enter the feeding cycle via cultivars grown on the said land (Lin et al., 1998).

A more common technique of clean-up is enhanced bioremediation, which includes the inclusion of certain microbial strains that are known to break down the contaminant. Together, bacteria and fungi may utilize a huge diversity of chemical compounds. However, some requirements must be satisfied for microbe-based bioremediation to be successful at a specific location. These include the chemical's accessibility or bioavailability; the existence of inducers to trigger the production of the required enzymes; and the capacity of the microorganisms with the requisite physiological activity to survive in that habitat. Numerous organic contaminants are resistant to break down and cannot serve as the exclusive source of carbon (Doty, 2008). The presence of these substrates is occasionally required for the expression of the genes since the contaminants are occasionally processed by enzymes together with other substrates. This criterion presents a challenge if the chemical that induces the reaction is also a dangerous pollutant like phenol. Additionally, the availability of adequate carbon and energy sources is necessary for bioremediation. It is frequently necessary to pump hundreds of volumes of feed supply, like molasses, through the area to promote microbial growth (Doty, 2008).

As autotrophic creatures, plants absorb a variety of hazardous substances, both natural and artificial, as a byproduct for which they have evolved a variety of detoxifying systems (Eapen et al., 2007; Van Aken, 2008). Plants have naturally occurring defensive mechanisms against the many allelochemicals emitted by adversarial species, such as bacteria, insects, and other plants, which are likely where the enzymes break down pollutants (Singer, 2006; Van Aken, 2008). This perspective gives rise to the idea of phytoremediation, where plants are used as organic, photovoltaic panels and treatment systems for cleaning up contamination in the environment (Pilon-Smits, 2005; Van Aken, 2008).

The method was first created to remove thrust metals from the soil, but it has subsequently been shown to effectively handle organic substances like explosives, polynuclear aromatics (PNAs), and chlorinated solvents (Pilon-Smits, 2005; Salt et al., 1998; Van Aken, 2008). In addition to removing toxins from soil, phytoremediation involves other procedures, including enzymatic breakdown, that may result in the detoxification of contaminants (Pilon-Smits, 2005; Dietz and Schnoor, 2001; Van Aken, 2008). Although showing considerable pledge, very sluggish expulsion levels and the possibility of harmful chemical build-up inside plants may have restricted the use of bioremediation (Eapen et al., 2007; Van Aken, 2008). Metal phytostabilization may simply consist of reducing discharge brought by above-ground foliage, preventing leaching through plant transpiration-generated upward water flow and reducing soil eroding through soil stability by plant roots (Vassilev et al., 2004; Martínez et al., 2006).

A developing technology called phytoremediation, which uses plants to remove, alter, clear out, or keep upright toxins such as hydrocarbons found in water, sediments, or soils, is both economical and ecologically benign (Cunningham et al., 1997; Cherian and Oliveira, 2005; Mello-Farias and Chaves, 2008). Due to its possible advantages over existing machinery like soil replacement, solidification, and cleansing techniques, this technology has gained increasing interest (Yang et al., 2005; Mello-Farias and Chaves, 2008). The benefits of phytoremediation over conventional bioremediation by microbes include the fact that plants, as autophytes with vast biomass, need little nutritional input and don't erode soil with water or wind, which prevents toxins from spreading (Pulford and Watson, 2003; Cherian and Oliveira, 2005; Mello-Farias and Chaves, 2008). Additionally, flora provides supplements to the microbes in the rhizosphere, fostering the development and conservation of a microbiological population for deeper pollutant reclamation (Cherian and Oliveira, 2005; Mello-Farias and Chaves, 2008). Utilizing both the transformation, bioavailability, and contaminant retention and decay capabilities of the overall plant system as well as the distinct, selective, and naturally occurring absorption capacities of plant root systems, In addition to being visually beautiful, phytoremediation is often ten times less expensive as opposed to alternative physical, biochemical, or thermal remediation processes since it is carried out in situ, is powered by sunlight, and requires little upkeep once it is established (Hooda, 2007).

2.2 RESURRECTING WASTE DISPOSAL SITES

2.2.1 PHYTOREMEDIATION

Phytoremediation is one of the best methods selected among the top ten genetic engineering techniques for the environmental remediation of new pollutants and nanoparticles (NPs) (Eapen et al., 2007). The elimination of environmental toxins by phytoremediation using hyperaccumulator plants is now broadly acknowledged as an environmentally acceptable and economically viable sustainable solution (Rai and Kim, 2020). Hyperaccumulators often assemble metal ions to more than 0.1% to 1% of the flora's dry mass (Meagher, 2000). When developing in their native environment, the hyperaccumulators of thrash metals/metalloids have a certain cutoff in dry weight biomass, such as > 100 g g^{-1} Cd, > 300 g g^{-1} Co/Cu/Cr, > 1000 g g^{-1} Ni/As/Pb, > 3000 g g^{-1} Zn, or > 10,000 g g^{-1} Mn (Reeves et al., 2018).

In aerial plant parts, general processes (separation, characterization, phytovolatilization, and phytotransformation) engaged in phytoremediation of organic/inorganic/evolving ecological pollutants of the main issue are shown in Figure 2.1. Transgenic phyto/rhizo-remediation for a variety of environmental pollutants involves an intricate process involving the connection between the root system with an endosymbiont microbiological community underground. This association is mediated by several enzymes, such as nitrate reductases, peroxidases, and oxidases (Lima et al., 2019; Tang et al., 2019).

To get over some obstacles to its quick adoption, hyperaccumulator-based phytotechnology can be further improved by genetic engineering (referred to as "gene remediation"). To further improve the effectiveness of phytotechnologies, understanding the molecular processes of transgenic plants is also essential (Rai, 2018a, 2018b). So, in the preceding several years, there has been an extensive survey regarding how to develop biotechnology through gene manipulation to speed up phytotechnology. To improve the chances for the genoremediation of environmental pollutants, gene regulation-generated transporters, enzymes, and molecular processes are being researched (Cherian and Oliveira, 2005). Transgenic hyperaccumulators can also be used to lessen the dangerous effects of environmental pollutants on the well-being of people (Kumar et al., 2015; Rai and Kim, 2020).

Rapid, sustainable remediation of many environmental toxins is intrinsically tied to bioremediation with the aid of hyperaccumulators, their genetic framework, and gene engineering. Concurrently, genetic modification remarkably changed the display of phenotype to raise the nutrient content in hybrid flora via means of bio-fortification, particularly in light of the context of significant minor components (Antosiewicz et al., 2014). Using bioengineering and crop modification based on new genes, it has been feasible to transform bioremediation and increase crop yield, enhance crop nutritive value (e.g., bio-fortification), and create biopharmaceuticals (e.g., plant biotechnology and Agrobacterium-mediated gene transfer) (Rai and Kim, 2020). All of these processes are inseparably linked to human health and welfare (Figure 2.1).

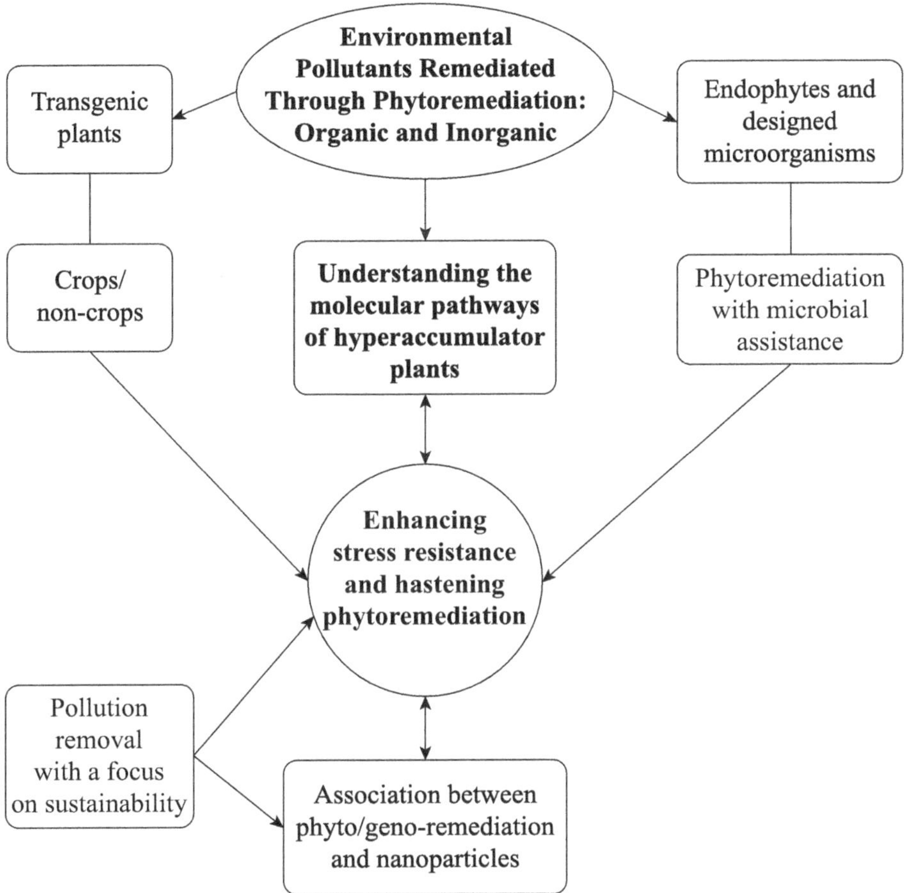

FIGURE 2.1 Environmental pollutants cleaned up by plants through molecular mechanisms

2.3 ORGANIC POLLUTANTS' MOLECULAR PROCESSES AND GENETIC TRANSFORMATION

It is widely acknowledged that organic pollutants contribute to the deterioration of the ecosystem. Unravelling the molecular and genetic pathways requires an understanding of the risks associated with organic pollutants. Organic xenobiotics can harm biotic food chain components by deteriorating the quality of the environment (Hussain et al., 2018; Rai et al., 2018a). According to Matteucci et al. (2015) and Hussain et al. (2018), organochlorides are highly constant in human surroundings, and polychlorinated biphenyl (PCB) is frequently utilized as a starting point for pesticides in the agricultural industry. According to a study, persistent organic pollutants are organic impurities with a particular fondness for phospholipids like lipophilics that are resistant to photochemical destruction. The "dirty dozen" group has a wide variety of permanence, toxicity, bioavailability, and amplification across several energy pyramids (Hussain et al., 2009; Martin et al., 2016; Fernandez-Cruz et al., 2017; Hussain et al., 2018). Similarly, a class of insecticides known as organophosphates (including paraoxon, parathion, and others) operates as neurotoxins, degrading acetylcholine esterase and causing bladder cancer. As a result, they pose substantial health hazards by interfering with regular nerve communication (Webster et al., 2002). Natural explosives and additional chlorinated compounds are deleterious to individual skin, liver, and immunological systems (Doty et al., 2017; Hussain et al., 2018).

Sadly, there isn't a biological transmitter for organic environmental pollutants inside plant cells. Instead, because organics and xenobiotics are man-made, their passive reception takes place. Hydrophobic organic pollutants can attach to the hemicellulose in the cell membrane and the lipid bilayers in the plasmalemma (Hussain et al., 2009). According to an anticipated index, the transpiration stream concentration factor, the pollutants could also be transferred to plant aerial parts and the channels of the transpiration stream (Cherian and Oliveira, 2005; Hussain et al., 2009). The organics can then be mineralized by rhizospheric microorganisms that are linked to particular plant species as accumulators, such as PNAs, PCBs, and petroleum hydrocarbons. Additionally, the importance of metabolic conversion (via nitroreductases, dehalogenases, laccases, peroxidases, etc.) in transforming dangerous xenobiotics into chemical compounds with comparatively lower toxicity cannot be overstated (Cherian and Oliveira, 2005).

Three stages can be distinguished in the conversion of organic materials: characterization/chemical changes through oxidations, reductions, and hydrolysis; recombination of foreign xenobiotics with sugars, glutathione, and amino acids; and categorization/compartmentation/segregation.

Explosive environmental contaminants are among the most dangerous environmental pollutants, and they are frequently found in military training and combat areas. Indian mallow (*Abutilon avicenna*), Indian joint vetch (*Aeschynomene indica*), and barnyard grass (*Echinochloa crusgalli*) were among the terrestrial plants for which the potency of bioremediation for the organic volatile trinitrotoluene (TNT) was witnessed at different levels (Lee et al., 2007). As an illustration, genetically engineered plants that had the old yellow enzyme (OYE3) gene from the fungus *Saccharomyces cerevisiae* inserted and expressed it more quickly progressed through TNT phytoremediation (Zhu et al., 2012). Some cultivated everlasting herbs (*Panicum virgatum* and *Agrostis stolonifera*, including in xplA and xplB vectors) also provided improved hexahydro-1, 3, 5-trinitro-1, and 3, 5-triazine (RDX) elimination (Zhang et al., 2017). Such results are crucial for averting deeper groundwater contamination brought on by RDX seeping from the soil. *Drosophila melanogaster*-based glutathione transferase (DmGSTE6) was also expressed in *Arabidopsis* (Table 2.1) to enhance the bioremediation potency of a fractious and potentially harmful volatile, 2,4,6-TNT (Tzafestas et al., 2017). A thorough examination of the molecular structure revealed that the flavodoxin-encoded reductase, or xplB, was coupled to the cytochrome P450-active xplA gene by a fusion with a flavodoxin redox partner (Rylott et al., 2011b). Genetically modified flora with xplA genes was sufficiently powerful for the phytoremediation of the RDX in polluted military sites, primarily in anaerobic environments (although the ultimate breakdown product lines can be changed if oxygen is available (Rylott et al., 2011a; Chatterjee et al., 2017).

2.3.1 ORGANIC POLLUTANT PHYTOREMEDIATION WITH MICROBIAL ASSISTANCE

A crucial part in the detoxification of organic xenobiotics is played by microbial consortiums connected to the underground biomass of plants. Even though their function only pertains to organic rhizome-/phyto-remediation instead of inorganics, microbes are critical in the breakdown of xenobiotics (Table 2.1). For instance, through heterologous overexpression in tobacco plants, glutathione S-transferase (GST) from the fungus *Trichoderma virens* increased the effectiveness of phytoremediation for refractory PAHs (such as Anthracene) (Dixit et al., 2011). If contrasted to transgenic hairy root technology, TPX1 (i.e., tobacco transgenic hairy roots) in combination with arbuscular mycorrhizal fungus also increased the bioremediation efficacy of phenol (Ibáñez et al., 2011). In this respect, powerful anti-oxidative enzyme networks that defend against phenol-induced oxidative stress can be credited to the transgenic tobacco hairy roots' high tolerance to hydroxybenzene.

Endophytic bacteria are a well-known essential part of the rhizomicrobiota that aids in the bioremediation of organic contaminants. To rectify petroleum hydrocarbons from soil polluted by diesel, four strains of endophytic bacteria (*Pseudomonas* sp., *Pantoea* sp., *Stenotrophomonas* sp., and *Flavobacterium* sp.) proved to be successful in reducing the harmful effects (Mitter et al., 2019). *Pseudomonas* sp. is one of such endophytic microorganisms, though was found to have a far greater impact on removing crude oil from soil using microbe-assisted phytoremediation. Several genes, including CYP153 and alkB, have been found to increase strain resistance in this setting of hydrocarbon clean-up (Mitter et al., 2019). Along with these genes, *Pseudomonas* sp. also has the genes Nah, Pah, and Phn that can improve phytoremediation that is supported by microbes (Abbasian et al., 2019; Mitter et al., 2019).

The most efficient and environmentally friendly method of enzymatic biodegrading organic materials is phytoremediation, which is engineered with microorganisms (such as plant-based endophytic and rhizospheric bacteria) (Hussain et al., 2018). Extra-chromosomal DNA, known as plasmids, was found to contain the genes for the enzymes responsible for remediation in Pseudomonas strains (Hussain et al., 2009, 2018). *Burkholderiacepacia*, a recombinant class of bacterium, may break down herbicides that are harmful to human health, like Agent Orange (Chauhan et al., 2008). The effectiveness of multiple gram-negative N-fixing symbiotic Rhizobiales (22 *Rhizobium* strains, including *Rhizobium phaseoli*) bacteria in the cleaning up of phenolics such as catechol and salicylic acid led to their initial identification (Teng et al., 2015). Their increased effectiveness was demonstrated by the bioremediation of PCBs (>70%) using a strain of *Sinorhizobium meliloti* genes.

Recent developments in molecular biotechnology have significantly altered the potential for phytoremediation of dangerous environmental toxins in light of omics techniques (e.g., environmental genomics, environmental transcriptomics, and metaproteomics) (Van Aken, 2011; Sarma et al., 2019). Cutting-edge research in this area can be seen in the *Dietzia* genome, which generates metabolites that give rise to PAH phytoremediation and contains the alkB gene, which codes for 1-hydroxylase, and the CYP153 gene, which codes for P450 alkane hydroxylase (Hernández-Vega et al., 2017; Sarma et al., 2019). Gene editing improves the plant–endophyte connection and the vascular tissues' and rhizosphere's stress resistance to organic compounds. Naphthalene, toluene, and trichloroethylene (TCE) phytoremediation were improved by the inoculation of genetically modified endophytes (Germaine et al., 2009; Ijaz et al., 2016; Feng et al., 2017). Additionally, this recombinant flora aids in the invasion of endophytes (Rasche et al., 2006; Feng et al., 2017). Additionally, several biomolecules, such as cyanides, cyanide hydratase, formamidase, nitrilase, nitrile hydratase, cyanide dioxygenase, cyanide monooxygenase, cyanate, and nitrogenase, are involved in the breakdown of cyanide. The potential for cyanide mycoremediation is projected to increase with additional developments in protein engineering and omics, though (Sharma et al., 2019).

2.3.2 Organic Contaminant Phyto/Genetically Based Remediation

When 2,3-dihydroxy biphenyl-1,2-dioxygenase (BphC.B) was overexpressed in alfalfa (through *Agrobacterium*-regulated transition, employing CaMV 35S as a promoter) and separated from a soil metagenomic bridge, the clean-up of combined chlorinated organics (i.e., 2,4-DCP/PCBs) was successfully completed (Wang et al., 2015). Additionally proven to provide PCB (e.g., dioxin) response in hybrid flora is the aryl hydrocarbon receptor, which is derived from mammals (Bao et al., 2013). Recombinant *Arabidopsis* demonstrated the co-expression of a retinoid X receptor and a pregnancy X receptor for coping with strain when exposed to non-dioxin-like PCBs. It's interesting to note that, in this model plant, the improved pectin synthesis was what truly provided the resistance. The phytoremediation capacity of PCB126 was also improved by GUS gene exchange (with glucuronidase activity) from *Solidago canadensis* into *Arabidopsis* (Shimazu et al., 2014).

Arabidopsis thaliana is a better plant for phytoremediation of persistent organic carcinogens like trichlorophenol (TCP), which has an antibacterial character and is a raw material for chlor-alkali agrochemicals. In this context, the overexpression of a protein similar to a disulfide isomerase, AtPDIL1–2, was credited as the biochemical pathway for improved phytoremediation of TCP by *Arabidopsis* (Peng et al., 2017). It was discovered that overexpressing the dehalogenase genes from the bacterium *Xanthobacter autotrophicus* enabled the phytoremediation of refractory 1,2-dichloroethane in hybrid tobacco. Similar to this, in a soiled setting, tolerance to TCP was created by *A. thaliana* overexpressing a cotton laccase gene to increase secretory laccase activity (James and Strand, 2009).

When organics like phenanthrene bioaccumulate in food crops like wheat, it can have a significant impact on crop production, food sustainability, and individual well-being (Rai and Kim, 2020; Shen et al., 2019). *Triticum aestivum* exposed to phenanthrene showed chlorosis and chloroplast distortion, according to molecular research. In this situation, phenanthrene treatment for wheat crops resulted in a complex up-regulation of 261 proteins, which caused thylakoid breakdown (Shen et al., 2019). Similar to that, the biotransformation of a xenobiotic (azoxystrobin) by maize root culture allowed researchers to examine the phytoremediation of the substance (Gautam et al., 2018). CsGSTUs (derived from *Citrus sinensis* or sweet

TABLE 2.1

Ecological pollutant clean-up via genetic engineering of genes and gene-derived products from microbes, plants, and mammals

Microbes/Higher plants/Mammals as an Origin of Gene	Gene/s	Selected flora	Enzymatic carrier expression for remediation	Phytoremediation effects	References
Streptomyces	CYP105 A1	*Nicotiana tabacum*	Cytochrome P450 monooxygenase	Improved stress tolerance to organic pollutants, such as sulfonylurea, and better phytoremediation than wild varieties	O'Keefe et al., (1994); Hussain et al. (2018)
Bacteria	AtzA	*Medicago sativa; N. tabacum*	Atrazine chlorohydrolase	Atrazine stress resistance and improved phytoremediation/photo-degradation	Wang et al. (2005); Hussain et al. (2018)
Pseudomonas pseudoalcaligenes	ophc2	*N. tabacum*	Organophosphoru s hydrolase (OPH)	Organophosphorus phytoremediation that is improved (methyl parathion)	Wang et al. (2008)
Bacillus subtilis	Protox	*Oryza sativa*	Protoporphyrinog en IX oxidase	Improved strain resistance and phytoremediation of the herbicide diphenyl ether	Jung et al. (2008)
Escherichia coli	gshI	*Populus xcanescens*	γ-Glutamylcysteine synthetase	Greater stress tolerance brought about by paraquat exposure due to increased GSH concentration	Gyulai et al. (2014); Ibañez et al. (2016)
Fusarium solani	lin A	*F. solani*	Dechlorinase	Clean-up of organic contaminants, such as dichloro-diphenyl trichloroethane (DDT), effectively	Mitra et al. (2001)
Trichoderma virens	GST	*N. tabacum*	Glutathione transferase	Anthracene phytoremediation with improved plants	Dixit et al. (2011)
Enterobacter cloacae	XplA, XplB	*Rhodococcus rhodrochrous 11Y*	Cytochrome P450; reductase	Improved RDX phytoremediation	Rylott et al. (2015); Chatterjee et al. (2017)
Saccharomyces cerevisiae	CUP1	Tobacco & cauliflower	Metallothionein for metal removal (Cd and Cu)	Significantly increased ability for Cu in tobacco and Cd in the cauliflower to be absorbed	Thomas et al. (2003); Cherian et al. (2012)

(Continued)

TABLE 2.1 (*Continued*)

Ecological pollutant clean-up via genetic engineering of genes and gene-derived products from microbes, plants, and mammals

Microbes/Higher plants/Mammals as an Origin of Gene	Gene/s	Selected flora	Enzymatic carrier expression for remediation	Phytoremediation effects	References
Caenorhabditis elegans	Phytochelat in synthase (PCS1)	*N. tabacum*	As (v)	Elevated As resistance/phytoremediation	Wojas et al. (2010)
Staphylococcus aureus	mer A	*Populus alba* × *Populus tremula var. glandulosa*	Mercuric ion reductase	Improved ionic Hg resistance including its transformation to the lesser dangerous version Hg (0)	Luo et al. (2016)
Pseudomonas fluorescens Sasm05	SaNramp1, SaIRT1	*Sedum alfredii*	The iron-regulated carrier and intrinsic immunity protein of macrophages	In an account of excessive expression of SaNramp1 as well as SaIRT1 during metal stress accelerated Zn phytoremediation	Wang et al. (2019)
Arabidopsis	PtUGT72B 1	*Populus trichocarpa* (Black cottonwood)	Glucosyl transferase	Utilizing plants to safely remove trichlorophenol (TCP)	Rylott et al. (2015)
Arabidopsis	AtGSTF2, AtGSTU1, AtGSTU24	*Arabidopsis* seedling	Glutathione S transferases	Enhancing remediation of a class of dangerous pesticides, such as acetochlor, metolachlor, and triazine, in *Arabidopsis* root cells through pesticide GSH conjugate interactions	Hussain et al. (2009)
Maize (*Zea mays*)	GST I	Tobacco	Glutathione S-transferase	The resulting transgenic plant has improved alachlor resistance	Hussain et al. (2009)
Homo sapiens	CYP1A1, CYP1A2	Tobacco	Cytochrome P450 is an enzyme with the capability for restoration in both plants and animals. Gene store exhibits a wide range of variation	N-dealkylation and deisopropylation for the elimination of metamitron and atrazine clean-up	Hussain et al. (2009)

Rabbit (*Oryctolag us cuniculus*)	Cytochrome P450 (2E1)	*Arabidopsis*	Cytochrome P450	Excellent removal of chlorinated hydrocarbons such as DDT and trichloroethylene (TCE)	Rylott et al. (2015)
Pig (*Sus scrofa*)	CYP2B22 and CYP2C49	Rice cultivars	Cytochrome P450	Improved herbicide remediation results from hastened metabolism	Hussain et al. (2009)
Malus domestica	SPDS1	*Pyrus communis*	Spermidine synthase	Cu phytoremediation is improved by MdSPDS1 overexpression and spermidine build-up	Luo et al. (2016)
Arabidopsis thaliana	HMA3; ABCC1; MTP3	*A. thaliana*	Metal-tolerant enzyme, ATP-binding cassette transporter, and heavy metal ATPase	Metal–ligand complex, which causes vacuolar sequestration and accelerated phytoremediation of Cd; MTP3 expression, which causes Zn bio-accumulation	Morel et al. (2009); Park et al. (2012)
Spinacia oleracea	Gene expressing Cysteine synthase	*N. tabacum*	Cysteine synthase	Increasing metals' resistance to stress for Cd, Se, and Ni	Kawashima et al. (2004)
O. sativa	NRAMP1	*A. thaliana*	Biological tolerance linked to the macrophages	Due to OsNRAMP1 overexpression and concurrent co-expression of the AtABCC1, AtABCC2, and AtHMA4 genes, there has been an extraordinary rise in As and Cd phytoremediation via xylem loading. This has led to simultaneous Fe and Mn oxidation	Luo et al. (2016)
Pteris vittate	ACR3	*A. thaliana*	Arsenate reductase	Enhanced stress resistance when subjected to the hazardous metalloid arsenic, leading to a rise in phytoremediation	Zhao et al. (2014)
Phragmites australis	PaPCS1	Tall Fescue	Phytochelatin synthase	Increased phytoremediation of Cd is caused by increased phytochelatin synthase build-up	Kumar et al. (2015)

orange) improved the phytoremediation/strain resistance potential of xenobiotics (fluorodifen, a diphenyl ether herbicide) against many abiotic conditions such as salt stress and water deficit when abundantly expressed in tobacco (Cicero et al., 2015). The GmGSTU4 isoenzyme from the versatile GST group, which is observed in agricultural crops like *Glycine max* (soybean) on *Agrobacterium*, was also discovered by (Benekos et al., 2010) to facilitate transitions in tobacco and boost the phytoremediation potency of specific herbicides such as chloroacetanilide and diphenyl. Linuron, a phenylurea herbicide utilized to regulate weeds in agroecosystems, is also renowned for having the ability to harm the environment and biota. However, the introduction of the human CYP1A2 gene into *A. thaliana* helped with its phytoremediation (Azab et al., 2018). Similar to this, it was demonstrated by (Duhoux et al., 2015) that hybrid flora with genes addressing GST can be used to remove herbicides from the environment. Biphenyl dioxygenase enzyme gene modifications can also effectively remove PCB congeners (of various chemical structures) (Hussain et al., 2018).

A biochemical organic byproduct called cyanide has considerable poisoning. High cyanide exposures have been demonstrated to be detrimental to reactive defenses. When these reactive pathways malfunction, it can disturb the overall redox homeostasis mechanisms since they are in charge of producing reactive oxygen species in plants under different abiotic stimuli (Kebeish et al., 2017). Cytochrome P450 monooxygenases are capable of cleaning up organic contaminants such as 9-methylanthracene, acenaphthalene, fluorine, and naphthalene (CYPs). This enzyme's capacity for phytoremediation has been improved by the precise changes made by gene modification (Chica et al., 2005). The most common application of CYP2E1 is the cultivation of transgenic plants for the clean-up of biodegradable toxins (Zhang et al., 2013). Additionally, it is shown that genetically modified endophytes, such as *Burkholderia cepacia* G4 and *Pseudomonas putida* VM1441 (pNAH7), can efficiently remediate organics such as toluene and naphthalene (Germaine et al., 2009).

Sequence-specific endonucleases can have a big impact in the field of genome editing by delivering desired features when organic and inorganic contaminants are remedied using plants. CRISPR, which stands for clustered regularly interspaced palindromic repeats, is a technology that can be used to potentiate the required features for phytoremediation (Basharat et al., 2018). Likewise, CRISPR may operate transcriptional activation or repression to control the genomic reprogramming of hyperaccumulator plants (CRISPRi and CRISPRa). This technique can significantly increase the ability of environmental pollutants to be cleaned up by plants (Basharat et al., 2018). Additionally, this biotechnological advancement in CRISPR DNA formatting can increase the potential outcome of renewables and phytomining, which are closely related to phytoremediation.

2.4 INORGANIC CONTAMINANT MOLECULAR PROCESSES AND GENETIC TRANSFORMATION

Acyl-CoA-binding protein (ACBP), a crucial peptide engaged in the molecular mechanisms of phytoremediation, was first identified as a neuropeptide that is present in prokaryotes and eukaryotes (Du et al., 2016). This housekeeping family peptide, ACBP, is crucial for plant development, including cuticle formation, germination, and seedling growth. It can also increase a plant's ability to withstand stress (Du et al., 2016). Understanding the molecular level foundation of phytoremediation, comparative proteome analysis can also be performed.

According to reports, the NRAMP family of proteins plays a significant part in the phytoremediation of metallic pollutants. For instance, after being overexpressed in A, there was a finding that the SaNramp6 cDNA obtained from *Sedum alfredii* conferred higher Cd tolerance in *A. thaliana* (Chen et al., 2017). Using phytotechnology, this chemical development can assist in the decontamination of inorganic pollutants in the form of transgenic plants based on SaNramp6. Additionally, elevated mRNA amounts of various genomic transporters, including ZIP2, ZIP6, NRAMP1.1, NRAMP 1.3, and hydrogen ATPases connected to cellular membranes, were found to be a mediator of the accelerated phytoremediation of transgenic poplar (Luo et al., 2016). Takahashi et al. (2014) also made an effort to create Anjana Dhan, an OsNRAMP5 RNAi-grown rice crop variety (A5i). To clarify the fate/phytoremediation mechanisms of Cd, these authors tracked its influence using positron-emitting tracer imaging methods. They demonstrated that OsNRAMP5 overexpression had no negative effects on biomass. The Cd values in A5i were notable in the shoots, suggesting a two-fold improvement over native flora.

Vegetation from marshes and marine ecosystems may indeed be helpful for heavy metal bioremediation (Singh and Rai, 2016; Ekperusi et al., 2019). However, research has seldom ever concentrated on molecular clarification or genetic engineering to improve their repair capacity. Duckweed's resistance to beta irradiation distress was also validated using cell wall remodelling, anti-oxidative reactions, and genetic repairing (Ekperusi et al., 2019). Duckweed, or Lemnaceae, is the smallest group of angiosperms, and sequencing its genome has shown size variations (ranging from 150 Mb in *Spirodela polyrhiza* to 1,881 Mb in *Wolffia arrhiza*) and significant changes in the biochemical processes that underlie phytoremediation (An et al., 2018). For instance, in this regard, the glutamate synthase gene family showed an increase in ammonia uptake (An et al., 2018). Additionally, the transcriptome evaluation of the *Spirodela*, *Landoltia*, and *Lemna* members of the duckweed family under various stresses demonstrated their aptitude for phytoremediation and bioenergy (An et al., 2018).

The nicotianamine synthase (SaNAS1) gene produces the non-protein amino acid nicotianamine, a metal chelator, which is essential for maintaining the metal homeostasis in the *S. alfredii* hyperaccumulator ecotype (Chen et al., 2019). Additionally, it was discovered in this study that SaNAS1 gene was substantially expressed in yeast mutants and *Arabidopsis* plants after it had been cloned. To improve the phytomediation of Cd and Zn, the SaNAS1 gene was overexpressed, which led to higher amounts of nicotianamin E (Chen et al., 2019). The NRAMP is an overexpressed form of the PaPCS/PaGCS (phytochelatin synthase/a glutamyl cysteine synthetase) gene from a well-known novel wetland flora; for example, *Phragmites australis* reed in grass-tall fescue (*Festuca arundinacea*) greatly improved Cd phytoremediation potential. High phytochelatin production was linked to this (Zhao et al., 2014).

Recent research on the biochemical and genomic potentials of Cd in the hyperaccumulator green food *Brassica rapa* (BrHMA3) (Zhang et al., 2019). The aforementioned study discovered that the BrHMA3 coding sequences controlled the redistribution of Cd from the base to the branches. It's fascinating to note that the BrHMA3 gene's promoter sequences did not play a role in the sustained relocation that resulted from the gene's overexpression in *Arabidopsis* and yeast. Moreover, BrHMA3 haplotypes were beneficial in molecular marker-assisted mating efforts to significantly lower the Cd mobility in succulent shoots, reducing the danger to human health from food intake (Zhang et al., 2019).

Accumulation of metallic pollutants in the vacuole, particularly As and Cd, is regarded to be the underlying biochemical pathway for bioremediation and efficient agricultural output (Tan et al., 2019). The vacuole is thought to represent the widest plant organelle in the advanced cell architecture of plants and is home to a large number of protein transporters for stress tolerance and phytoremediation when exposed to environmental contaminants, including copper transporter COPT5, multidrug, and toxic compound extrusion, transporters, vacuolar sugar transporters (AtSuc4), and vacuolar iron transporter (Tan et al., 2019).

2.5 PHYTOREMEDIATION OF SOIL WITH VARIOUS POLLUTANTS (BOTH ORGANIC AND INORGANIC)

To increase a wild plant species' tolerance to stress caused by contamination exposure, phytoremediation is used to overexpress genes from hyper-accumulating species of wild plants in non-accumulating wild plant species (Antosiewicz et al., 2014). The phytoremediation of organic/inorganic toxins is explored as a few power-challenging, environmentally benign, and economically viable solutions accompanying the appearance of novel living apparatus including genomic modifications (Teng et al., 2015; Rai and Kim, 2020). So, as the withdrawal of organics and inorganics separately, phytoremediation is extensively acknowledged as an economical grassy method (Yadav et al., 2018).

To identify phytotechnological solutions for areas contaminated with complex circumstances and various environmental toxins, numerous different investigations have been conducted. The development of the genoremediation of various pollutants also depends on the microbiological variety's function, perhaps on its own or in complex connection with plants (microbiome). Additionally, specific biological strategies, such as N_2-fixing rhizobia combined with a variety of advantageous microbial consortiums, can work in concert to enhance the capability for phytoremediation against environmental contaminants that are both organic and inorganic (Teng et al., 2015; Basu et al., 2018). In summary, there is a strong need for additional biotechnological surveys in the area of the general clean-up of numerous pollutants using a microbiome. This issue of numerous pollutants has been covered in several researches. GST and human cytochrome P450 2E1 (CYP2E1) genes were found to offer improved phytoremediation ability to heterogeneous environmental pollutants like Hg along with TCE in altered alfalfa plants by (Zhang et al., 2013). Combining the transcription of the GST gene with the *Agrobacterium*-mediated conversion of CYP2E1 enhanced the plant's ability to remove both inorganic toxins like Cd and composite organic pollutants like TCE (Zhang and Liu, 2011).

The ability to limit the harm caused by various environmental pollutants at the cellular, metabolic, and genetic domains, genetic amplifications in transformants was tested. As a result, these initiatives will provide a boost for the promotion of phytoremediation. The stress tolerance strategies of transgenic such as rice are concentrated in the root/rhizome areas because they prefer to decrease metals being transferred in shoots. The goal of molecular breeders and molecular designers is to increase the ultimate load resistance as a result, allowing the phytoremediation machinery to proliferate successfully in the root or rhizosphere areas of food crops (Luo et al., 2018; Rai and Kim, 2020). For such reason, elevated levels of metal-binding transporter proteins and enzymes are primarily intended to minimize oxidative damage. Additionally, metabolic engineering allows transgenic crop lines to maintain adequate amounts of crucial minerals and nutrients (Yang et al., 2018). With the use of microbe-assisted phytoremediation, the genoremediation attempts of environmental toxins are carefully coordinated and facilitated (Cai et al., 2019).

2.6 UTILIZING A TRANSGENIC STRATEGY TO RECLAIM OR RESTORE MINE WASTES

One strategy for contemporary plant rejuvenation and bioremediation projects is the use of transgenic methods. Primarily, there must already be a conversion procedure for the species of plants, as developing one could hold for a year or longer. Several appropriate species include yellow poplar (*Liriodendron tulipifera*), Indian mustard (*B. juncea*), poplar (*Populus* spp.), and cordgrass (*Spartina* spp) (Pilon-Smits and Pilon, 2002). In addition to being incredibly helpful for abiotic stress tolerance, genetically engineered flora can be exploited as reservoirs for novel genetic constitutions. Genetic engineering uses a wide range of sophisticated procedures, yet its fundamental ideas are surprisingly straightforward. However, vital information includes the control of gene regulation, the physicochemical operations, as well as the security of the genome and genetic products that will be used. The effective execution of the procedure of genetic modification necessitates several phases, including nucleotide (DNA/RNA) isolation, genetic copying (synthesis of DNA fragments, linking to a carrier, multiplication in a host organism, including picking of the required sequence), genome architecture, and encapsulation. The said DNA fragments will either enhance or terminate the expression of the inserted gene. It is possible to design and package genes by, among many other things, adding additional promoters, selectable marker genes, reporter genes, gene controller sections, introns, and organelle-restricted sequences. To enable the production of numerous copies of the gene module, the desired gene is encapsulated and then given to bacteria. The DNA isolated from the microbial colonies is subsequently used to convert plant cells by particle bombardment. *Agrobacterium tumefaciens*, a soil bacterium, is used to alter plants. The left and right margins of a binary vector should be utilized to replicate the entire gene package between each one. This will enable the *Agrobacterium* to be processed in a way that really only allows the transfer-DNA (T-DNA) to be integrated inside the plant genome (Figure 2.2). Sarangi et al. (2009) highlighted the use of transgenic plants in phytoremediation to eradicate heavy metal toxins from

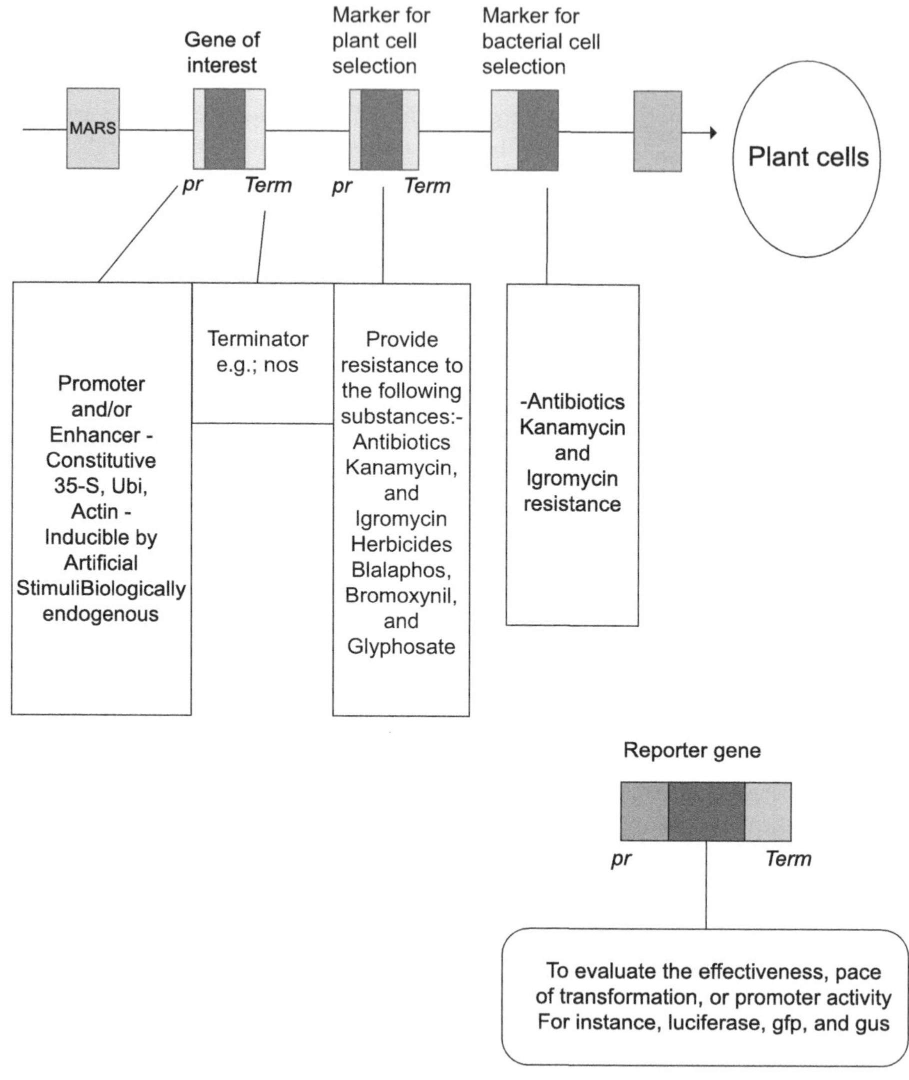

FIGURE 2.2　　Gene architecture for genetic modification

the ecosystem, such as arsenic and chromium. The plant species have traits that are advantageous, including the capacity to gather metals out of above-ground regions, resistance to the concentration of gathered metals, rapid progression, large output, widely dispersed extensive root network, and ease of reaping for phytoremediation studies. The potency of inherently altered species of plants to accumulate or detoxify mercury, cadmium, lead, selenium, and arsenic has been described in various papers (Pilon-Smits et al., 2000; Dhankher et al., 2002; Kawashima et al., 2004). A methodology was created by Samantaray et al. (2001a, 2001b) for chromium and nickel-tolerant *Echinochloa colona* cell lines. Chromium and nickel-rich mining waste are used to rejuvenate the resistant cell lines into full seedlings, which are then cultivated. This methodology will aid in choosing plants that can tolerate metals for phytoremediation operations.

2.7 TECHNIQUES FOR UTILIZING TRANSGENIC PLANTS

One conceivable method to boost the nutrient content of organisms that are strong gatherers is to employ seedlings that are resistant to metals through a transgenic approach. Enhancing metal accumulation and/or resistance in organisms with higher yields seems to be another tactic. Knowing the routes and genes involved can help us speed up plant processes that currently slow down phytoremediation of a component (such as root influx, root–shoot dislocation, sequestration in certain tissues or sub-cell components, solubilization, and plant–microbe interplay). There have been well over 400 species of biological hyperaccumulators documented (McGrath and Zhao, 2003). They include cheap biomass plants. Using genetic manipulation strategies has become available. *T. caerulescens* has been one of the most widely studied hyperaccumulators. The generation of biomass and the bioconcentration effectiveness are two crucial elements that determine how effective a plant is at extracting phytochemicals. Despite being excellent choices for phytoremediation, hyperaccumulators make it difficult to integrate the necessary genes or hyperaccumulation features into large biomass plants. Additionally, the possibility for remediation of genes acquired from bacteria and overexpressed in plant systems has been revealed. At the molecular level, several trace-element detoxification mechanisms in bacteria have been genetically and functionally identified. The generation of transgenics utilized in phytoremediation projects has received widespread reporting (De Souza et al., 2002; Krämer and Chardonnens, 2001). In transgenic plants, certain genes or features can be altered to increase resistance and phytoextraction capability. Several plant species have recently been designated as hyperaccumulators (Reeves, 1992). Many trees can successfully absorb metal since they have deeper root systems and produce more biomass (Greger and Landberg, 1999). The foundation for effective in situ restoration is higher efficiency and increased absorption and transportation of contaminants to the harvestable biomass (Lasat, 1999). A few woody species were also utilized in soil phytoremediation (Shaekh et al., 2013). There has been new research on the capacity of forest trees for the phytoremediation of metal-contaminated soils (Lingua et al., 2005; Shaekh et al., 2013). Many times, it seems like a clone- or hybrid-specific tolerance to metals instead of species-specific tolerance (Punshon and Dickinson, 1999). As they have previously been examined for in vivo and in vitro studies on metal tolerance (Lingua et al., 2005), poplars are particularly well-suited for remediation objectives. Thus, the utilization of willows and poplars in phytoextraction has received a lot of research attention. High zinc and copper concentrations were tested for their impact on polyamine production and accumulation in *Populus alba* through in vitro investigations. Zn was observed to be transiently hazardous at 0.51 mM concentration, and progressively harmful at 2–4 mM according to leaf indications, the rise of adventitious root development, and ethylene biosynthesis. For the purpose of revegetating coal and chromium mine overburden, (Rout et al., 1999) created metal-resistant cell lines for generating metal-tolerant plants. In the past few years, the search for naturally existing flora that may collect harmful metals from the soil has been ongoing. The effectiveness of hybrid poplar plants in in vitro tests for a bioremediation strategy was revealed (Song et al., 2003). The baker's yeast (*S. cerevisiae*) ABC transporter YCF1 is responsible for the transfer of hazardous substances such as toxic metals inside the vacuole. It has been shown that YCF1-transcription enhances *A. thaliana*'s tolerance to Cd and Pb by enhancing the sequestration of those metals in vacuoles. Comparable to *Arabidopsis* plants, the expression of YCF1 also improved the Cd sensitivity of poplar. The YCF1-transgenic poplar plants outperformed the non-transgenic plants in growth on agar plates featuring 0.1 and 0.3 mM of CdCl2. Through field tests, Song et al. (2003) also made a comparison between transgenic and non-transgenic plants. In the mining waste region, poplar plants—both transgenic and non-transgenic—were grown, and the amount of metal accumulated by various plant components was assessed. The survivability, according to their research, exhibited notable variation in the form of the build-up of heavy metals from the mining waste and the neighbourhood. The development of phytoremediation methods for Se and Hg employing technological methods that leverage the transformation of the metals to little harmful and explosive forms has been the subject of the efforts of several research organizations. According to Zwolak and Zaporowska (2012), selenium (Se) is a vital element for a variety of organisms. At higher quantities, it turns extremely poisonous. Balestrazzi et al. (2009) found that the hybrid white poplar (*P. alba L*) plants did not survive at 0.1 mM $CuCl_2$, whereas the marker-free hybrid plants did. The transgenic plants produced an MT-like protein from the *Pisum sativum* gene PsMTA1.

A short time ago, researchers have been looking for naturally occurring or inherently altered flora that can take harmful metals extracted out of the soil, physiologically transform them into aeriform species inside the plant, and then release them within the atmosphere. It is known as phytovolatilization. The first account of higher plants releasing volatile Se compounds was studied (Lewis et al., 1966). Brassicaceae can release up to 40 g of selenium each day 21 in a variety of gaseous molecules (Terry et al., 1992). *Typha latifolia* L. is also beneficial for Se assemblage (Pilon-Smits et al., 1999). The bacterial organomercurial lyase (MerB) and mercuric reductase (MerA) genes have been infused into the genetic material of *Nicotiana tabacum* L. and *A. thaliana* L. (Heaton et al., 1998; Rugh et al., 1998). The aforementioned plants take up radical mercury (Hg II) and methyl mercury (Me Hg) from the soil and expel volatile mercury (Hg O) into the environment from their leaves (Heaton et al., 1998). Inside the atmosphere, the phytovolatilization of Se and Hg has a number of benefits (De Souza et al., 2000). Dimethyl selenide is one-sixth to one-fifth as hazardous as the inorganic forms of selenium that are present in the soil. Due to the removal of the inorganic forms of Se and Hg and the likelihood that the aeriform species will not be re-deposited at or adjacent to the site, volatilization of these components is also a sustainable site remedy (Atkinson et al., 1990; Heaton et al., 1998). The additional advantages of this remediation technique are low site disturbance, reduced erosion, and a lack of necessity to dispose of attenuated flora waste (Heaton et al., 1998; Rugh, 2001). Hg (O) introduction to the environment would not have a substantial impact on the climatic pool (Heaton et al., 1998; Rugh et al., 2000). Contrary to certain clean-up methods, after adulterants have been eliminated through volatilization, control over their movement to other locations is lost.

2.8 CONCLUSION

To shorten the development period of phytotechnologies, research activities in molecular biology, omics, and transgenics must be intensified. To increase the success of phytotechnologies, their co-benefits such as bioenergy, biopolymers, biofuels, biofertilizers, phytomining, and phytosynthesis of NPs need to be extensively researched and publicized. After the phytoremediation process is finished, the issue of the safe and sustainable disposal of contaminant-loaded biomass must be appropriately addressed. The environment can be protected from the subsequent release or leakage of sequestered metals or organics by properly disposing of and using biomass. The speed of phytotechnology for environmental decontamination has accelerated due to significant advancements in molecular biology and gene modification. However, further development is required to continue the ongoing environmental clean-up. Likewise, with a further understanding of influx/stress coping systems and fundamental metabolic knowledge, phytoremediation of pollutants according to hereditary human resources (HR) technologies may be enhanced additionally. A specific technique for utilizing HR technology in macrophytes for increased genoremediation is the use of constructed wetlands.

REFERENCES

Abbasian, F., Lockington, R., Megharaj, M., & Naidu, R. (2019). A review on the genetics of aliphatic and aromatic hydrocarbon degradation. *Applied Biochemistry and Biotechnology*, *178*(2), 224–250.

Ahmed, T., Shahid, M., Azeem, F., Rasul, I., Shah, A. A., Noman, M., . . . & Muhammad, S. (2018). Biodegradation of plastics: Current scenario and future prospects for environmental safety. *Environmental Science and Pollution Research*, *25*(8), 7287–7298.

Alexander, M. (1999). *Biodegradation and bioremediation.* Gulf Professional Publishing.

An, D., Li, C., Zhou, Y., Wu, Y., & Wang, W. (2018). Genomes and transcriptomes of duckweeds. *Frontiers in Chemistry*, *6*, 230.

Antosiewicz, D. M., Barabasz, A., & Siemianowski, O. (2014). Phenotypic and molecular consequences of overexpression of metal-homeostasis genes. *Frontiers in Plant Science*, *5*, 1–7.

Atkinson, R., Aschmann, S. M., Hasegawa, D., Thompson-Eagle, E. T., & Frankenberger Jr, W. T. (1990). Kinetics of the atmospherically important reactions of dimethyl selenide. *Environmental Science & Technology*, *24*(9), 1326–1332.

Awasthi, S. K., Sarsaiya, S., Kumar, V., Chaturvedi, P., Sindhu, R., Binod, P., . . . & Awasthi, M. K. (2022). Processing of municipal solid waste resources for a circular economy in China: An overview. *Fuel*, *317*, 123478.

Azab, E., Kebeish, R., & Hegazy, A. K. (2018). Expression of the human gene CYP1A2 enhances tolerance and detoxification of the phenylurea herbicide linuron in Arabidopsis thaliana plants and Escherichia coli. *Environmental Pollution*, *238*, 281–290.

Balestrazzi, A., Botti, S., Zelasco, S., Biondi, S., Franchin, C., Calligari, P., . . . & Carbonera, D. (2009). Expression of the PsMT A1 gene in white poplar engineered with the MAT system is associated with heavy metal tolerance and protection against

8-hydroxy-2′-deoxyguanosine mediated-DNA damage. *Plant Cell Reports, 28*(8), 1179–1192.

Bao, L., Gao, C., Li, M., Chen, Y., Lin, W., Yang, Y., . . . & Wang, J. (2013). Biomonitoring of non-dioxin-like polychlorinated biphenyls in transgenic Arabidopsis using the mammalian pregnane X receptor system: A role of pectin in pollutant uptake. *Plos One, 8*(11), e79428.

Basharat, Z., Novo, L. A., & Yasmin, A. (2018). Genome editing weds CRISPR: What is in it for phytoremediation? *Plants, 7*(3), 1–8.

Basu, S., Rabara, R. C., Negi, S., & Shukla, P. (2018). Engineering PGPMOs through gene editing and systems biology: A solution for phytoremediation? *Trends in Biotechnology, 36*(5), 499–510.

Baudouin, C., Charveron, M., Tarroux, R., & Gall, Y. (2002). Environmental pollutants and skin cancer. *Cell Biology and Toxicology, 18*(5), 341–348.

Benekos, K., Kissoudis, C., Nianiou-Obeidat, I., Labrou, N., Madesis, P., Kalamaki, M., . . . & Tsaftaris, A. (2010). Overexpression of a specific soybean GmGSTU4 isoenzyme improves diphenyl ether and chloroacetanilide herbicide tolerance of transgenic tobacco plants. *Journal of Biotechnology, 150*(1), 195–201.

Cai, C., Zhao, M., Yu, Z., Rong, H., & Zhang, C. (2019). Utilization of nanomaterials for in-situ remediation of heavy metal (loid) contaminated sediments: A review. *Science of the Total Environment, 662*, 205–217.

Chatterjee, S., Deb, U., Datta, S., Walther, C., & Gupta, D. K. (2017). Common explosives (TNT, RDX, HMX) and their fate in the environment: Emphasizing bioremediation. *Chemosphere, 184*, 438–451.

Chauhan, A., Oakeshott, J. G., & Jain, R. K. (2008). Bacterial metabolism of polycyclic aromatic hydrocarbons: Strategies for bioremediation. *Indian Journal of Microbiology, 48*(1), 95–113.

Chen, S., Han, X., Fang, J., Lu, Z., Qiu, W., Liu, M., . . . & Zhuo, R. (2017). Sedum alfredii SaNramp6 metal transporter contributes to cadmium accumulation in transgenic Arabidopsis thaliana. *Scientific Reports, 7*(1), 1–13.

Chen, S., Zhang, M., Feng, Y., Sahito, Z. A., Tian, S., & Yang, X. (2019). Nicotianamine Synthase Gene 1 from the hyperaccumulator Sedum alfredii Hance is associated with Cd/Zn tolerance and accumulation in plants. *Plant and Soil, 443*(1), 413–427.

Cherian, S., & Oliveira, M. M. (2005). Transgenic plants in phytoremediation: Recent advances and new possibilities. *Environmental Science & Technology, 39*(24), 9377–9390.

Cherian, S., Weyens, N., Lindberg, S., & Vangronsveld, J. (2012). Phytoremediation of trace element—Contaminated environments and the potential of endophytic bacteria for improving this process. *Critical Reviews in Environmental Science and Technology, 42*(21), 2215–2260.

Chica, R. A., Doucet, N., & Pelletier, J. N. (2005). Semi-rational approaches to engineering enzyme activity: Combining the benefits of directed evolution and rational design. *Current Opinion in Biotechnology, 16*(4), 378–384.

Cicero, L. L., Madesis, P., Tsaftaris, A., & Piero, A. R. L. (2015). Tobacco plants over-expressing the sweet orange tau glutathione transferases (CsGSTUs) acquire tolerance to the diphenyl ether herbicide fluorodifen and to salt and drought stresses. *Phytochemistry, 116*, 69–77.

De Souza, M. P., Pickering, I. J., Walla, M., & Terry, N. (2002). Selenium assimilation and volatilization from selenocyanate-treated Indian mustard and musk grass. *Plant Physiology, 128*(2), 625–633.

De Souza, M. P., Pilon-Smits, E. A. H., & Terry, N. (2000). The physiology and biochemistry of selenium volatilization by plants. In: *Phytoremediation of toxic metals: Using plants to clean up the environment* (pp. 171–190). John Wiley & Sons.

Dhankher, O. P., Li, Y., Rosen, B. P., Shi, J., Salt, D., Senecoff, J. F., . . . & Meagher, R. B. (2002). Engineering tolerance and hyperaccumulation of arsenic in plants by combining arsenate reductase and γ-glutamylcysteine synthetase expression. *Nature Biotechnology, 20*(11), 1140–1145.

Dietz, A. C., & Schnoor, J. L. (2001). Advances in phytoremediation. *Environmental Health Perspectives, 109*(suppl 1), 163–168.

Dixit, P., Mukherjee, P. K., Sherkhane, P. D., Kale, S. P., & Eapen, S. (2011). Enhanced tolerance and remediation of anthracene by transgenic tobacco plants expressing a fungal glutathione transferase gene. *Journal of Hazardous Materials, 192*(1), 270–276.

Doty, S. L. (2008). Enhancing phytoremediation through the use of transgenics and endophytes. *New Phytologist, 179*(2), 318–333.

Doty, S. L., Freeman, J. L., Cohu, C. M., Burken, J. G., Firrincieli, A., Simon, A., . . . & Blaylock, M. J. (2017). Enhanced degradation of TCE on a superfund site using endophyte-assisted poplar tree phytoremediation. *Environmental Science & Technology, 51*(17), 10050–10058.

Du, Z. Y., Arias, T., Meng, W., & Chye, M. L. (2016). Plant acyl-CoA-binding proteins: An emerging family involved in plant development and stress responses. *Progress in Lipid Research, 63*, 165–181.

Duhoux, A., Carrère, S., Gouzy, J., Bonin, L., & Délye, C. (2015). RNA-Seq analysis of ryegrass transcriptomic response to an herbicide inhibiting acetolactate-synthase identifies transcripts linked to non-target-site-based resistance. *Plant Molecular Biology, 87*(4), 473–487.

Eapen, S., Singh, S., & D'souza, S. F. (2007). Advances in the development of transgenic plants for remediation of xenobiotic pollutants. *Biotechnology Advances, 25*(5), 442–451.

Ekperusi, A. O., Sikoki, F. D., & Nwachukwu, E. O. (2019). Application of common duckweed (*Lemna minor*) in phytoremediation of chemicals in the environment: State and future perspective. *Chemosphere, 223*, 285–309.

Fernandez-Cruz, T., Martinez-Carballo, E., & Simal-Gandara, J. (2017). Perspective on pre- and post-natal agro-food exposure to persistent organic pollutants and their effects on quality of life. *Environment International, 100*, 79–101.

Feng, N. X., Yu, J., Zhao, H. M., Cheng, Y. T., Mo, C. H., Cai, Q. Y., . . . & Wong, M. H. (2017). Efficient phytoremediation of organic contaminants in soils using plant–endophyte partnerships. *Science of the Total Environment, 583*, 352–368.

Gautam, M., Elhiti, M., & Fomsgaard, I. S. (2018). Maize root culture as a model system for studying azoxystrobin biotransformation in plants. *Chemosphere, 195*, 624–631.

Germaine, K. J., Keogh, E., Ryan, D., & Dowling, D. N. (2009). Bacterial endophyte-mediated naphthalene phytoprotection and phytoremediation. *FEMS Microbiology Letters, 296*(2), 226–234.

Greger, M., & Landberg, T. (1999). Use of willow in phytoextraction. *International Journal of Phytoremediation, 1*(2), 115–123.

Gyulai, G., Bittsánszky, A., Szabó, Z., Waters Jr, L., Gullner, G., Kampfl, G., . . . & Kőmíves, T. (2014). Phytoextraction potential of wild type and 35S-GSH I transgenic poplar trees (*Populus× Canescens*) for environmental pollutants herbicide paraquat, salt sodium, zinc sulfate and nitric oxide in vitro. *International Journal of Phytoremediation, 16*(4), 379–396.

Heaton, A. C., Rugh, C. L., Wang, N. J., & Meagher, R. B. (1998). Phytoremediation of mercury-and methylmercury-polluted soils using genetically engineered plants. *Journal of Soil Contamination, 7*(4), 497–509.

Hernández-Vega, J. C., Cady, B., Kayanja, G., Mauriello, A., Cervantes, N., Gillespie, A., . . . & Colón-Carmona, A. (2017). Detoxification of Polycyclic Aromatic Hydrocarbons (PAHs) in Arabidopsis thaliana involves a putative flavonol synthase. *Journal of Hazardous Materials, 321*, 268–280.

Hooda, V. (2007). Phytoremediation of toxic metals from soil and waste water. *Journal of Environmental Biology, 28*(2), 367.

Hussain, I., Aleti, G., Naidu, R., Puschenreiter, M., Mahmood, Q., Rahman, M. M., . . . & Reichenauer, T. G. (2018). Microbe and plant assisted-remediation of organic xenobiotics and its enhancement by genetically modified organisms and recombinant technology: A review. *Science of the Total Environment, 628*, 1582–1599.

Hussain, S., Siddique, T., Arshad, M., & Saleem, M. (2009). Bioremediation and phytoremediation of pesticides: Recent advances. *Critical Reviews in Environmental Science and Technology, 39*(10), 843–907.

Ibañez, S., Talano, M., Ontañon, O., Suman, J., Medina, M. I., Macek, T., & Agostini, E. (2016). Transgenic plants and hairy roots: Exploiting the potential of plant species to remediate contaminants. *New Biotechnology, 33*(5), 625–635.

Ibáñez, S. G., Medina, M. I., & Agostini, E. (2011). Phenol tolerance, changes of antioxidative enzymes and cellular damage in transgenic tobacco hairy roots colonized by arbuscular mycorrhizal fungi. *Chemosphere, 83*(5), 700–705.

Ijaz, A., Imran, A., Anwar ul Haq, M., Khan, Q. M., & Afzal, M. (2016). Phytoremediation: Recent advances in plant-endophytic synergistic interactions. *Plant and Soil, 405*(1), 179–195.

Jadaun, J. S., Bansal, S., Sonthalia, A., Rai, A. K., & Singh, S. P. (2022). Biodegradation of plastics for sustainable environment. *Bioresource Technology*, 126697.

James, C. A., & Strand, S. E. (2009). Phytoremediation of small organic contaminants using transgenic plants. *Current Opinion in Biotechnology, 20*(2), 237–241.

Jung, H. I., Kuk, Y. I., Back, K., & Burgos, N. R. (2008). Resistance pattern and antioxidant enzyme profiles of protoporphyrinogen oxidase (PROTOX) inhibitor-resistant transgenic rice. *Pesticide Biochemistry and Physiology, 91*(1), 53–65.

Kawashima, C. G., Noji, M., Nakamura, M., Ogra, Y., Suzuki, K. T., & Saito, K. (2004). Heavy metal tolerance of transgenic tobacco plants over-expressing cysteine synthase. *Biotechnology Letters, 26*(2), 153–157.

Kebeish, R., El-Ayouty, Y., El-Naggar, A., & Saleh, A. M. (2017). Synchronous overexpression of glutathione-S-transferase and cyanidase maintains the redox homeostasis and improves cyanide remediation capacity in tobacco. *Environmental and Experimental Botany, 137*, 220–227.

Khan, S., Ali, S. A., & Ali, A. S. (2022). Biodegradation of Low-Density Polyethylene (LDPE) by mesophilic fungus 'Penicillium citrinum' isolated from soils of plastic waste dump yard, Bhopal, India. *Environmental Technology*, 1–15.

Knasmüller, S., Gottmann, E., Steinkellner, H., Fomin, A., Pickl, C., Paschke, A., . . . & Kundi, M. (1998). Detection of genotoxic effects of heavy metal contaminated soils with plant bioassays. *Mutation Research/Genetic Toxicology and Environmental Mutagenesis, 420*(1–3), 37–48.

Krämer, U., & Chardonnens, A. (2001). The use of transgenic plants in the bioremediation of soils contaminated with trace elements. *Applied Microbiology and Biotechnology, 55*(6), 661–672.

Kumar, S., Dubey, R. S., Tripathi, R. D., Chakrabarty, D., & Trivedi, P. K. (2015). Omics and biotechnology of arsenic stress and detoxification in plants: Current updates and prospective. *Environment International, 74*, 221–230.

Lasat, M. M. (1999). Phytoextraction of metals from contaminated soil: A review of plant/soil/ metal interaction and assessment of pertinent agronomic issues. *Journal of Hazardous Substance Research, 2*(1), 5.

Lee, I., Baek, K., Kim, H., Kim, S., Kim, J., Kwon, Y., . . . & Bae, B. (2007). Phytoremediation of soil co-contaminated with heavy metals and TNT using four plant species. *Journal of Environmental Science and Health Part A, 42*(13), 2039–2045.

Lewis, B. G., Johnson, C. M., & Delwiche, C. C. (1966). Release of volatile selenium compounds by plants: Collection procedures and preliminary observations. *Journal of Agricultural and Food Chemistry, 14*(6), 638–640.

Lima, L. R., Silva, H. F., Brignoni, A. S., Silva, F. G., Camargos, L. S., & Souza, L. A. (2019). Characterization of biomass sorghum for copper phytoremediation: Photosynthetic response and possibility as a bioenergy feedstock from contaminated land. *Physiology and Molecular Biology of Plants, 25*(2), 433–441.

Lin, C. F., Lo, S. S., Lin, H. Y., & Lee, Y. (1998). Stabilization of cadmium-contaminated soils using synthesized zeolite. *Journal of Hazardous Materials, 60*(3), 217–226.

Lingua, G., Castiglione, S., Todeschini, V., Franchin, C., Fossati, T., Peterson, E. A., . . . & Berta, G. (July 2005). Selection of elite poplar clones for phytoremediation of soil contaminated by heavy metals in field and glass-house experiments. In: *Proceedings of the XVII International Botanical Congress, Vienna* (Vol. 624). S. Biondi.

Luo, J. S., Huang, J., Zeng, D. L., Peng, J. S., Zhang, G. B., Ma, H. L., . . . & Gong, J. M. (2018). A defensin-like protein drives cadmium efflux and allocation in rice. *Nature Communications, 9*(1), 1–9.

Luo, Z. B., He, J., Polle, A., & Rennenberg, H. (2016). Heavy metal accumulation and signal transduction in herbaceous and woody plants: Paving the way for enhancing phytoremediation efficiency. *Biotechnology Advances, 34*(6), 1131–1148.

Martin, B. C., George, S. J., Price, C. A., Shahsavari, E., Ball, A. S., Tibbett, M., & Ryan, M. H. (2016). Citrate and malonate increase microbial activity and alter microbial community composition in uncontaminated and diesel-contaminated soil microcosms. *Soil, 2*(3), 487–498.

Martínez, M., Bernal, P., Almela, C., Vélez, D., García-Agustín, P., Serrano, R., & Navarro-Aviñó, J. (2006). An engineered plant that accumulates higher levels of heavy metals than Thlaspi caerulescens, with yields of 100 times more biomass in mine soils. *Chemosphere, 64*(3), 478–485.

Matteucci, F., Ercole, C., & Del Gallo, M. (2015). A study of chlorinated solvent contamination of the aquifers of an industrial area in central Italy: A possibility of bioremediation. *Frontiers in Microbiology, 6*, 924.

McGrath, S. P., & Zhao, F. J. (2003). Phytoextraction of metals and metalloids from contaminated soils. *Current Opinion in Biotechnology, 14*(3), 277–282.

Meagher, R. B. (2000). Phytoremediation of toxic elemental and organic pollutants. *Current Opinion in Plant Biology, 3*(2), 153–162.

Mello-Farias, P. C., & Chaves, A. L. S. (2008). Biochemical and molecular aspects of toxic metals phytoremediation using transgenic plants. *Transgenic Approach in Plant Biochemistry and Physiology*, 253–266.

Mitra, J., Mukherjee, P. K., Kale, S. P., & Murthy, N. B. K. (2001). Bioremediation of DDT in soil by genetically improved strains of soil fungus Fusarium solani. *Biodegradation, 12*(4), 235–245.

Mitter, E. K., Kataoka, R., de Freitas, J. R., & Germida, J. J. (2019). Potential use of endophytic root bacteria and host plants to degrade hydrocarbons. *International Journal of Phytoremediation, 21*(9), 928–938.

Morel, M., Crouzet, J., Gravot, A., Auroy, P., Leonhardt, N., Vavasseur, A., & Richaud, P. (2009). AtHMA3, a P1B-ATPase allowing Cd/Zn/co/Pb vacuolar storage in Arabidopsis. *Plant Physiology, 149*(2), 894–904.

O'Keefe, D. P., Tepperman, J. M., Dean, C., Leto, K. J., Erbes, D. L., & Odell, J. T. (1994). Plant expression of a bacterial cytochrome P450 that catalyzes activation of a sulfonylurea pro-herbicide. *Plant Physiology, 105*(2), 473–482.

Park, J., Song, W. Y., Ko, D., Eom, Y., Hansen, T. H., Schiller, M., . . . & Lee, Y. (2012). The phytochelatin transporters AtABCC1 and AtABCC2 mediate tolerance to cadmium and mercury. *The Plant Journal, 69*(2), 278–288.

Peng, R. H., Qiu, J., Tian, Y. S., Gao, J. J., Han, H. J., Fu, X. Y., . . . & Yao, Q. H. (2017). Disulfide isomerase-like protein AtPDIL1–2 is a good candidate for trichlorophenol phytodetoxification. *Scientific Reports, 7*(1), 1–10.

Pilon-Smits, E. A. (2005). Phytoremediation. *Annual Review of Plant Biology, 56*, 15–39.

Pilon-Smits, E. A., Hwang, S., Mel Lytle, C., Zhu, Y., Tai, J. C., Bravo, R. C., . . . & Terry, N. (1999). Overexpression of ATP sulfurylase in Indian mustard leads to increased selenate uptake, reduction, and tolerance. *Plant Physiology, 119*(1), 123–132.

Pilon-Smits, E. A., & Pilon, M. (2002). Phytoremediation of metals using transgenic plants. *Critical Reviews in Plant Sciences, 21*(5), 439–456.

Pilon-Smits, E. A., Zhu, Y. L., Sears, T., & Terry, N. (2000). Overexpression of glutathione reductase in *Brassica juncea*: Effects on cadmium accumulation and tolerance. *Physiologia Plantarum, 110*(4), 455–460.

Pulford, I. D., & Watson, C. (2003). Phytoremediation of heavy metal-contaminated land by trees—A review. *Environment International, 29*(4), 529–540.

Punshon, T., & Dickinson, N. (1999). Heavy metal resistance and accumulation characteristics in willows. *International Journal of Phytoremediation, 1*(4), 361–385.

Rai, P. K. (2018a). Heavy metal phyto-technologies from Ramsar wetland plants: Green approach. *Chemistry and Ecology, 34*(8), 786–796.

Rai, P. K., & Kim, K. H. (2020). Invasive alien plants and environmental remediation: A new paradigm for sustainable restoration ecology. *Restoration Ecology, 28*(1), 3–7.

Rasche, F., Velvis, H., Zachow, C., Berg, G., Van Elsas, J. D., & Sessitsch, A. (2006). Impact of transgenic potatoes expressing anti-bacterial agents on bacterial endophytes is comparable with the effects of plant genotype, soil type, and pathogen infection. *Journal of Applied Ecology, 43*(3), 555–566.

Reeves, R. D. (1992). The hyperaccumulation of nickel by serpentine plants. In: *The vegetation of ultramafic (serpentine) soils* (pp. 253–277). Intercept Ltd.

Reeves, R. D., Baker, A. J., Jaffré, T., Erskine, P. D., Echevarria, G., & van der Ent, A. (2018). A global database for plants that hyperaccumulate metal and metalloid trace elements. *New Phytologist, 218*(2), 407–411.

Rodriguez-Hernandez, M. C., Bonifas, I., Torre, M. C. A., Flores-Flores, J. L., Bañuelos-Hernández, B., & Patiño-Rodríguez, A. (2015). Increased accumulation of cadmium and lead under Ca and Fe deficiency in *Typha latifolia*: A study of Two-Pore Channel (TPC1) gene responses. *Environmental and Experimental Botany., 115*, 38–48.

Rout, G. R., Samantaray, S., & Das, P. (1999). In vitro selection and biochemical characterisation of zinc and manganese adapted callus lines in Brassica spp. *Plant Science, 146*(2), 89–100.

Rugh, C. L. (2001). Mercury detoxification with transgenic plants and other biotechnological breakthroughs for phytoremediation. *In Vitro Cellular & Developmental Biology-Plant, 37*(3), 321–325.

Rugh, C. L., Bizily, S. P., & Meagher, R. B. (2000). Phytoreduction of environmental mercury pollution. In: *Phytoremediation of toxic metals: Using plants to clean-up the environment* (pp. 151–170). John Wiley and Sons.

Rugh, C. L., Senecoff, J. F., Meagher, R. B., & Merkle, S. A. (1998). Development of transgenic yellow poplar for mercury phytoremediation. *Nature Biotechnology, 16*(10), 925–928.

Rylott, E. L., Budarina, M. V., Barker, A., Lorenz, A., Strand, S. E., & Bruce, N. C. (2011a). Engineering plants for the phytoremediation of RDX in the presence of the co-contaminating explosive TNT. *New Phytologist, 192*(2), 405–413.

Rylott, E. L., Jackson, R. G., Sabbadin, F., Seth-Smith, H. M., Edwards, J., Chong, C. S., . . . & Bruce, N. C. (2011b). The explosive-degrading cytochrome P450 XplA: Biochemistry, structural features and prospects for bioremediation. *Biochimica et Biophysica Acta (BBA)-Proteins and Proteomics, 1814*(1), 230–236.

Rylott, E. L., Johnston, E. J., & Bruce, N. C. (2015). Harnessing microbial gene pools to remediate persistent organic pollutants using genetically modified plants—A viable technology? *Journal of Experimental Botany, 66*(21), 6519–6533.

Salt, D. E., Smith, R. D., & Raskin, I. (1998). Phytoremediation. *Annual Review of Plant Biology, 49*(1), 643–668.

Samantaray, S., Rout, G. R., & Das, P. (2001a). Induction, selection and characterization of Cr and Ni-tolerant cell lines of Echinochloa colona (L.) link in vitro. *Journal of Plant Physiology, 158*(10), 1281–1290.

Samantaray, S., Rout, G. R., & Das, P. (2001b). Heavy metal and nutrient concentration in soil and plants growing on a metalliferous chromite minespoil. *Environmental Technology, 22*(10), 1147–1154.

Sarangi, B. K., Kalve, S., Pandey, R. A., & Chakrabarti, T. (2009). Transgenic plants for phytoremediation of arsenic and chromium to enhance tolerance and hyperaccumulation. *Transgenic Plant Journal, 3*(1), 57–86.

Sarma, H., Nava, A. R., & Prasad, M. N. V. (2019). Mechanistic understanding and future prospect of microbe-enhanced phytoremediation of polycyclic aromatic hydrocarbons in soil. *Environmental Technology & Innovation, 13*, 318–330.

Shaekh, M. P. E., Mondol, A., Islam, M. M., Kabir, A. S., Saleh, M. A., Uddin, M. S., . . . & Ekram, A. E. (2013). Isolation, characterization and identification of an antagonistic bacterium from Penaeus monodon. *IJSER, 4*(10), 254–261.

Sharma, M., Akhter, Y., & Chatterjee, S. (2019). A review on remediation of cyanide containing industrial wastes using biological systems with special reference to enzymatic degradation. *World Journal of Microbiology and Biotechnology, 35*(5), 1–14.

Shen, Y., Li, J., Gu, R., Zhan, X., & Xing, B. (2019). Proteomic analysis for phenanthrene-elicited wheat chloroplast deformation. *Environment International, 123*, 273–281.

Shimazu, S., Ohta, M., & Ashida, H. (2014). Application of lipid extracts from Solidago canadensis to phytomonitoring of PCB126 in transgenic Arabidopsis plants. *Science of the Total Environment, 491*, 240–245.

Singh, E., Kumar, A., Mishra, R., You, S., Singh, L., Kumar, S., & Kumar, R. (2021). Pyrolysis of waste biomass and plastics for production of biochar and its use for removal of heavy metals from aqueous solution. *Bioresource Technology, 320*, 124278.

Singh, M. M., & Rai, P. K. (2016). A microcosm investigation of Fe (iron) removal using macrophytes of ramsar lake: A phytoremediation approach. *International Journal of Phytoremediation, 18*(12), 1231–1236.

Song, W. Y., Ju Sohn, E., Martinoia, E., Jik Lee, Y., Yang, Y. Y., Jasinski, M., . . . & Lee, Y. (2003). Engineering tolerance and accumulation of lead and cadmium in transgenic plants. *Nature Biotechnology, 21*(8), 914–919.

Takahashi, R., Ishimaru, Y., Shimo, H., Bashir, K., Senoura, T., Sugimoto, K., . . . & Nakanishi, H. (2014). From laboratory to field: OsNRAMP5-knockdown rice is a promising candidate for Cd phytoremediation in paddy fields. *PLoS One*, *9*(6), e98816.

Tan, D., Wang, Y., Tong, Y., & Chen, G. Q. (2021). Grand challenges for industrializing polyhydroxyalkanoates (PHAs). *Trends in Biotechnology*, *39*(9), 953–963.

Tan, X., Li, K., Wang, Z., Zhu, K., Tan, X., & Cao, J. (2019). A review of plant vacuoles: Formation, located proteins, and functions. *Plants*, *8*(9), 1–11.

Tang, L., Hamid, Y., Gurajala, H. K., He, Z., & Yang, X. (2019). Effects of CO_2 application and endophytic bacterial inoculation on morphological properties, photosynthetic characteristics and cadmium uptake of two ecotypes of Sedum alfredii Hance. *Environmental Science and Pollution Research*, *26*(2), 1809–1820.

Teng, Y., Wang, X., Li, L., Li, Z., & Luo, Y. (2015). Rhizobia and their bio-partners as novel drivers for functional remediation in contaminated soils. *Frontiers in Plant Science*, *6*(32), 1–11.

Terry, N., Carlson, C., Raab, T. K., & Zayed, A. M. (1992). *Rates of selenium volatilization among crop species* (Vol. 21, No. 3, pp. 341–344). American Society of Agronomy, Crop Science Society of America, and Soil Science Society of America.

Thakur, S., Chaudhary, J., Singh, P., Alsanie, W. F., Grammatikos, S. A., & Thakur, V. K. (2022). Synthesis of bio-based monomers and polymers using microbes for a sustainable bioeconomy. *Bioresource Technology*, *344*, 126156.

Thomas, J. C., Davies, E. C., Malick, F. K., Endreszl, C., Williams, C. R., Abbas, M., . . . & Murray, K. S. (2003). Yeast metallothionein in transgenic tobacco promotes copper uptake from contaminated soils. *Biotechnology Progress*, *19*(2), 273–280.

Van Aken, B. (2008). Transgenic plants for phytoremediation: Helping nature to clean up environmental pollution. *Trends in Biotechnology*, *26*(5), 225–227.

Van Aken, B. (2011). *Transgenic plants and associated bacteria for phytoremediation of organic pollutants*. Temple University, Elsevier.

Vassilev, A., Schwitzguébel, J. P., Thewys, T., Van Der Lelie, D., & Vangronsveld, J. (2004). The use of plants for remediation of metal-contaminated soils. *The Scientific World Journal*, *4*, 9–34.

Wang, L., Samac, D. A., Shapir, N., Wackett, L. P., Vance, C. P., Olszewski, N. E., & Sadowsky, M. J. (2005). Biodegradation of atrazine in transgenic plants expressing a modified bacterial atrazine chlorohydrolase (atzA) gene. *Plant Biotechnology Journal*, *3*(5), 475–486.

Wang, X., Wu, N., Guo, J., Chu, X., Tian, J., Yao, B., & Fan, Y. (2008). Phytodegradation of organophosphorus compounds by transgenic plants expressing a bacterial organophosphorus hydrolase. *Biochemical and Biophysical Research Communications*, *365*(3), 453–458.

Wang, Q., Ye, J., Wu, Y., Luo, S., Chen, B., Ma, L., . . . & Yang, X. (2019). Promotion of the root development and Zn uptake of Sedum alfredii was achieved by an endophytic bacterium Sasm05. *Ecotoxicology and Environmental Safety*, *172*, 97–104.

Wang, Y., Ren, H., Pan, H., Liu, J., & Zhang, L. (2015). Enhanced tolerance and remediation to mixed contaminates of PCBs and 2, 4-DCP by transgenic alfalfa plants expressing the 2, 3-dihydroxybiphenyl-1, 2-dioxygenase. *Journal of Hazardous Materials*, *286*, 269–275.

Webster, L. R., McKenzie, G. H., & Moriarty, H. T. (2002). Organophosphate-based pesticides and genetic damage implicated in bladder cancer. *Cancer Genetics and Cytogenetics*, *133*(2), 112–117.

Wojas, S., Clemens, S., Skłodowska, A., & Antosiewicz, D. M. (2010). Arsenic response of AtPCS1-and CePCS-expressing plants—Effects of external As (V) concentration on as-accumulation pattern and NPT metabolism. *Journal of Plant Physiology*, *167*(3), 169–175.

Yadav, K. K., Gupta, N., Kumar, A., Reece, L. M., Singh, N., Rezania, S., & Khan, S. A. (2018). Mechanistic understanding and holistic approach of phytoremediation: A review on application and future prospects. *Ecological Engineering, 120*, 274–298.

Yang, M., Lu, K., Zhao, F. J., Xie, W., Ramakrishna, P., Wang, G., . . . & Lian, X. (2018). Genome-wide association studies reveal the genetic basis of ionomic variation in rice. *The Plant Cell, 30*(11), 2720–2740.

Yang, X., Feng, Y., He, Z., & Stoffella, P. J. (2005). Molecular mechanisms of heavy metal hyperaccumulation and phytoremediation. *Journal of Trace Elements in Medicine and Biology, 18*(4), 339–353.

Yatoo, A. M., Hamid. B., Sheikh, T. A., Ali, S., Bhat, S. A., Ramola, S., Baba, Z. A., & Kumar, S. (2024). Global perspective of municipal solid waste and landfill leachate: Generation, composition, ecotoxicity, and sustainable management strategies. *Environmental Science and Pollution Research. 31*(16), 23363–23392.

Zhang, L., Routsong, R., Nguyen, Q., Rylott, E. L., Bruce, N. C., & Strand, S. E. (2017). Expression in grasses of multiple transgenes for degradation of munitions compounds on live-fire training ranges. *Plant Biotechnology Journal, 15*(5), 624–633.

Zhang, L., Tsui, T. H., Loh, K. C., Dai, Y., & Tong, Y. W. (2021). Effects of plastics on reactor performance and microbial communities during acidogenic fermentation of food waste for production of volatile fatty acids. *Bioresource Technology, 337*, 125481.

Zhang, W. H., Sun, R. B., Xu, L., Liang, J. N., Wu, T. Y., & Zhou, J. (2019). Effects of micro-/nano-hydroxyapatite and phytoremediation on fungal community structure in copper contaminated soil. *Ecotoxicology and Environmental Safety, 174*, 100–109.

Zhang, X., Ma, D., Lv, J., Feng, Q., Liang, Z., Chen, H., & Feng, J. (2022). Food waste composting based on patented compost bins: Carbon dioxide and nitrous oxide emissions and the denitrifying community analysis. *Bioresource Technology, 346*, 126643.

Zhang, Y., & Liu, J. (2011). Transgenic alfalfa plants co-expressing Glutathione S-Transferase (GST) and human CYP2E1 show enhanced resistance to mixed contaminates of heavy metals and organic pollutants. *Journal of Hazardous Materials, 189*(1–2), 357–362.

Zhang, Y., Liu, J., Zhou, Y., Gong, T., Wang, J., & Ge, Y. (2013). Enhanced phytoremediation of mixed heavy metal (mercury)—Organic pollutants (trichloroethylene) with transgenic alfalfa co-expressing glutathione S-transferase and human P450 2E1. *Journal of Hazardous Materials, 260*, 1100–1107.

Zhao, C., Xu, J., Li, Q., Li, S., Wang, P., & Xiang, F. (2014). Cloning and characterization of a Phragmites australis Phytochelatin Synthase (PaPCS) and achieving Cd tolerance in tall fescue. *PLoS One, 9*(8), e103771.

Zhou, Y., Kumar, M., Sarsaiya, S., Sirohi, R., Awasthi, S. K., Sindhu, R., . . . & Awasthi, M. K. (2022). Challenges and opportunities in bioremediation of micro-nano plastics: A review. *Science of the Total Environment, 802*, 149823.

Zhu, B., Peng, R. H., Fu, X. Y., Jin, X. F., Zhao, W., Xu, J., . . . & Yao, Q. H. (2012). Enhanced transformation of TNT by *Arabidopsis* plants expressing an old yellow enzyme. *PloS One, 7*(7), e39861.

Zwolak, I., & Zaporowska, H. (2012). Selenium interactions and toxicity: A review. *Cell Biology and Toxicology, 28*(1), 31–46.

3 Role of Algae in Bioremediation of Waste Disposal Sites

Shagufta Iqbal, Khair Ul Nisa, Najeebul Tarfeen, Humaira, Saba Wani, Qadrul Nisa, and Shafat Ali

3.1 INTRODUCTION

Waste products, which can range from nuclear waste to untreated sewage, have always been a major problem. Increased emissions of contaminants, particularly hazardous metals, into the environment are mostly due to industrialization. Heavy metals, having specific gravity ≥ 5 g cm^3, have generated massive concern amongst environmentalists as they exhibit a profound propensity to bioaccumulate, remaining persistent, and both human health and environmental quality may suffer as a result of their accumulation in the environment (Ouyang et al., 2018). It is obvious that hazardous metal poisoning of soil and water sources harms the environment, and the accumulation of toxic metals in humans is responsible for serious consequences including abnormal growth and development, renal malfunction, certain kind of cancers and mental retardation (Bauddh and Korstad, 2022). The use of soils polluted by heavy metals for crop production becomes the cause of the entrance of pollutants into the food chain. Additionally, toxic metal contamination reduces an area's productivity (Gosavi et al., 2004). Hazardous heavy elements are extremely toxic to all components of aquatic communities as well. Water contamination occurs mainly through the use of fertilizers used in agricultural operations, and several other carcinogenic and immunotoxic compounds from growing industries pollute water bodies to a great extent. The contaminated water, used for irrigation of crops, is responsible for its accumulation in the food products.

Mostly disposal of solid wastes in the past would be done by digging a hole and dumping the waste followed by its filling. Although this method appears to be a suitable option, the leakage of toxic materials from these "dig-and-dump" sites into water sources and other potential habitats has raised a serious concern, thereby urging researchers to find a more convenient way out for the appropriate disposal of waste materials along with the remediation of waste disposal sites. Conventional techniques, such as electrochemical treatment, precipitation and ion exchange, are often expensive for large projects; rigorous oversight and ongoing supervision are required; and, consequently, there is still a need for a workable and cost-effective solution. The elimination of environmental contaminants using live organisms (mainly bacteria)

DOI: 10.1201/9781003359326-3

or conversion of them into less dangerous or non-hazardous compounds, a process known as bioremediation, has in the recent past gained significant attention (Leung, 2004). Due to their capacity for adsorption, microorganisms provide an alternative to traditional physical/chemical methods for eliminating heavy metals. By destroying or reducing the concentration of hazardous waste utilizing microbes or other living entities like plants, bioremediation is essentially a biological treatment system for eliminating pollutants and decontaminating damaged or contaminated sites (Dwivedi, 2012; Sarmah and Rout, 2020).

Utilizing plants to absorb hazardous substances from soil and water, a process called phytoremediation, has various benefits, including economic viability, simple management, cheap maintenance and installation costs (Yan et al., 2020). Moreover, this process is environmentally friendly, reducing the amount of toxins that are exposed to the ecosystem, and may be applied over a large area and quickly disposed of. On top of this it stabilizes heavy metals to prevent metal leaching and erosion while lowering the likelihood that pollutants will spread. Releasing a number of organic substances into the soil may also help increase soil fertility. It is a technique for cleaning up contaminated soil and water by allowing plants, algae or fungi to absorb heavy metals.

Lately, algae have gained attention as possible candidates for bioremediation for their ability to accumulate dissolved metals, thus offering an eco-friendly, cost-effective and potential replacement for traditional procedures for the decontamination of polluted sites, having a vital role in "green clean" technologies (Nhat et al., 2018).

3.2 ALGAE AS BIO-REMEDIATORS

Non-vascular plants, known as algae, range in size from tiny unicellular or colonial phytoplanktons to enormous multicellular macroalgae (Anbuchezhian et al., 2015). Ecologically, algae are known to have a worldwide distribution, practically occurring in every liveable environment. They have the potential to thrive in a multitude of environments, including various types of wastewater, by producing a wide array of bioactive substances and cellular processes (Marella et al., 2019). Being primarily photosynthetic, algae are the major organic compound manufacturers and play a significant part in the food chain. Besides this, their immense capability to sorb metals has made them potential contributors to the bioremediation of polluted sites, particularly water bodies (Figure 3.1) (Kaur and Bhatnagar, 2002).

3.2.1 DEGRADATION OF HEAVY METALS USING ALGAE

Phycoremediation, the bioremediation of algae, has become a practical technique for eliminating heavy metals from wastewater (Koul et al., 2022). Algae primarily consume pollutants as food; for instance, they require nitrates, phosphates and specific heavy metals for normal growth and sustenance. Some heavy metals, such as cobalt (Co), copper (Cu), iron (Fe), manganese (Mn), molybdenum (Mo), nickel (Ni), vanadium (V) and Zinc (Zn), are nutritionally important at lower levels and are considered essential for plants as well as for algae growth and metabolism (Sunda, 1989), whereas aluminium (Al), arsenic (As), cadmium (Cd), chromium (Cr), lead (Pb),

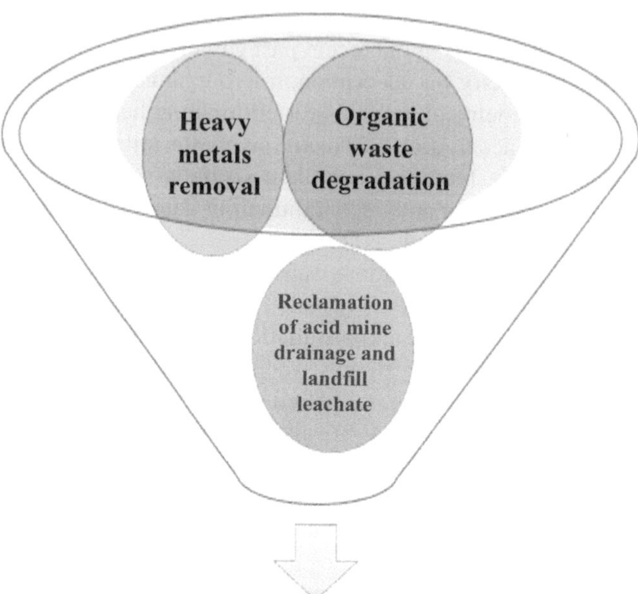

FIGURE 3.1 Algal-based bioremediation system for the remediation of contaminated sites

mercury (Hg), silver (Ag) and tin (Sn) are non-essential metals providing no significant contribution to biological processes (Priya et al., 2014). Nevertheless, excessive exposure to heavy metals may have harmful effects; therefore, attempts to remediate heavy metal contamination require the deployment of large-scale measures so that the threat to ecological and human health is reduced. Due to the hazard posed by bioaccumulation and biomagnification within the food chain, the physiological and metabolic features of aquatic organisms can be altered by heavy metals, which reduce productivity (Kobielska et al., 2018). One of the most pervasive pollutants in aquatic ecosystems is heavy metal. Additionally, since the start of the Industrial Revolution, heavy metal pollution of aquatic habitats has increased.

Wastewater remediation using algae has been promoted for the removal of heavy metals, herbicides and other waste products (Marella et al., 2018; Baghour, 2019; Tiwari and Marella, 2020). Considering their capacity to eliminate hazardous compounds from the environment, macro- and microalgae have recently attracted a lot of attention for usage in many applications. A brief description of various algal species as potential degraders of toxic heavy metals is shown in Table 3.1. Through internal and extracellular assimilation systems, phytoplankton controls the levels of trace metal ions in water. The primary population of phytoplankton, diatom algae, performs a critical role in the aquatic ecosystem by removing heavy metals from the polluted environment and rendering them innocuous so that the degraded environment can return to its former state (Marella et al., 2020). Metals such as Cd, Co, Cu, Hg, Ni, Pb and Zn are concentrated in green algae's polyphosphate bodies, which enable these freshwater unicellular algae to store nutrients, providing a nutrient storage pool and a component of the detoxification process. Other metals significant for algal growth include iron and chromium for photosynthesis and metabolism, respectively. Metal ions are taken up by algae that depend on their ability of biosorption and bioaccumulation. Sequentially, the process of physical adsorption is followed by chemisorption where ions are absorbed into the cytoplasm from the surface of the cell. This phenomenon facilitates the pollution removal from contaminated places, thereby restoring these sites and maintaining the surroundings. Heavy metal ion biosorption is mostly caused by the algal cell wall components with a strong metal affinity, such as alginates and sulphated polysaccharides (Davis et al., 2003). Heavy metal ions are allowed to adhere to functional groups on the cell surface by fucoidan and alginate, the two most important functional groups required for heavy metal biosorption (Kumar et al., 2015). Due to phytochelatins and metallothioneins, which result in metal-chelated complexes, heavy metals can be collected and stored in the cytoplasm and storage vacuoles of diatoms as insoluble salts (Perrein-Ettajani et al., 1999). *Skeletonema costatum* displayed a heavy metal-chelated complex transport and storage mechanism in response to cadmium (Cd) and copper (Cu) stress, together with increased vacuole storage (Nassiri et al., 1997)

For removing metal, *Chlorella* and *Scenedesmus* are considered the ideal microalgae. It was revealed that, when the media's phosphorous content was raised, the alga *Scenedesmus obliquus* began to acquire several metals such as Cd and Zn. *Chlamydomonas reinhardtii, Chlorella vulgaris, Chlorella miniata, Chlorella salina, Chlorella sorokiniana, Scenedesmus abundans, Scenedesmus quadricauda, Scenedesmus subspicatus, Porphyridium purpureum, Phaeodactylum tricornutum, Spirogyra* spp., *Spirulina platensis, Stichococcus bacillaris* and *Stigeoclonium tenue* have been studied for their capacity to take in metals (Perales-Vela et al., 2006; Brinza et al., 2007). Low quantities of strontium have been observed to be absorbed by the marine green microalgae *Platymonas subcordiformis* (Mei et al., 2006). It has been discovered that metals can bind to peptides produced by microalgae, causing the development of organometallic structures that are then placed inside vacuoles to manage the quantity of heavy metals in the cytoplasm, thus eradicating their possible negative effects (Hernández-Ávila et al., 2017). The ability of algae, especially microalgae (living as well as nonliving biomass) to remove metals, has been reviewed along with several mechanisms for absorbing heavy metals (Mondal et al., 2019; Marella et al., 2020; Bauddh and Korstad, 2022). These microorganisms have been discovered to be highly appropriate for recovering metals and capable of achieving notable percent removals (Monteiro et al., 2012). To purge wastewater from heavy metal contamination, the use of diatom algal-mediated technology has proved to be extremely promising and economical.

Biofilms can be formed by bacterial and microalgal aggregates, which makes it easier for them to cohabit in harmony. Heavy metal removal is made easier by concentrating them into biomass using the algal biofilm-based remediation technology (Adey et al., 2011).

3.2.2 DEGRADATION OF ORGANIC POLLUTANTS USING ALGAE

Toxic organic substances are discharged into the environment because of industrial waste, agricultural activities or poor waste management procedures. The buildup of organic environmental toxins can result in major issues that have a detrimental impact on the stability of many aquatic ecosystems and have negative environmental and health repercussions. In aquatic ecosystems, algae play a key role in the bio-monitoring and management of organic pollutants. The three types of algae showing promise in the bioremediation process include cyanobacteria, which are blue-green algae, macroalgae and microalgae, which are typically green. Numerous researches have examined how specific microalgae and cyanobacteria species can be used to phytoremediate organic pollutants in aquatic environments (Gonçalves, 2021). Since microalgae play a crucial role in carbon dioxide fixation, they have been utilized in wastewater bioremediation, thereby reducing the amount of carbon dioxide in the atmosphere, cleaning up contaminated ecosystems, preventing global warming and so on (Ahmad et al., 2022), and they have become a potentially affordable substitute for physical–chemical therapies. Furthermore, due to their role in bioremediation, they are highly suitable for environmental sustainability (Ellis et al., 2012).

Hydrocarbons can be detrimental if they enter the food chain as some of the most long-lasting organic pollutants, like polycyclic aromatic hydrocarbons (PAHs) and synthetic chemicals polychlorinated biphenyls, possess carcinogenic properties (IARC, 1983). These organic contaminants can be removed from ecosystems by algae in three distinct methods: the electrostatic attraction and complexation of PAHs to the cell wall active groups of algae; bioaccumulation inside of the cells; and enzymatic oxidation–reduction processes involving redox enzymes to change organic pollutants into useful substances (Mona et al., 2011). The primary mechanism is fuel combustion, which results in the production of PAHs, which have high boiling temperatures, very poor water solubility and great stability, and are poisonous (Lima et al., 2005). Due to their hydrophobicity and limited water solubility, these pentacyclic molecules are extremely resistant and can survive in the environment for a longer duration (Cerniglia, 1992). Hong et al. (2008) have examined how phenanthrene (PHE) and fluoranthene, two prevalent PAHs, are built up and biodegraded by the diatoms using two algal models; *Nitzschia* sp. and *S. costatum*. *Chlamydomonas* sp., *C. miniata*, *C. vulgaris*, *Scenedesmus platydiscus*, *Scenedesmus quadricola*, *Selenestrum capricornutum* and *Synechocystis* sp. have been studied for their efficacy in removing pyrene (Lei et al., 2002).

It has been proposed that it may be feasible to employ microalgae to provide specialized bacteria with the oxygen they require to digest harmful pollutants such as organic solvents, phenolics and PAHs (Muñoz and Guieysse, 2006). It has been demonstrated that *Chlorella sorokiniana* and *Pseudomonas migulae* strains can form an algal–bacterial consortium that can biodegrade PHE dissolved in silicone oil or tetradecane at a rate of 200–500 mg/l without the need for an external oxygen supply, indicating that the biosurfactants released by microalgae may improve the breakdown of PHE even further (Munoz et al., 2003). In the past few years, the interaction between microalgae and bacteria, particularly the biodegradation and biosorption mechanisms, has received extensive study and research (Mondal et al., 2019).

TABLE 3.1
Summary of some potential algal species reported to effectively degrade toxic pollutants

Pollutant	Algal species	Reference
Bisphenol	*Monoraphidium braunii*	Gattullo et al. (2012)
Benzene, toluene, naphthalene, phenanthrene, pyrene	*Selenastrum capricornutum*	Pathak et al. (2018)
Phenanthrene	Consortium of *Chlorella sorokiniana* and *Pseudomonas migulae*	Munoz et al. (2003)
Mesotrione	*Pediastrum tetras, Ankistrodesmus fusiformis Amphora coffeaeformis*	Valiente Moro et al. (2012
Cu (II), Co (II), Ni (II), Zn (II)	*Ulva reticulat*	Senthilkumar et al. (2006)
Cu (II)	*Posidonia oceanic*	Izquierdo et al. (2010)
Pb (II), Cu (II), Cr (III)	*Gelidium* sp.	Vilar et al. (2008)
Hg	*Aphanothece flocculosa* and *Spirulina platensis*	Sood et al. (2015)
Mn, Co, Pb, Cu	*Anabaena, subcylindrical* and *Nostoc muscorum*	Dönmez et al. (1999)
Cr	*Gloeocapsa* and *Lyngbya*	Shukla et al. (2017)
Cr	*Chroococcus* sp. and *Nostoc calcicola*	Zhai et al. (2017)
Trace element	*Spirulina* sp.	Oukarroum (2016)
Cd and Zn	*N. rivularis* and *Nostoc linckia*	Liu et al. (2002)
Acid mine drainage effluents	Blue-green algae cyanobacterial mat	Phillips et al. (1995)
Acid mine drainage effluents (Fe, Cu, Pb)	*Spirulina* sp.	Rose et al. (1998)
Cu	*Scenedesmus acuminatus*	Hamed et al. (2017)
Zn^{2+}	*C. sorokiniana*	Hamed et al. (2017)
17β-estradiol, 17α-ethinylestradiol, salicylic acid	*Nannochloris* sp.	Bai and Acharya (2019)
Cu	*Desmodesmus communis* and *Monoraphidium pusillum*	Novák et al. (2020)
Cu	*Scenedesmus* sp.	Cheng et al. (2020)
Ni, V, As, Cd	*Enteromorpha intestinalis, Rhizoclonium riparium, Cystoseira myrica* and *Sargassum boveanum*	Haghshenas et al. (2020)
Cu, Cr, Cd	*Chlorella vulgaris*	Mubashar et al. (2020)
Cu	*Desmodesmus* sp.	Liu et al. (2021)
Cd, Cu	*Chlorella sorokiniana*	León-Vaz et al. (2021)
Cd	*Chlorella pyrenoidosa* and *Scenedesmus acutus*	Chandrashekharaiah et al. (2021)
Cd, Cr	*Parachlorella kessleri Bh-2*	Bauenova et al. (2021)
Cd, Cr	*Dunaliella salina*	Elleuch et al. (2021)
Zn	*Dunaliella* sp. AL-1	Elleuch et al. (2021)
Cr	*Chlorella thermophile*	Majhi and Samantaray (2021)
Cd	*Chlorella vulgaris* and *Coelastrella* sp.	Plöhn et al. (2021)

TABLE 3.1 (*Continued*)
Summary of some potential algal species reported to effectively degrade toxic pollutants

Pollutant	Algal species	Reference
Cr	*Chlamydomonas* sp.	Ayele et al. (2021)
Cd, Zn	*Sargassum polycystum*	Jayakumar et al. (2021)
Cr	*Chlamydomonas moewusii, Auxenochlorella pyrenoidosa* and *Scenedesmus* sp.	Venkatesan and Sathiavelu (2022)
Pb (II)	*Dunaliella salina*	Ziaei et al. (2022)
Cd, Cr, Cu	*Chlorella pyrenoidosa*	Kothari et al. (2022)
Co (II), Methylene blue	*Sargassum latifolium*	Moussa et al. (2022)
Pb (II)	*Phormidium* sp., *Monoraphidium* sp. and *Uronema* sp.	Lakmali et al. (2022)
2,2-*bis* (4-Chlorophenyl)-1,1,1-trichloroethane and 2,2-*bis* (4-chlorophenyl)-1,1,1-dichloroethylene	*Stenotrophomonas* sp.	Xie et al. (2022)
Bisphenol	*Chlorella pyrenoidosa*	Fu et al. (2023)

Bioremediation of explosives using algae has been reviewed (Chekroun et al., 2014). Trinitrotoluene (TNT), an old explosive that has been used extensively globally, is a common contaminant at numerous military installations (Glass, 2005). Earlier research by Cruz-Uribe and Rorrer (2006) claimed that *Portieria hornemannii*, a marine red alga, can eliminate TNT from saltwater. According to Cruz-Uribe et al. (2007), the conversion of TNT to 2-amino-4,6-dinitrotoluene and 4-amino-2,6-dinitrotoluene by green alga *Acrosiphonia coalita*, *Portieria hornemannii* and the red algae *Porphyra yezoensis* resulted in a 20% elimination of TNT from seawater. For bacteria to use TNT as a source of nitrogen, in *Enterobacter cloacae* strain PB2, an enzyme called pentaerythritol tetranitrate reductase must be produced (Dhankher et al., 2012). Using this bacteria, different microalgal species could form a consortium to degrade related chemicals.

As a result of the extensive usage of agrochemicals, such as fertilizers and insecticides, in agriculture, the world's food supply has been guaranteed, but a variety of environmental contaminants including lindane, phenol, dichloro-diphenyl trichloroethane and naphthalene are discharged due to agricultural runoff. These chemicals' endurance in animal fatty acids is boosted by their lipophilic character. *Scenedesmus bijugatus*, *Phormidium tenue*, *Synechococcus elongatus* and *Nostoc linckia* have been reported to degrade, over the course of 30 days, the organic phosphorus insecticides monocrotophos and quinalphos (Megharaj et al., 1987). *Cabomba aquatic*, *Elodea canadensis* and *Leman minor* were tested by Olette et al. (2008) for their ability to absorb and eliminate copper sulphate and dimethomorph (fungicides) as well as flazasulfuron (herbicide). *Scenedesmus quadricauda* has recently been shown to be more efficient at removing the herbicide isoproturon and fungicides dimethomorph and pyrimethanil from their medium (Olette et al., 2010). *Chlamydomonas reinhardtii*, a green alga, demonstrated a remarkable capacity for the rapid absorption and degradation of the herbicide prometryne (Jin et al., 2012).

Genetic engineering has lately been used to improve the bioremediation of organic contaminants (Kaur and Bhatnagar, 2002). It has been effectively attempted to use transgenic technology to create plants with bacterial or animal xenobiotic-degrading genes for in situ bioremediation, heavy metal deposition, as well as the breakdown of diverse xenobiotics, notably explosives and hydrocarbons (Suresh and Ravishankar, 2004). A new generation of genetically modified organisms, consisting of transgenic plants and related microorganisms, is capable of treating water and soil polluted with organic contaminants in an effective and environmentally acceptable manner (Aken et al., 2010, Chekroun et al., 2014).

3.3 RECLAMATION OF ACID MINE DRAINAGE USING ALGAE

Acid mine drainage (AMD) is mostly caused by the discharge of acidic water from operating or abandoned mines that contain sulphide, pyrite and other harmful, toxic heavy metals responsible for significant global environmental concerns. Loss of bio-diversity and degraded aquatic ecosystems are some of the longer-lasting effects. The literature has offered a thorough explanation of the causes of AMD and the various pathways that lead to AMD origin (Johnson and Hallberg, 2005; Kalin et al., 2006; Rambabu et al., 2020). An effective course of AMD treatment and maintenance is crucial to mitigate the effects of the condition and improve ecological sustainability. AMD treatment may not be successful using conventional techniques; thus, efficient techniques for capturing and sequestering heavy metals from wastewater are necessary for environmental remediation and avoiding downstream contamination. A modern and appealing biological technique assuring to provide a viable method of treating AMD is bioremediation using algae strains since it is cost-effective and highly successful at removing metal and sulphates (Punia, 2019). A variety of algae strains have been used to study the bioremediation of acidic-rich streams including *Anabaena, Chlamydomonas, Chlorella, Cladophora, Oscillatoria, Phaeodactylum, Scenedesmus* and *Spirulina* sp., and algae-based treatment can remove a consider-able proportion of heavy metals from AMD (Dean et al., 2019). In general, rather than acting as direct absorbers, algae are most frequently used as phycoremedia-tors because of their ability to increase alkalinity. The metabolism of algal biomass results in high alkalinity, which balances the acidity of the drain stream and facili-tates metal precipitation. By generating inorganic bicarbonate salts, several species of algae produce alkaline substances on a constant basis that work to regulate the acidity of the AMD. Since the early 1990s, AMD or other contaminated wastewa-ters have been cleansed of heavy metals using algae. Specific algae species, includ-ing *Lepocinclis* sp., *Klebsormidium* sp. and *Spirulina* sp., have been investigated for their potential contribution to the bioremediation of AMD (Das et al., 2009). Additionally, making use of fungi and algae as a consortium in AMD has been sug-gested. According to a 2011 study on the algal biosorption of heavy metals in AMD from coal mining in Brazil, algal biomass can accumulate heavy metals constitut-ing about 6.3% of the biomass (Freitas et al., 2011). At least 14 algae species have been investigated for their capability to absorb heavy metal elements from AMD and other contamination sources (Ben and Baghour, 2013). The employment of specific types of microalgae for the bioremediation of AMD has recently been reviewed (Bwapwa et al., 2017).

3.4 ROLE OF ALGAE IN LANDFILL LEACHATE TREATMENT

The application of the algal community to treat a variety of hazardous toxicants associated with landfill leachate has been reported in the literature. Generally, the pre-treatment or diluted leachate treatment was usually achieved using this green technology; however, the use of algae for the biological treatment is the major focus of research as it can accomplish dual goals, that is, providing the best source of algal biomass and byproducts while also treating landfill leachate, thereby fulfilling sustainable energy goal (Dogaris et al., 2019). As described earlier, algae have the ability to grow in a toxic environment and can utilize the same nutrient source and simply biotransform them into less toxic substances. Landfill leachate is reported to act as an excellent carbon, nitrogen and phosphorus source and to allow algal cells to bioassimilate to fulfil their metabolic requirements (Cai et al., 2013). Different studies have documented the employment of microalgae in the cultivation of closed photobioreactors or open systems including ponds and raceways to appropriately treat landfill leachate (Zittelli et al., 2013). For example, the study conducted by Borowitzka and Moheimani (2013) excellently reviews the usage of algae for large-scale open cultivation for landfill leachate treatment. Several small-scale landfill leachate treatment methods are documented in the literature. As an illustration, a study conducted by Casazza and Rovatti (2018) reported the usage of the chlorophyta green alga *C. vulgaris* to treat ammonia, nitrate and nitrite levels from landfill leachate. Similarly, in another study, approximately 100% of the ammonia, 27% of the nitrate and 100% of the phosphate were estimated to be eliminated by *C. vulgaris* from the landfill leachate effluent. Furthermore, it was concluded that the addition of phosphorous enhances the capacity of *C. vulgaris* for growth and nutrient removal. In another study, Desai (2016) concluded that the multispecies removal efficiencies were effective in treating raw and pretreated landfill leachate at various dilutions for the treatment of nitrogenous waste effluents. Additionally, a variety of species like *Chlamydomonas*, *Oscillatoria* and *Scenedesmus* are reported as excellent players in the removal of landfill leachate effluents in biological management setups (Cheng and Tian, 2013; Nordin et al., 2017).

3.5 CONCLUSION

The excessive deposition of organic and heavy metals in the environment entails detrimental impacts on both human health and the ecosystem, as well as severely influences the integrity of many aquatic ecosystems. Utilizing algae to eliminate, decompose, or render harmless these toxic contaminants has recently attracted more attention as this approach appears more productive in contrast to conventional methods. Due to their widespread availability, low cost, superior metal removal effectiveness and eco-friendliness, microalgae are increasingly being used in heavy metal and other noxious waste phycoremediation. Lately, genetic engineering has provided a method for promoting better organic pollutant absorption and bioremediation as well as increased algal tolerance to these contaminants.

REFERENCES

Adey, W. H., Kangas, P. C., & Mulbry, W. (2011). Algal turf scrubbing: Cleaning surface waters with solar energy while producing a biofuel. *Bioscience, 61*, 434–441.

Ahmad, A., Banat, F., Alsafar, H., & Hasan, S. W. (2022). Algae biotechnology for industrial wastewater treatment, bioenergy production, and high-value bioproducts. *Science of The Total Environment, 806*, 150585.

Aken, B. V., Correa, P. A., & Schnoor, J. L. (2010). Phytoremediation of polychlorinated biphenyls: New trends and promises. *Environmental Science & Technology, 44*, 2767–2776.

Anbuchezhian, R., Karuppiah, V., & Li, Z. (2015). Prospect of marine algae for production of industrially important chemicals. In: *Algal biorefinery: An integrated approach* (pp. 195–217). Springer.

Ayele, A., Suresh, A., Benor, S., & Konwarh, R. (2021). Optimization of chromium (VI) removal by indigenous microalga (Chlamydomonas sp.)-based biosorbent using response surface methodology. *Water Environment Research, 93*(8), 1276–1288.

Baghour, M. (2019). Algal degradation of organic pollutants. In: Martínez, L., Kharissova, O., & Kharisov, B. (Eds.), *Handbook of ecomaterials* (pp. 1–22). Springer.

Bai, X., & Acharya, K. (2019). Removal of seven Endocrine Disrupting Chemicals (EDCs) from municipal wastewater effluents by a freshwater green alga. *Environmental Pollution, 247*, 534–540.

Bauddh, K., & Korstad, J. (2022). Phycoremediation: Use of algae to sequester heavy metals. *Hydrobiology, 1*(3), 288–303.

Bauenova, M. O., Sadvakasova, A. K., Mustapayeva, Z. O., Kokociński, M., Zayadan, B. K., Wojciechowicz, M. K., . . . & Allakhverdiev, S. I. (2021). Potential of microalgae Parachlorella kessleri Bh-2 as bioremediation agent of heavy metals cadmium and chromium. *Algal Research, 59*, 102463.

Ben, C. K., & Baghour, M. (2013). The role of algae in phytoremediation of heavy metals: A review. *Journal of Materials and Environmental Science, 4*(6), 873e880.

Borowitzka, M. A., & Moheimani, N. R. (2013). Open pond culture systems. In: *Algae for biofuels and energy* (pp. 133–152). Springer.

Brinza, L., Dring, M. J., & Gavrilescu, M. (2007). Marine micro and macroalgal species as biosorbents for heavy metals. *Environmental Engineering & Management Journal, 6*, 237–251.

Bwapwa, J. K., Jaiyeola, A. T., & Chetty, R. (2017). Bioremediation of acid mine drainage using algae strains: A review. *South African Journal of C9hemical Engineering, 24*, 62–70.

Cai, T., Park, S. Y., & Li, Y. (2013). Nutrient recovery from wastewater streams by microalgae: Status and prospects. *Renewable and Sustainable Energy Reviews, 19*, 360–369.

Casazza, A. A., & Rovatti, M. (2018). Reduction of nitrogen content in landfill leachate using microalgae. *Desalination and Water Treatment, 127*, 71–74.

Cerniglia, C. E. (1992). Biodegradation of polycyclic aromatic hydrocarbons. *Biodegradation, 3*, 351–368.

Chandrashekharaiah, P. S., Sanyal, D., Dasgupta, S., & Banik, A. (2021). Cadmium biosorption and biomass production by two freshwater microalgae Scenedesmus acutus and Chlorella pyrenoidosa: An integrated approach. *Chemosphere, 269*, 128755.

Chekroun, K. B., Sánchez, E., & Baghour, M. (2014). The role of algae in bioremediation of organic pollutants. *Journal Issues ISSN, 2360*, 8803.

Cheng, H. X., & Tian, G. M. (2013). Preliminary evaluation of a newly isolated microalga Scenedesmus sp. CHX1 for treating landfill leachate. In *2013 third international conference on intelligent system design and engineering applications* (pp. 1057–1060). IEEE.

Cheng, P., Osei-Wusu, D., Zhou, C., Wang, Y., Xu, Z., Chang, T., & Huo, S. (2020). The effects of refractory pollutants in swine wastewater on the growth of Scenedesmus sp. with biofilm attached culture. *International Journal of Phytoremediation, 22*(3), 241–250.

Cruz-Uribe, O., & Rorrer, G. L. (2006). Uptake and transformation of 2,4,6-trinitrotolune (TNT) from seawater by microplantlet suspension cultures of the marine red macroalga *Portieria hornemannii. Biotechnology and Bioengineering, 93*, 401–412.

Cruz-Uribe, O., Cheney, D. P., & Rorrer, G. L. (2007). Comparison of TNT removal from seawater by three marine macroalgae. *Chemosphere, 67*(8), 1469–1476.

Das, B. K., Roy, A., Koschorreck, M., Mandal, S. M., Potthoff, K. W., & Bhattachary, J. (2009). Occurrence and role of algae and fungi in acid mine drainage environment with special reference to metals and sulfate immobilization. *Water Research, 43*(4), 883e894.

Davis, T. A., Volesky, B., & Mucci, A. (2003). A review of the biochemistry of heavy metal biosorption by brown algae. *Water Research, 37*, 4311.

Dean, A. P., Hartley, A., McIntosh, O. A., Smith, A., Feord, H. K., Holmberg, N. H., . . . & Pittman, J. K. (2019). Metabolic adaptation of a Chlamydomonas acidophila strain isolated from acid mine drainage ponds with low eukaryotic diversity. *Science of the Total Environment, 647*, 75–87.

Desai, M. (2016). The efcacy of using the microalgae Chlorella sp. for the treatment of hazardous landfll leachate [Doctoral dissertation]. https://pdfs.semanticscholar.org/b9fc/4 23f34e61a488ced4a5e54b9ec967b2f4150

Dhankher, O. P., Pilon-Smits, E. A., Meagher, R. B., & Doty, S. (2012). Biotechnological approaches for phytoremediation. In: *Plant biotechnology and agriculture* (pp. 309–328). Academic Press.

Dogaris, I., Loya, B., Cox, J., & Philippidis, G. (2019). Study of landfill leachate as a sustainable source of water and nutrients for algal biofuels and bioproducts using the microalga Picochlorum oculatum in a novel scalable bioreactor. *Bioresource Technology, 282*, 18–27.

Dönmez, G. Ç., Aksu, Z., Öztürk, A., & Kutsal, T. (1999). A comparative study on heavy metal biosorption characteristics of some algae. *Process Biochemistry, 34*(9), 885–892.

Dwivedi, S. (2012). Bioremediation of heavy metal by algae: Current and future perspective. *Journal of Advanced Laboratory Research in Biology, 3*(3), 195–199.

Elleuch, J., Amor, F. B., Chaaben, Z., Frikha, F., Michaud, P., Fendri, I., & Abdelkafi, S. (2021). Zinc biosorption by Dunaliella sp. AL-1: Mechanism and effects on cell metabolism. *Science of the Total Environment, 773*, 145024.

Elleuch, J., Hmani, R., Drira, M., Michaud, P., Fendri, I., & Abdelkafi, S. (2021). Potential of three local marine microalgae from Tunisian coasts for cadmium, lead and chromium removals. *Science of the Total Environment, 799*, 149464.

Ellis, J. T., Hengge, N. N., Sims, R. C., & Miller, C. D. (2012). Acetone, butanol, and ethanol production from wastewater algae. *Bioresource Technology, 111*, 491–495.

Freitas, A. P. P., Schneider, I. A. H., & Schwartzbold, A. (2011). Biosorption of heavy metals by algal communities in water streams affected by the acid mine drainage in the coal-mining region of Santa Catarina state. *Brazil. Minerals Engineering, 24*, 1215e1218.

Fu, W., Yang, Y., & Song, D. (2023). Enhanced degradation of bisphenol A: Influence of optimization of removal, kinetic model studies, application of machine learning and microalgae-bacteria consortia. *Science of the Total Environment, 858,* 159876.

Gattullo, C. E., Bährs, H., Steinberg, C. E., & Loffredo, E. (2012). Removal of bisphenol a by the freshwater green alga Monoraphidium braunii and the role of natural organic matter. *Science of the Total Environment, 416,* 501–506.

Glass, D. J. (2005). Commercial use of Genetically Modified Organisms (GMOs) in bioremediation and phytoremediation. In: Fingerman, M., & Nagabhushanam, R. (Eds), *Bioremediation of aquatic and terrestrial ecosystems* (pp. 41–96). Science Publishers, Inc.

Gonçalves, A. L. (2021). The use of microalgae and cyanobacteria in the improvement of agricultural practices: A review on their biofertilising, biostimulating and biopesticide roles. *Applied Sciences, 11,* 871.

Gosavi, K., Sammut, J., Gifford, S., & Jankowski, J. (2004). Macroalgal biomonitors of trace metal contamination in acid sulfate soil aquaculture ponds. *Science of the Total Environment, 324,* 25–39.

Haghshenas, V., Kafaei, R., Tahmasebi, R., Dobaradaran, S., Hashemi, S., Sahebi, S., . . . & Ramavandi, B. (2020). Potential of green/brown algae for monitoring of metal (loid) s pollution in the coastal seawater and sediments of the Persian Gulf: Ecological and health risk assessment. *Environmental Science and Pollution Research, 27,* 7463–7475.

Hamed, S. M., Selim, S., Klöck, G., & AbdElgawad, H. (2017). Sensitivity of two green microalgae to copper stress: Growth, oxidative and antioxidants analyses. *Ecotoxicology and Environmental Safety, 144,* 19–25.

Hamed, S. M., Zinta, G., Klöck, G., Asard, H., Selim, S., & AbdElgawad, H. (2017). Zinc-induced differential oxidative stress and antioxidant responses in Chlorella sorokiniana and Scenedesmus acuminatus. *Ecotoxicology and Environmental Safety, 140,* 256–263.

Hernández-Ávila, J., Salinas-Rodríguez, E., Cerecedo-Sáenz, E., Reyes-Valderrama, I., Arenas-Flores, A., Román-Gutiérrez, A. D., & Rodríguez-Lugo, V. (2017). Diatoms and their capability for heavy metal removal by cationic exchange. *Metals, 7,* 169.

Hong, Y. W., Yuan, D. X., Lin, Q. M., & Yang, T. L. (2008). Accumulation and biodegradation of phenanthrene and fluoranthene by the algae enriched from a mangrove aquatic ecosystem. *Marine Pollution Bulletin, 6,* 1400–1405.

IARC (International Agency for Research on Cancer). (1983). In: IARC (Ed.), *IARC monographs on the evaluation of the carcinogenic risk of chemicals to humans: Polynuclear aromatic compounds part I.* IARC Press.

Izquierdo, M., Gabaldón, C., Marzal, P., & Álvarez-Hornos, F. J. (2010). Modeling of copper fixed-bed biosorption from wastewater by Posidonia oceanica. *Bioresource Technology, 101*(2), 510–517.

Jayakumar, V., Govindaradjane, S., Rajamohan, N., & Rajasimman, M. (2021). Biosorption potential of brown algae, Sargassum polycystum, for the removal of toxic metals, cadmium and zinc. *Environmental Science and Pollution Research,* 1–14.

Jin, Z. P., Luo, K., Zhang, S., Zheng, Q., & Yang, H. (2012). Bioaccumulation and catabolism of prometryne in green algae. *Chemosphere, 87,* 278–284.

Johnson, D. B., & Hallberg, K. B. (2005). Acid mine drainage remediation options: A review. *Science of the Total Environment, 338*(1–2), 3–14.

Kalin, M., Fyson, A., & Wheeler, W. N. (2006). The chemistry of conventional and alternative treatment systems for the neutralization of acid mine drainage. *Science of the Total Envionment, 366*(2–3), 395–408.

Kaur, I., & Bhatnagar, A. K. (2002). Algae-dependent bioremediation of hazardous wastes. *Progress in Industrial Microbiology, 36,* 457–516.

Kobielska, P. A., Howarth, A. J., Farha, O. K., & Nayak, S. (2018). Metal—Organic frameworks for heavy metal removal from water. *Coordination Chemistry Reviews, 358,* 92–107.

Kothari, R., Pandey, A., Ahmad, S., Singh, H. M., Pathak, V. V., Tyagi, V. V., . . . & Sari, A. (2022). Utilization of Chlorella pyrenoidosa for remediation of common effluent treatment plant wastewater in coupling with Co-relational study: An experimental approach. *Bulletin of Environmental Contamination and Toxicology,* 1–11.

Koul, B., Sharma, K., & Shah, M. P. (2022). Phycoremediation: A sustainable alternative in Wastewater Treatment (WWT) regime. *Environmental Technology & Innovation, 25,* 102040.

Kumar, K. S., Dahms, H. U., Won, E. J., Lee, J. S., & Shin, K. H. (2015). Microalgae—A promising tool for heavy metal remediation. *Ecotoxicology and Environmental Safety, 113,* 329–352.

Lakmali, W. M., Athukorala, A. S. N., & Jayasundera, K. B. (2022). Investigation of Pb (II) bioremediation potential of algae and cyanobacteria strains isolated from polluted water. *Water Science and Engineering, 15*(3), 237–246.

Lei, A. P., Wong, Y. S., & Tam, N. F. Y. (2002). Removal of pyrene by different microalgal species. *Water Science and Technology, 46*(11–12), 195–201.

León-Vaz, A., León, R., Giráldez, I., Vega, J. M., & Vigara, J. (2021). Impact of heavy metals in the microalga Chlorella sorokiniana and assessment of its potential use in cadmium bioremediation. *Aquatic Toxicology, 239,* 105941.

Leung, M. (2004). Bioremediation: Techniques for cleaning up a mess. *Bio Teach Journal, 2,* 18–22.

Lima, A. L. C., Farrington, J. W., & Reddy, C. M. (2005). Combustion-derived polycyclic aromatic hydrocarbons in the environment: A review. *Environmental Forensics, 6,* 109–131.

Liu, L., Lin, X., Luo, L., Yang, J., Luo, J., Liao, X., & Cheng, H. (2021). Biosorption of copper ions through microalgae from piggery digestate: Optimization, kinetic, isotherm and mechanism. *Journal of Cleaner Production, 319,* 128724.

Liu, Y., Yang, S. F., Tan, S. F., Lin, Y. M., & Tay, J. H. (2002). Aerobic granules: A novel zinc biosorbent. *Letters in Applied Microbiology, 35*(6), 548–551.

Majhi, P., & Samantaray, S. M. (2021). Bio-reduction of hexavalent chromium by an indigenous green alga and its impact on the germination of rice seed in chromium enriched environment. *Bioremediation Journal, 25*(2), 128–147.

Marella, T. K., Datta, A., Patil, M. D., Dixit, S., & Tiwari, A. (2019). Biodiesel production through algal cultivation in urban wastewater using algal floway. *Bioresource Technology, 280,* 222–228.

Marella, T. K., Parine, N. R., & Tiwari, A. (2018). Potential of diatom consortium developed by nutrient enrichment for biodiesel production and simultaneous nutrient removal from waste water. *Saudi Journal of Biological Sciences, 25,* 704–770.

Marella, T. K., Saxena, A., & Tiwari, A. (2020). Diatom mediated heavy metal remediation: A review. *Bioresource Technology, 305,* 123068.

Megharaj, M., Venkateswarlu, K., & Rao, A. S. (1987). Metabolism of monocrotophos and quinalphos by algae isolated from soil. *Bulletin of Environmental Contamination and Toxicology, 39,* 251–256.

Mei, L. I., Xitao, X., Renhao, X. U. E., & Zhili, L. I. U. (2006). Effects of strontium-induced stress on marine microalgae Platymonas subcordiformis (Chlorophyta: Volvocales). *Chinese Journal of Oceanology and Limnology, 24,* 154–160.

Mona, S., Kaushik, A., & Kaushik, C. P. (2011). Biosorption of reactive dye by waste biomass of *Nostoc linckia. Ecological Engineering, 37,* 1589–1594.

Mondal, M., Halder, G., Oinam, G., Indrama, T., & Tiwari, O. N. (2019). Bioremediation of organic and inorganic pollutants using microalgae. In: *New and future developments in microbial biotechnology and bioengineering* (pp. 223–235). Elsevier.

Monteiro, C. M., Castro, P. M., & Malcata, F. X. (2012). Metal uptake by microalgae: Underlying mechanisms and practical applications. *Biotechnology Progress, 28*(2), 299–311.

Moussa, Z., Ghoniem, A. A., Elsayed, A., Alotaibi, A. S., Alenzi, A. M., Hamed, S. E., . . . & Saber, W. I. (2022). Innovative binary sorption of Cobalt (II) and methylene blue by Sargassum latifolium using Taguchi and hybrid artificial neural network paradigms. *Scientific Reports, 12*(1), 18291.

Mubashar, M., Naveed, M., Mustafa, A., Ashraf, S., Shehzad Baig, K., Alamri, S., . . . & Kalaji, H. M. (2020). Experimental investigation of Chlorella vulgaris and Enterobacter sp. MN17 for decolorization and removal of heavy metals from textile wastewater. *Water, 12*(11), 3034.

Muñoz, R., & Guieysse, B. (2006). Algal—Bacterial processes for the treatment of hazardous contaminants: A review. *Water Research, 40*(15), 2799–2815.

Munoz, R., Guieysse, B., & Mattiasson, B. (2003). Phenanthrene biodegradation by an algal-bacterial consortium in two-phase partitioning bioreactors. *Applied Microbiology and Biotechnology, 61*(3), 261–267.

Nassiri, Y., Mansot, J., Wéry, J., Ginsburger-Vogel, T., & Amiard, J. (1997). Ultrastructural and electron energy loss spectroscopy studies of sequestration mechanisms of Cd and Cu in the marine diatom Skeletonema costatum. *Archives of Environmental Contamination and Toxicology, 33*, 147–155.

Nhat, P. V. H., Ngo, H. H., Guo, W. S., Chang, S. W., Nguyen, D. D., Nguyen, P. D., . . . & Guo, J. B. (2018). Can algae-based technologies be an affordable green process for biofuel production and wastewater remediation? *Bioresource Technology, 256*, 491–501.

Nordin, N., Yusof, N., & Samsudin, S. (2017). Biomass production of Chlorella sp., Scenedesmus sp., and Oscillatoria sp. in nitrified landfill leachate. *Waste and Biomass Valorization, 8*(7), 2301–2311.

Novák, Z., Harangi, S., Baranyai, E., Gonda, S., B-Béres, V., & Bácsi, I. (2020). Effects of metal quantity and quality to the removal of zinc and copper by two common green microalgae (Chlorophyceae) species. *Phycological Research, 68*(3), 227–235.

Olette, R., Couderchet, M., Biagianti, S., & Eullaffroy, P. (2008). Toxicity and removal of pesticides by selected aquatic plants. *Chemosphere, 70*, 1414–1421.

Olette, R., Couderchet, M., Biagianti, S., & Eullaffroy, P. (2010). Fungicides and herbicide removal in Scenedesmus cell suspensions. *Chemosphere, 79*, 117–123.

Oukarroum, A. (2016). Alleviation of metal-induced toxicity in aquatic plants by exogenous compounds: A mini-review. *Water, Air, & Soil Pollution, 227*(6), 1–9.

Ouyang, W., Wang, Y., Lin, C., He, M., Hao, F., Liu, H., & Zhu, W. (2018). Heavy metal loss from agricultural watershed to aquatic system: A sciento metrics review. *Science of the Total Environment, 37*, 208–220.

Pathak, B., Gupta, S., & Verma, R. (2018). Biosorption and biodegradation of Polycyclic Aromatic Hydrocarbons (PAHs) by microalgae. In: *Green adsorbents for pollutant removal* (pp. 215–247). Springer.

Perales-Vela, H. V., Peña-Castro, J. M., & Cañizares Villanueva, R. O. (2006). Heavy metal detoxification in eukaryoticmicroalgae. *Chemosphere, 64*, 1–10.

Perrein-Ettajani, H., Amiard, J., Haure, J., & Renaud, C. (1999). Effets des métaux (Ag, Cd, Cu) sur la composition biochimique et compartimentation de ces métaux chez deux microalgues Skeletonema costatum et Tetraselmis suecica. *Canadian Journal of Fisheries and Aquatic Sciences, 56*, 1757–1765.

Phillips, P., Bender, J., Simms, R., Rodriguez-Eaton, S., & Britt, C. (1995). Manganese removal from acid coal-mine drainage by a pond containing green algae and microbial mat. *Water Science and Technology, 31*(12), 161–170.

Plöhn, M., Escudero-Onate, C., & Funk, C. (2021). Biosorption of Cd (II) by Nordic microalgae: Tolerance, kinetics and equilibrium studies. *Algal Research, 59*, 102471.

Priya, M., Gurung, N., Mukherjee, K., & Bose, S. (2014). Microalgae in removal of heavy metal and organic pollutants from soil. In: *Microbial biodegradation and bioremediation* (pp. 519–537). Elsevier.

Punia, A. (2019). Innovative and sustainable approach for phytoremediation of mine tailings: A review. *Waste Disposal & Sustainable Energy, 1*(3), 169–176.

Rambabu, K., Banat, F., Pham, Q. M., Ho, S. H., Ren, N. Q., & Show, P. L. (2020). Biological remediation of acid mine drainage: Review of past trends and current outlook. *Environmental Science and Ecotechnology, 2*, 100024.

Rose, P. D., Boshoff, G. A., Van Hille, R. P., Wallace, L. C. M., Dunn, K. M., & Duncan, J. R. (1998). An integrated algal sulphate reducing high rate ponding process for the treatment of acid mine drainage wastewaters. *Biodegradation, 9*(3), 247–257.

Sarmah, P., & Rout, J. (2020). Role of algae and cyanobacteria in bioremediation: Prospects in polyethylene biodegradation. In: *Advances in cyanobacterial biology* (pp. 333–349). Academic Press.

Senthilkumar, R., Vijayaraghavan, K., Thilakavathi, M., Iyer, P. V. R., & Velan, M. (2006). Seaweeds for the remediation of wastewaters contaminated with zinc (II) ions. *Journal of Hazardous Materials, 136*(3), 791–799.

Shukla, R., Pandey, A. K., & Mishra, K. N. (2017). The efficacy of modified cyanobacterial biomass to remove Cr (VI) ions from aqueous solution. *Indian Journal of Scientific Research*, 31–34.

Sood, A., Renuka, N., Prasanna, R., & Ahluwalia, A. S. (2015). Cyanobacteria as potential options for wastewater treatment. In: *Phytoremediation* (pp. 83–93). Springer.

Sunda, W. G. (1989). Trace metal interactions with marine phytoplankton. *Biological Oceanography, 6*, 411–442.

Suresh, B., & Ravishankar, G. A. (2004). Phytoremediation—A novel and promising approach for environmental clean-up. *Critical Reviews in Biotechnology, 24*, 97–124.

Tiwari, A., & Marella, T. K. (2020). Algal biomass: Potential renewable feedstock for biofuel production. In: *Substrate analysis for effective biofuels production* (pp. 1–32). Springer.

Valiente Moro, C., Bricheux, G., Portelli, C., & Bohatier, J. (2012). Comparative effects of the herbicides chlortoluron and mesotrione on freshwater microalgae. *Environmental Toxicology and Chemistry, 31*(4), 778–786.

Venkatesan, P., & Sathiavelu, M. (2022). Effective kinetic modeling and phycoremediation of Cr (IV) ions from tannery effluent by using microalgae—Chlamydomonas moewusii, Auxenochlorella pyrenoidosa, Scenedesmus sp. *Bioremediation Journal*, 1–20.

Vilar, V. J., Botelho, C. M., & Boaventura, R. A. (2008). Lead uptake by algae Gelidium and composite material particles in a packed bed column. *Chemical Engineering Journal, 144*(3), 420–430.

Xie, H., Liu, R., Xu, Y., Liu, X., Sun, F., Ma, Y., & Wang, Y. (2022). Effect of in situ bioremediation of soil contaminated with DDT and DDE by stenotrophomonas sp. strain DXZ9 and ryegrass on soil microorganism. *Microbiology Research, 13*(1), 64–86.

Yan, A., Wang, Y., Tan, S. N., Mohd Yusof, M. L., Ghosh, S., & Chen, Z. (2020). Phytoremediation: A promising approach for revegetation of heavy metal-polluted land. *Frontiers in Plant Science, 11*, 359.

Zhai, J., Li, X., Li, W., Rahaman, M. H., Zhao, Y., Wei, B., & Wei, H. (2017). Optimization of biomass production and nutrients removal by Spirulina platensis from municipal wastewater. *Ecological Engineering, 108*, 83–92.

Ziaei, S., Ahmadzadeh, H., & Es'haghi, Z. (2022). Dynamic removal of Pb (II) by live Dunaliella salina: A competitive uptake and isotherm model study. https://doi.org/10.21203/rs.3.rs-2226836/v1

Zittelli, G. C., Biondi, N., Rodolfi, L., & Tredici, M. R. (2013). Photobioreactors for mass production of microalgae. *Handbook of Microalgal Culture: Applied Phycology and Biotechnology*, 225–266.

4 Heavy Metals, Powerful Environmental Poisons, from Waste Disposal Site

Their Environmental Effects and Remediation

Neesa Majeed, Burhan Hamid,
Tanveer Ahmad Mir, Ali Mohd Yatoo,
and R. Z. Sayyed

4.1 INTRODUCTION

The industrialization process has advanced rapidly during the past 100 years. As a result, there is now more stipulation for the reckless utilization of natural resources on the earth, which has created a major concern of environmental degradation throughout the world (Gautam et al., 2016). Numerous pollutants such as radioactive isotopes, organo-metallic compounds, gaseous pollutants, nanoparticles, inorganic ions, and organic pollutants have crudely damaged the ecosystem (Beasley et al., 2012). Accumulation of toxic and transition metals results in environmental degradation through pollut-ants from the quickly developing industrial districts, the dumping of metal wastes, petroleum refining and paints, and so on. The sustainability of the environment is also threatened by the use of fertilizers and pesticides (Yatoo et al., 2021, 2022), sewage sludge, wastewater irrigation, spills of petrochemicals, coal combustion byproducts, and air deposition from smelting (Nath et al., 2019). Heavy metals are usually metals having a specific density > 5 g/cm^3 and negatively influence the surroundings and liv-ing creatures (Duruibe et al., 2007). The deposition of heavy metals in the environment comes from a variety of natural processes like forest fires, weathering of minerals, and volcanic activities, and by human activities through mining, the use of pesticides, com-bustion of fossil fuel, waste disposal, fertilizer industries, and so on (Zhang and Wang, 2020; Martins et al., 2016; Dixit et al., 2015). The mining and industrial sectors are the main sources responsible for the contamination of the soil, groundwater, and air with heavy metals. Pollution of heavy metals is caused by the application of sewage irriga-tion, industrial waste, and sludge farms due to the acceleration of urbanization and industry as well as the discharge of various types of exhaust gases (Wan et al., 2016; Rezania et al., 2016; Huang et al., 2016; Chandrasekaran et al., 2015).

DOI: 10.1201/9781003359326-4

In the environment heavy metals are present in various oxidation states and have different toxicity levels. These metals produce toxic effects by interfering with biological macromolecules. There are potentially severe hazards in living organisms because of the increased exposition of heavy metals (Sarma et al., 2017). Release of heavy metals into the soil could have an adverse impact on human health and the ecosystem. Absorption of these toxic compounds in living organisms occurs through different mechanisms including oral ingestion, inhalation, skin contact, absorption, and soil diet (Jaishankar et al., 2014). It is estimated that more than 20 million hectares of land around the world is polluted by various heavy metals such as Zn, Hg, Pb, Ni, As, Cr, Se, Cd, Cu, and Co, with current soil concentrations above regulatory standards (Wuana and Okieimen, 2011). Accumulation of heavy metals in the human body can occur through the intake of different foods such as vegetables grown in contaminated soils or fish or oysters polluted through the food chain, which can lead to serious health issues such as cancer (Dixit et al., 2015; Ali and Khan, 2018). Heavy metal toxicity has also a detrimental effect on various physiological activities of the plant including interference in the electron transport chain, photosynthesis, respiration, and cell division (Jadia and Fulekar, 2009; Pourrut et al., 2011).

Presently, heavy metal polluted areas have become severe environmental issues, and there is an imperative need to treat these heavy metal ions. There are numerous methods that rely on chemical and electrochemical procedures that have been used to remediate the polluted areas including ultrafiltration, ion exchange, electrodialysis, and chemical precipitation (Crini and Lichtfouse, 2018). Among them, the extensive method used for heavy metal detoxification is chemical precipitation, which is less expensive and effective (Azimi et al., 2017). However, chemical precipitation has various drawbacks, including the emergence of low-density slurry, secondary contaminants, and difficulties removing large areas of heavy metal ions. The ion exchange method has high efficacy removal of heavy metals, fast kinetics, and high treatment capacity (Zamri et al., 2017). Moreover, mainly ion exchange methods should be utilized in an acidic environment that minimizes its implementations. High-performance membrane filtration alternatives are available for heavy metal contamination; however, membrane material manufacture is highly intricate (Bhattacharjee et al., 2021). Conventional remediation methods have many limitations such as they are expensive for implementation and produce toxic chemicals which are mutagenic and less eco-friendly (Qazilbash et al., 2004). Therefore, a highly efficient, economical, and simple method that is eco-friendly is crucially required. Currently, bioremediation methods rely on microbes because of their high capability, low price, and environmental friendliness, particularly at low metal levels (Yin et al., 2019).

4.2 SOURCES OF HEAVY METALS

Heavy metals have been naturally available in the earth's crust since its creation, but their enormous use has resulted in the accumulation of these metallic compounds within aquatic as well as terrestrial environments (Gautam et al., 2016). Anthropogenic activities are considered the primary cause of heavy metal pollution; these activities include smelting, metal mining, foundries, and metals leaching from a variety of sources, such as road construction, runoffs, waste dumps, automobiles, livestock manure, and landfills (Masindi and Muedi, 2018; Briffa et al., 2020). The use of fertilizers, insecticides, and herbicides in agricultural fields is considered the secondary cause of heavy metal contamination (Alloway, 2013). Natural processes including geological weathering, volcanic activity, soil erosion, and evaporation of metals from soil can also lead to a rise in heavy metal pollution (Nachana, 2019). The various sources of heavy metals in the environment are given in Table 4.1.

Effluents that are directly discharged from tannery industries have become a growing environmental issue. The wastewater from tanneries contains a complex mixture of inorganic and organic compounds (Yusif et al., 2016). Effluent from a tannery is extremely polluted as it carries unevenly suspended solids, sulphite, manganese (Mn), copper (Cu), chromium (Cr), cadmium (Cd), and sulphate (Zahid et al., 2006; Mondal et al., 2005). Chrome liquor and powder are also used in the tannery industry, which has extremely toxic heavy metals (Cr^{6+}) that cause water pollution (Bashi, 2021). Lead (Pb) and cadmium (Cd), which are highly poisonous to humans, animals, and plants, are major contaminants of the environment and cause damage to the structure of DNA, cell membranes, and enzymes. The effluent from the dye industry is also the main source of heavy metal toxicity (Shifaw, 2018; Ayangbenro and Babalola, 2017; Sarker et al., 2015). The heavy metals in textile dye effluents including zinc, lead, and cadmium present either as complex metals or in free ionic form, usually originating from the dyeing process, are highly carcinogenic (Basha and Rajaganesh, 2014). Azo dye, which is the main pollutant, contributes up to 70% of the textile and paper industry. Industries involved in electrochemical reaction, anodizing–cleaning, milling, and etching create active heavy metals like Pt, Pb, Cd,

---

TABLE 4.1
Sources of heavy metal pollution

Heavy metals	Sources	References
Chromium (Cr)	Chromium steel, coal burning, fertilizers, metal plating tanneries, oil well drilling, and pigment oxidants	Ghani and Ghani (2011)
Arsenic (As)	Fuel burning, thermal power plants, natural/geogenic processes, and smelting operations	Medfu Tarekegn et al. (2020)
Copper (Cu)	Fungicides, batteries, electronic equipment, mining, insecticides, and catalysts	Gossuin and Vuong (2018); Batool et al. (2017)
Iron (Fe)	Mining, coal seams	Rajak et al. (2020)
Lead (Pb)	Metal plating and finishing operations, wastes from battery industries, factory chimneys, additives in gasoline and pigments, soil wastes, and smelting of ores	Yeboah et al. (2019)
Mercury (Hg)	Agriculture industry, pulp and paper preservatives, pharmaceuticals, industries producing chlorine and caustic soda	Morais et al. (2012)
Cadmium (Cd)	Smelting, volcanic eruptions, mining, tobacco smoking, weathering, fertilizers manufacturing, and municipal waste incineration	Bhat et al. (2019)

and Cr. Heavy metal wastes, such as Pb, Ni, and Sn, can be produced during the fabrication of printed circuit boards (Veit et al., 2005). The major sources of mercury pollution include incineration, municipal wastewater, and discharge of industrial wastewater (Chen et al., 2012). Inorganic pigment manufacturing and petroleum refining can generate chromium and cadmium contamination. These anthropogenic activities are primarily responsible for contamination of the environment with heavy metals, which intimidate human health (Vareda et al., 2019).

4.3 TOXIC EFFECTS OF HEAVY METALS

4.3.1 EFFECTS OF HEAVY METALS ON ENVIRONMENT

Heavy metal toxicity refers to a metal's ability to have negative impacts on the environment and is usually determined by the effective concentration and availability of heavy metals (Rasmussen et al., 2000). Heavy metal contamination has turned into a serious menace to human beings and ecosystems in the environment (Igiri et al., 2018; Okolo et al., 2016; Hrynkiewicz and Baum, 2014; Deepa and Suresha, 2014). Due to a lack of environment management schemes, most industries discharge raw wastes directly into the environment, resulting in major environmental and public health problems (Martins et al., 2016). Heavy metal influences the drinking water, food quality, land use, and food chain, which poses a danger to humans and the ecosystem (Adhikari et al., 2022). The toxicity of heavy metals involves various mechanisms, including disrupting the function of enzymes, interfering with DNA synthesis, generation of reactive oxygen species, destroying ion regulation, and disrupting protein synthesis. Heavy metal can change the conformation of enzymes, thus inhibiting vital enzymatic activity (Gauthier et al., 2014). Besides, cell surface adhesion heavy metals enter cells through transmembrane carriers and ion channels resulting in ion imbalance (Chen et al., 2020). Plants are among the organisms most impacted by heavy metal poisoning, as their normal physiological functions are significantly hampered. Through various laboratory experiments, it has been shown that elevated levels of heavy metals have a deleterious impact on plant activities such as electron transport chain, photosynthesis, respiration, and cell division (Pourrut et al., 2011; Jadia and Fulekar, 2009). Furthermore, due to high metal toxicity in plant cell, enzymes of cytoplasm are inhibited, which disrupts cell structures by oxidative stress, resulting in reduced plant metabolism and growth (Gaur et al., 2014; Chibuike and Obiora, 2014). Various toxic effects of different heavy metals are mentioned in Table 4.2.

4.3.2 EFFECT OF HEAVY METALS ON SOIL

Soil contamination by heavy metals is becoming a worrying issue in industrialized places of the world. Pollution due to heavy metals influences the size, composition, and activity of microbial flora in addition to the quality and yield of plants (Alsherif et al., 2021). As a result, these metals are considered the major cause of soil deterioration. The primary causes of soil pollution include Pb, Zn, Cr, Cd, Cu, and Ni-like metals, which have a significant effect on clay content, pH, organic matter, soil biochemistry, and soil biology (Qin et al., 2021). By changing the microbial community that aids in the manufacturing of certain enzymes, they indirectly influence the enzymatic performance of soil (Wang et al., 2021). Additionally, they damage the soil microbiota by altering important microbial functions and lowering biotic activity and population. Increased bacterial and fungal populations' tolerance is one of the long-term impacts; for instance, arbuscular mycorrhizal (AM) fungi are crucial for the restoration of contaminated ecosystems (Meena et al., 2020). Due to heavy metals, bacterial species have decreased, whereas actinomycetes have grown in proportion (Selim et al., 2021). Additionally, in contaminated soils, it has been found that there is a decrease in biomass and diversity of bacterial populations (Burges et al., 2020). To maintain soil structure, recycle plant nutrients, detoxify hazardous substances, and manage pests, soil bacteria are essential. Analysis of soil organisms in ecosystems subjected to chronic heavy metal pollution is required (Rajmohan et al., 2020).

4.3.3 Effect of Heavy Metals on Human Health

Heavy metals have the ability to accumulate inside the living tissues of humans and can cause various serious health implications (Engwa et al., 2019). The potential for biomagnification is greatly increased when the heavy metals are delivered to the carnivores through the food chain. They enhance the concentration of reactive oxygen species, which are harmful to fish species and other aquatic animals (Meitha et al., 2020). One of the main ways that humans are exposed to heavy metals is through plants, which absorb them in high quantities. Consuming foods grown on polluted soil carries a significant danger (Rehman et al., 2021). If the metals are not metabolized and instead build up in soft tissues, they become poisonous to the organism. After several years of exposure, symptoms start to appear in chronic toxic cases (Johnston and Strobel, 2020). For instance, the liver, kidneys, placenta, lungs, bones, and brain are among the target organs in Cd poisoning. Depending upon the exposure level, symptoms may include vomiting, dyspnoea, nausea, cramping in the abdomen, and muscle weakness. Exposure to extremes causes pulmonary oedema and mortality. Emphysema, alveolitis, and bronchiolitis are pulmonary and renal consequences that can result from the sub-chronic inhalation of cadmium and its derivatives (Jaiswal et al., 2018).

The itai-itai sickness in Japan brought on by Cd exposure has drawn interest from all around the world. Clinical diseases including cerebrovascular infarction, anosmia, heart failure, proteinuria, malignancies, cataracts, emphysema, and osteoporosis are all linked to lower or higher levels of Cd (Barinova et al., 2020). However, if Zn is taken orally, it is rather harmless, but in high levels it can lead to dysfunctions that can affect growth and reproduction. Kidney failure, anaemia, icterus, bloody urine, vomiting, and diarrhoea are all symptoms of zinc poisoning (Shelar et al., 2021). Copper serves as an electron donor or acceptor in metalloenzymes, which makes it a crucial component of mammalian nutrition (Perea-García et al., 2022). A variety of negative consequences, however, might result from exposure to higher amounts. Some people may be more susceptible to elevated Cu consumption because of a sickness or genetic predisposition. High intakes may result in irritation and severe mucosal corrosiveness, extensive capillary damage, renal and hepatic damage, and depression-induced central nervous system irritation. The kidney and liver may also have necrotic changes, as well as severe gastrointestinal discomfort (Jaiswal et al., 2018).

Mercury is entirely hazardous since it serves no known purpose in the biochemistry or physiology of humans. Inorganic Hg causes congenital malformations, spontaneous abortion, and gastrointestinal conditions such as corrosive hematochezia and esophagitis (Bhoot and Bhoot, 2019). Lead is hazardous to human physiology and the nervous system. Acute Pb poisoning causes damage to the brain, kidney, liver, and reproductive systems, which eventually culminates in illness and death. Pb is incredibly hazardous even at very low concentrations. Teratogenic effects are another result. Pb poisoning causes cardiovascular problems, chronic and acute central and peripheral nervous system damage, and suppression of haemoglobin formation. The chemical form of arsenic consumed affects the harmful effects (Singh et al., 2018). Arsenic causes protein coagulation, creates coenzyme complexes, and restricts the generation of adenosine triphosphate (ATP) during respiration. In its oxidation stages, it is carcinogenic, and prolonged exposure results in death (Kakar et al., 2022). Guillain-Barre syndrome, a condition in which the PNS is attacked by the immune system and results in nerve inflammation and muscular paralysis, is also brought on by arsenic poisoning (Florian et al., 2021). Some of the common effects of heavy metals on human beings are mentioned in Figure 4.1.

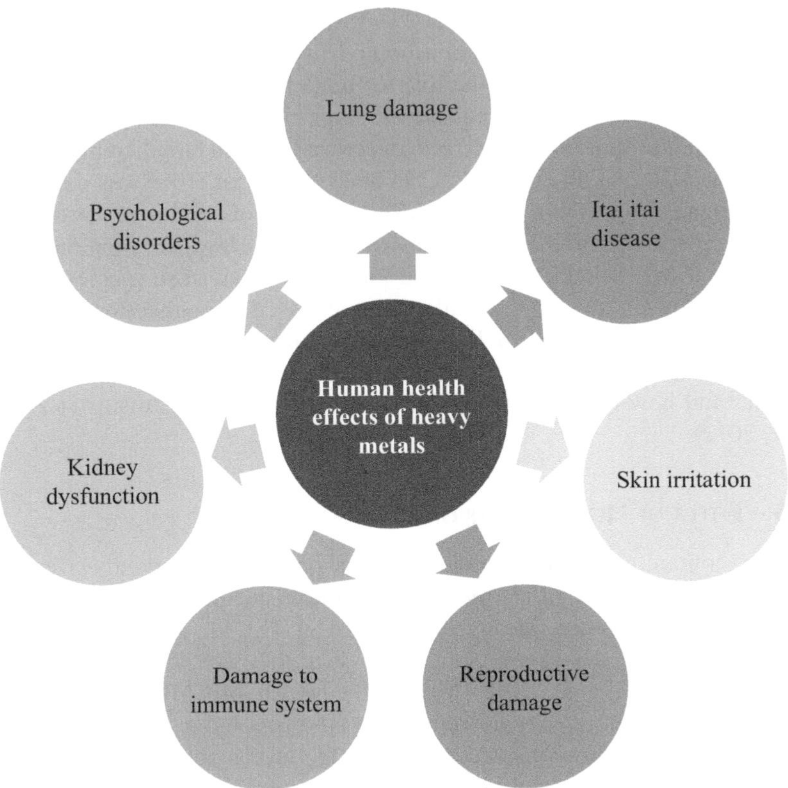

FIGURE 4.1   Some commonly known negative effects of heavy metals on humans

4.3.4   Effect of Heavy Metals on Aquatic Environment

Fish suffer significant consequences from heavy metals in aquatic habitats. These heavy metals cause high oxidative activity in aquatic organisms and are surprisingly persistent and toxic in small amounts (Vardhan et al., 2019). They are therefore very important in terms of ecotoxicology. Due to their inability to biodegrade, these metals remain in the environment forever. The balance of aquatic climates is harmed by contamination of rivers by heavy metals, and a variety of aquatic creatures decline in proportion to the level of pollution (Jaiswal et al., 2018). The number of heavy metals released into aquatic ecosystems is often restricted to particles, which eventually settle into sediments (Shah, 2021). In aquatic climates, surface sediments are important metal and other toxicant repositories. They can be inhabited by water plants and other macrophytes. Most trace metals that enter aquatic environments eventually settle in the lowest sediments, where they have a deleterious impact on microorganisms. Heavy metals initially trigger the formation of certain fish enzymes that change metabolism, which then shows up as necrosis, cellular toxicity, and eventually tissue death (Renu et al., 2021). Mercury (Hg) is a significant contaminant which poses harmful effects on aquatic creatures. For instance, the toxic chemical substance present in fish flesh is methyl mercury, which is produced in marine sediments by microorganisms methylating organic mercury (Al-Sulaiti et al., 2022). Fish are exposed to pollutants through their gills, skin, water, food, and nonfood particles among other ways (Alti et al., 2022). Blood carries them into the liver or bone for storage and then transformation once they have been absorbed. If the liver changes pollutants, they are kept there and can be either expelled in bile or returned to the bloodstream to be eliminated by the gills or kidneys or also can be stored in fat. To assess the ecological effects of metal pollution in streams, several species of benthic macroinvertebrate with varying sensitivity to contaminants are present. However, even though invertebrates are a significant source of food for moving-water fish species, the impact on living organisms by heavy metals is not assessed in connection to their value as food (Clements et al., 2000). Most salmonid species that are important for recreation or trade depend on drift-prone macroinvertebrates; hence, it is essential to understand how heavy metal pollution impacts these organisms (Kappes and Haase, 2012).

4.3.5   Effect of Heavy Metals on Plants

Efflux of contaminants into the environment has grown markedly over the past several decades because of changes in agricultural methods, escalating urbanization, and industrialization. Heavy metals like Se, Hg, Pb, As, Pb, and Cd do not alter the physiological makeup of plants; hence, they are not necessary for plant development (Jaiswal et al., 2018). Metals like Ni, Zn, Mo, Mn, Cu, Fe, and Co, on the other hand, are necessary for regular metabolism and development but can be harmful when concentrations exceed acceptable levels (Rodríguez-Álvarez et al., 2021). Heavy metals also contaminate the food chain that plants absorb via their roots, which is extremely dangerous for both animal and human health. These substances' uptake and accumulation in plant tissue are influenced by conditions including moisture, pH, organic matter, and nutrient availability. The accumulation of heavy metals in plants also depends upon plant species and the effectiveness of metal assimilation, which is governed by either soil-to-plant transfer factors or plant consumption (Pehoiu et al., 2020). When the Pb level is high in the soil, it can cause

TABLE 4.2
Toxicity of heavy metals on humans, plants, and microorganisms

Metal	Effect on humans and plants	Effect on microorganisms	References
Lead (Pb)	Disturbs plant physiological processes by the production of ROS, damages chlorophyll, and reduces seed germination; excess exposure causes short-term memory loss, impaired development, cardiovascular disease in children	Destroys DNA and protein, inhibiting enzyme activity, disrupting cell membrane and causing oxidative phosphorylation	Fashola et al. (2020); Igiri et al. (2018)
Chromium (Cr)	Reduces the growth of plants, buds, roots, shoots, and leaves, fall of hair, nose infection, and issues with breathing	Lag phase elongation, inhibits growth and hindrance in the uptake of oxygen	Fu and Xi (2020); Igiri et al. (2018)
Mercury (Hg)	Damages lungs, brain, and kidney, visual disorders, autoimmune disorders, disrupts the functions of mitochondria	Causes protein denaturation, destroys cell membrane and enzyme inhibition	Han et al. (2022); Ahmad and Mahmood (2019); Igiri et al. (2018)
Cadmium (Cd)	Affects calcium regulation, carcinogenic and mutagenic; damages lungs and bones, and causes muscle twitching	Damages nucleic acid and protein and causes hindrance in cell division and transcription	Igiri et al. (2018); Dhaliwal et al. (2020)
Nickel (Ni)	Inhibits growth and development; causes leaf chlorosis; affects fertility; causes lung, nose, and throat cancer; induces allergic skin diseases	Disrupts cell membrane, causes oxidative stress, and inhibits enzyme activity	Dhaliwal et al. (2020); Igiri et al. (2018)
Aluminium (Al)	Impairs the functions of central nervous system and leads to the loss of coordination and memory; damages bones and brain; causes anaemia and impaired iron absorption	Stabilizes superoxide radicals, which damages DNA; inactivates enzymes such as phosphoxidase, phosphatase, phosphodiesterase, and hexokinase	Krewski et al. (2007); Jaishankar et al. (2014)

plants to develop short brown roots, dark green leaves, less foliage, and older leaves to droop (Ghag et al., 2012). Nitrogen fixation failure by leguminous plants, chlorosis, lower production, poor nutrient absorption, and metabolic abnormalities are all effects of phytotoxicity (Mondal et al., 2021). A greater Pb content eventually hinders seed germination. The harmful effects of metal binding, leaching, chelation, and Pb buildup by microorganisms are all neutralized by prolonged incubation (Singh and Kalamdhad, 2011).

4.4 REMEDIATION TECHNIQUES TO REMOVE HEAVY METALS FROM THE ENVIRONMENT

To address the issue of heavy metal pollution, various techniques and strategies have been used. Depending on the site's features, the pollutants to be eliminated, the concentration of contaminants, and the intended use of the polluted medium, the best remediation technique must be chosen (Li et al., 2019). There are physical, chemical, and biological techniques for the remediation of heavy metals (Figure 4.2).

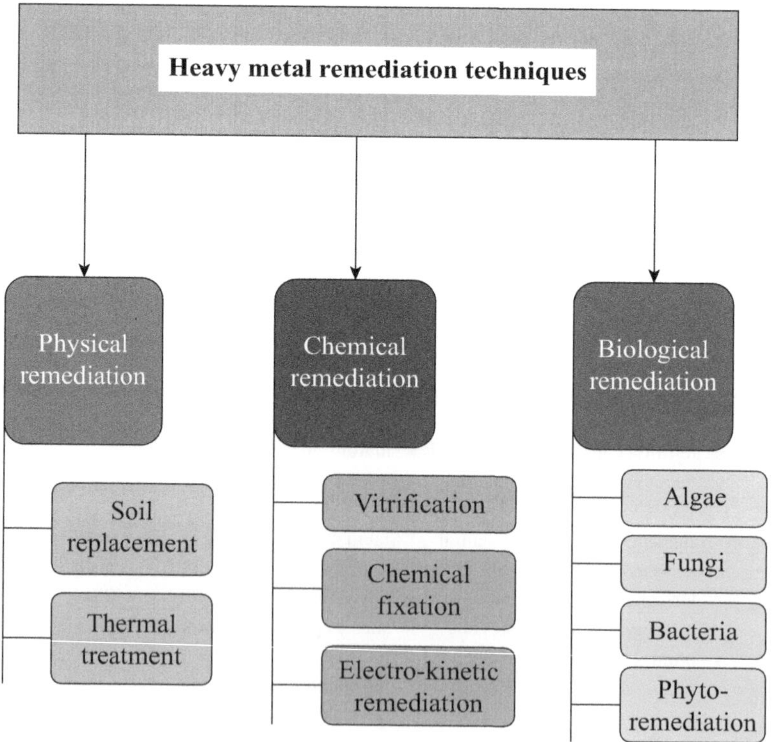

FIGURE 4.2   Different physico-chemical and biological heavy metal remediation techniques

4.4.1 Physical Remediation Methods

Physical methods including soil replacement, containment, isolation techniques, and thermal treatment are used in the physical remediation process to reverse or stop environmental harm.

4.4.1.1 Soil Replacement

Landfilling, encapsulation, and surface capping are examples of procedures used in soil replacement, which entails utilizing a substantial volume of clean soil to cover or mix with contaminated soil (Liu et al., 2018). Due to its high labour and expense, soil removal is best suited for severely polluted soil that is in a small region. It may also efficiently reduce the pollutant concentration (Ghosh et al., 2011). To stop further dispersion from the site, pollutants can be separated and limited by establishing barrier walls. For capping, horizontal and vertical containment, impermeable physical barriers constructed of grout, bentonite, cement, and steel are employed. The practice of soil isolation or confinement was utilized to minimize heavy metal migration into groundwater rather than as a direct remedial procedure (Jankaite and Vasarevičius, 2005).

4.4.2 Thermal Treatment

Based on the contaminant's volatility, the thermal treatment of soil is carried out by heating its subsurface for remediation. The principal heating strategies are conductive heating, electrical resistive heating, radio-frequency heating, and heating based on steam. Although it has been demonstrated that this method may successfully remove toxins with high vapour pressure, like mercury, it may significantly alter the characteristics of soil (Huan et al., 2017). The carrier gas or vacuum-negative pressure is used to collect and remove the volatile heavy metals. Thermal desorption of heavy metals takes place at high temperatures (320–560°C) and low temperatures (90–320°C), depending on the temperature. The USA established commercial services and employed this technique for mercury collecting on-site repairs. However, the use of these devices in soil remediation is constrained by considerations including expensive equipment and extended desorption times (Lianwen et al., 2018).

4.5 CHEMICAL REMEDIATION METHODS

Chemical techniques can minimize the heavy metal quantity which is accessible to plants. It can be done in many ways, such as by creating insoluble complexes with metals or by changing the pH of soil, which causes precipitation of metals (Sarwar et al., 2017).

4.5.1 Vitrification

To decompose or volatilize organic materials, the process of vitrification entails raising the temperature of soil between 1400 and 2000°C (Fatdillah and Pauzi, 2018). The product of pyrolysis is extracted from exhaust gas, which is created by a treatment system and is used in this way to make steam. Energy may be provided for ex-situ reclamation either by the burning of fossil fuels or through direct heating with the use of microwaves, electrodes, and plasma (Ossai et al., 2020). During in situ remediation, electrodes can be placed directly in contaminated soil to provide heat. This method is quite effective in getting rid of heavy metals. However, this technology is exceedingly costly and has only a restricted range of applications because of its high complexity and increased energy needed for fusion (Lee et al., 2013).

4.5.2 CHEMICAL LEACHING

The chemical leaching process involves washing polluted soil with water, chemicals, and other fluids that can quickly remove contaminants from the polluted area (Yang et al., 2010). Through precipitation, ion exchange, chelation, and adsorption, heavy metals from soil have been moved to the liquid phase where they were later determined from the infiltration. Surfactants, chelating agents, and inorganic fluent make up most of the infiltration. Tokunaga and Hakuta (2002) tested hydrochloric acid, phosphoric acid, sulphuric acid, hydrofluoric acid, and nitric acid as extractants for removing contaminants from soil. They recovered metals at various concentrations from contaminated soil.

4.5.3 CHEMICAL FIXATION

Chemical fixation, which helps to prevent the movement of heavy metals into the water sources, plants, and other environmental media, results in the remediation of soil (Jinadasa et al., 2016) by adding materials or reagents to the contaminated soil to create compounds that are soluble in a moderate manner (Huang et al., 2016; Huan et al., 2017). As a result, stability is achieved by chemical fixation rather than decontamination, which involves the transfer of metal into an inert state.

4.5.4 ELECTROKINETIC REMEDIATION

Electrokinetic remediation is a technique in which high voltage is used to establish an electric field gradient to remove contaminants from polluted areas (Luo et al., 2004). This procedure involves the electromigration, electroosmotic flow, and electrophoresis of charged contaminants to the poles. This technique benefits soil with low permeability, is inexpensive, and is simple to install and use while preserving the soil's natural composition and safeguarding the ecotype (Xu et al., 2006). However, because this technique couldn't regulate the pH of the soil, treatment effectiveness was poor. The most current techniques include using an ion exchange membrane to adjust the pH of the soil to promote migration or adding a buffer solution to the soil to alter pH (Fasani et al., 2017).

4.6 BIOREMEDIATION

Bioremediation is a technique for removing pollutants from the environment. It makes use of various biological mechanisms found in plants and microbes to eliminate harmful contaminants from the ecosystem and restore it to its natural state (Ayangbenro and Babalola, 2017). The fundamental principle of bioremediation is to lower the solubility of environmental toxins by redox processes, altering pH, and adsorption of contaminants from polluted sites (Jain and Arnepalli, 2019). Redox reactions involve the chemical transformation of toxic heavy metal contaminants into harmless compounds that are more stable, immobile, or inactive (Gadd, 2010; Rajapaksha et al., 2013; Tandon and Singh, 2016). The efficiency of bioremediation is determined by various factors, including the type of organisms used, the degree of pollution in that ecosystem, and the surrounding environment at the contaminated site (Azubuike et al., 2016).

The bioremediation process can be accomplished using various microorganisms, which rely on their metabolic potential to detoxify environmental contaminants and convert them to harmless form via redox reactions. This can also be achieved by using different plant species (phytoremediation) which bind, remove, and remediate contaminants in the environment (Tak et al., 2013). Phytoremediation generates a considerable amount of biomass that is high in heavy metals. However, it is becoming a major issue to dispose of these toxic plants (Liu et al., 2016). He et al. (2019), Rezvani et al. (2012), and Cestone et al. (2010) reported that the paucity of research work on heavy metal disposal has increased biomass produced by phytoremediation that eventually hampered its widespread use and is responsible for causing secondary pollution in the environment. Presently, there is a dire need for a bioremediation technique which is simple, eco-friendly, cost-effective, and highly

---

TABLE 4.3
Difference between in situ and ex situ bioremediation

In situ bioremediation	Ex situ bioremediation	References
It is a contaminated environment cleanup process that involves augmenting microbes with nutrients to stimulate their ability to eliminate contaminants from soil, as well as degrading contaminants from the environment by adding new microbes.	It is the process of transferring a polluted sample from its original site to other places for treatment, which is determined by the cost of treatment, the depth and type of contamination, the geographic location, and the geology of the polluted site.	Rayu et al. (2012); Azubuike et al. (2016); Mani and Kumar (2014)
The treatment of polluted soil was found to be quite efficient for in situ bioremediation approaches. It is less expensive because polluted soil is not evacuated from the contaminated site.	This method involves removing soil before treating it; it is faster, easier to maintain, and allows the treatment of a wide range of pollutants and soils.	Thakare et al. (2021)
It has the disadvantages of being less manageable and taking a longer time to decontaminate.	Packaging, transportation, and maintenance costs are some of the drawbacks of ex-situ bioremediation.	Thakare et al. (2021)
This type of remediation involves the following techniques: phytoremediation, biosparging, bioaugmentation, and so on.	Bioreactors, land farming, and biopiles are the methods involved in ex-situ bioremediation.	Thakare et al. (2021)

efficient. Microorganisms, unlike plants and animals, can withstand environmental stress by mutating and evolving quickly. Heavy metal ions present in an active form can affect the metabolic activity and diversity of microbes, and, as a result, microbes can develop various resistance mechanism systems to overwhelm the stress caused by toxic heavy metal ions (Yin et al., 2019). Bioremediation involves two different techniques that are used to remove contaminants from a polluted environment: in situ bioremediation and ex-situ bioremediation (Thakare et al., 2021). Table 4.3 shows the difference between in situ and ex-situ bioremediation.

4.6.1 Algae

Algae are a type of photosynthetic organisms that consume less amount of nutrients and produce a huge quantity of biomass in comparison to other microorganisms (Medfu Tarekegn et al., 2020). Algae have recently been reported to have a huge capacity for absorbing heavy metal ions. Various peptides that are produced by algae can accumulate heavy metal ions, which can prevent algae from toxicity of heavy metal ions. Heavy metals accumulate more in dead algal cells than in active cells. Several pre-treatments improve the ability of algae to absorb metals. The most cost-effective approach to activating algal biomass is to use calcium chloride as a pre-treatment (de Souza Coracao et al., 2020). According to Freundlich and Langmuir adsorption model, *Fucusvesiculosus* was reported as an efficient adsorbent used for Pb (II) while *Sargassum* has the capacity to remove Cu (II) from water (Barquilha et al., 2017). (Joo et al., 2021) employed that dead *Chlorella vulgaris* cells can remove ions of Pb, Cd, and Cu from various aqueous solutions under different pHs, contact time, and biosorbent dosage conditions. This research suggests that *C. vulgaris* biomass is an extremely efficient biosorbent for the removal of Cd, Cu, and Pb with an efficiency of 95.5%, 97.7%, and 99.4%, respectively, and therefore could be considered a bioremediation agent.

4.6.2 Fungi

Heavy metal-rich settings can support the growth of fungi, which can also absorb micronutrients and heavy metals. Because of their great potential for metal absorption, fungi are being extensively used to absorb heavy metal ions. Adsorption of heavy metals is aided by glucuronic acid, phosphate, chitin–chitonase complex, and polysaccharides found in fungal cells. Fungi, such as *Trichodermaaureoviride*, *Aspergillus niger*, *Penicillium* species, *T. harzianum*, and *T. virens* can be used as effective bioaccumulator for removing Cr and Cd from polluted soil (Martins et al., 2016). Heavy metals can be removed from the contaminated environment by fungi through different mechanisms (intracellular sequestration and extracellular sequestration) including extracellular sequestration, intracellular sequestration, and integrating the metals into their spores and mycelium. *A. niger*s is thought to be the perfect biosorbent for removing Pb (II). Adsorption capacity and selectivity of fungal species to heavy metal are influenced by various functional groups and several kinds of ionizable sites, including sulphhydryl, hydroxyl, phosphate amine, and carboxyl groups (Iram et al., 2015). Removal oCf opper ions from wastewater was examined by using the biomass of *Saccharomyces cerevisiae* (Amirnia et al., 2015). Copper, zinc, and cadmium pollution can also be remediated by *S. cerevisiae* in a hyaline environment, and the adsorption capability can be improved using sodium chloride (Li et al., 2018).

4.6.3 BACTERIA

Bacteria are among the most common microorganisms on the earth that are present in a wide range of environments (Nekouei et al., 2015). They are utilized for bio-recovery and reduction of heavy metals because of their advantages, such as rapid growth rate, small size, and easy cultivation (Yin et al., 2019; Thakare et al., 2021). Because bacteria include functional groups like sulphate, amino, and carboxyl groups, heavy metal ions may become adsorbed on the bacterial cell surface. With the help of these functional groups, heavy metal ions can interact and build up with powerful adsorption capacities (Yin et al., 2016). To improve the process of biosorption, the changes in physical, chemical, and bioreactor configuration must be responded to efficiently by microbial cells (Ayangbenro and Babalola, 2017). Several microbial species such as *Micrococcus* sp., *Bacillus* sp., *Pseudomonas* sp., and *Flavobacterium* sp. were used to examine various heavy metals; their surface-to-volume ratios and active chemisorptions sites on the cell surface contribute to their high biosorption capacities (Dey et al., 2021). Mercury-resistant strain, *Pseudomonas aeroginosa* can preferentially adsorb mercury ions with a maximal adsorption ability of about 180 mg/g (Yin et al., 2016). Mercury ions are taken with high affinity by cysteine-rich transport proteins, which are rich in sulfhydryl groups (Li et al., 2018). *Arthobacterviscosus* biomass has an efficient recovery, high adsorption capacity, and regeneration ability for Cr (IV) and therefore reduces Cr(IV) to Cr(III) in aqueous solution by both living and dead cells (Hlihor et al., 2017). After adsorption the heavy metal ions can be metabolically transferred into living bacterial cells, and modification in the redox state of heavy metals can minimize their toxicity (Zhang and Wang, 2020).

4.6.4 PHYTOREMEDIATION

To clean up polluted environments, a series of strategies known as phytoremediation use various plant species to degrade, immobilize, and minimize toxins in the environment brought on by human activity (Mukhopadhyay and Maiti, 2010). Several studies suggested that the application of chelating agents, fertilizer, organic amendments, and adjusting pH might increase the adsorption and bioavailability of metals by plants. For the rehabilitation of polluted soil, phytoremediation has recently attracted a lot of interest (Huang et al., 2016). A polluted region is treated with plants as part of phytoremediation to get rid of the contaminants. The fundamental idea behind phytoremediation is to either absorb contaminants and store them in the plant's stems and leaves or break them down into less hazardous forms through the roots of plants (Kaur et al., 2018). As a result, it is meant to be a substitute method of getting rid of, or more specifically, reducing, hazardous contaminants from the environment (Yadav and Srivastava, 2014). Many plant species, referred to as hyper-accumulators, have the capacity to collect significant quantities of heavy metals (Memon and Schroder, 2009). These plants had a remarkable capacity to absorb metal that was crucial for phyto-extraction, and they could withstand high metal concentrations (Yadav and Srivastava, 2014). The Euphorbiaceae, Violaceae, Poaceae, Lamiaceae, Caryophyllaceae, Flacourtaceae, Brassicaceae, Fabaceae, and Asteraceae families are mostly home to the four hundred hyper-accumulator plants that are currently known. The choice of the plant depends also upon seed supply, the capacity of the plant to establish itself and thrive in contaminated soil, and its ability to draw metals from the soil to its root biomass. The capacity of plants to bioaccumulate heavy metals from soil and water has been discussed in several researches. Studies have demonstrated that using phytoremediation technology is an option to decontaminate regions that are affected by heavy metals and may also be utilized as an excellent approach to remediate the environment (Bolan et al., 2014).

4.7 CONCLUSION

Heavy metals are poisonous, enduring, and irreversible contaminants that pose a grave threat to the environment. When heavy metals are used carelessly to boost soil productivity, they build up in the soil and have an adverse impact on the plants and soil. In rivers, seas, and oceans, toxic waste from industry is dumped, harming aquatic ecosystems. As a result, fish and other aquatic species have a variety of deformities. Through their roots, plants may absorb heavy metal ions from the soil, which can affect leaf growth, photosynthesis, and transpiration. By consuming plants or seafood, people consume these accumulated poisons. Neurogenic and teratogenic consequences can result from toxic doses. Due to its importance to public health, this serious issue must be given immediate attention. Generally, several approaches have been devised for heavy metal decontamination. Conventional approaches such as physical and chemical methods for restoring and cleaning up heavy metal-contaminated soils have significant drawbacks in terms of cost, alteration of the native microflora and soil qualities, and potential for additional contamination issues. In contrast, the solution that bioremediation offers is more effective. Without requiring costly equipment or specific maintenance for hazardous places, it is economical and ecologically beneficial.

REFERENCES

Adhikari, S., Marcelo-Silva, J., Rajakaruna, N., & Siebert, S. J. (2022). Influence of land use and topography on distribution and bioaccumulation of potentially toxic metals in soil and plant leaves: A case study from Sekhukhuneland, South Africa. *Science of the Total Environment, 806*, 150659.

Ahmad, S., & Mahmood, R. (2019). Mercury chloride toxicity in human erythrocytes: Enhanced generation of ROS and RNS, hemoglobin oxidation, impaired antioxidant power, and inhibition of plasma membrane redox system. *Environmental Science and Pollution Research, 26*(6), 5645–5657.

Ali, H., & Khan, E. (2018). Bioaccumulation of non-essential hazardous heavy metals and metalloids in freshwater fish: Risk to human health. *Environmental Chemistry Letters, 16*(3), 903–917.

Alloway, B. J. (2013). Sources of heavy metals and metalloids in soils. In: *Heavy metals in soils* (pp. 11–50). Springer.

Alsherif, E. A., Al-Shaikh, T. M., Almaghrabi, O., & AbdElgawad, H. (2021). High redox status as the basis for heavy metal tolerance of sesuviumportulacastrum L. inhabiting contaminated soil in Jeddah, Saudi Arabia. *Antioxidants, 11*(1), 19.

Al-Sulaiti, M. M., Soubra, L., & Al-Ghouti, M. A. (2022). The causes and effects of mercury and methylmercury contamination in the marine environment: A review. *Current Pollution Reports*, 1–24.

Amirnia, S., Ray, M. B., & Margaritis, A. (2015). Heavy metals removal from aqueous solutions using Saccharomyces cerevisiae in a novel continuous bioreactor—Biosorption system. *Chemical Engineering Journal, 264*, 863–872.

Ayangbenro, A. S., & Babalola, O. O. (2017). A new strategy for heavy metal polluted environments: A review of microbial biosorbents. *International Journal of Environmental Research and Public Health, 14*(1), 94.

Azimi, A., Azari, A., Rezakazemi, M., & Ansarpour, M. (2017). Removal of heavy metals from industrial wastewaters: A review. *ChemBioEng Reviews, 4*(1), 37–59.

Azubuike, C. C., Chikere, C. B., & Okpokwasili, G. C. (2016). Bioremediation techniques—Classification based on site of application: Principles, advantages, limitations and prospects. *World Journal of Microbiology and Biotechnology, 32*(11), 1–18.

Barinova, G. M., Gaeva, D. V., & Krasnov, E. V. (2020). Hazardous chemicals and air, water, and soil pollution and contamination. *Good Health and Well-Being*, 255–266.

Barquilha, C. E. R., Cossich, E. S., Tavares, C. R. G., & Silva, E. A. (2017). Biosorption of nickel (II) and copper (II) ions in batch and fixed-bed columns by free and immobilized marine algae Sargassum sp. *Journal of Cleaner Production, 150*, 58–64.

Basha, S. A., & Rajaganesh, K. (2014). Microbial bioremediation of heavy metals from textile industry dye effluents using isolated bacterial strains. *Intenational Journal Current Microbiology Applied Science, 3*, 785–794.

Bashi, N. F. (2021). Chromium (VI) removal methods from effluents—A review article. *University of Khartoum Engineering Journal, 11*(2).

Batool, S., Idrees, M., Hussain, Q., & Kong, J. (2017). Adsorption of copper (II) by using derived-farmyard and poultry manure biochars: Efficiency and mechanism. *Chemical Physics Letters, 689*, 190–198.

Beasley, V. R., & Levengood, J. M. (2012). Principles of ecotoxicology. In: *Veterinary toxicology* (pp. 831–855). Elsevier Inc.

Bhat, S. A., Hassan, T., & Majid, S. (2019). Heavy metal toxicity and their harmful effects on living organisms—A review. *International Journal of Medical Science and Diagnosis Research, 3*(1).

Bhattacharjee, T., Islam, M., Chowdhury, D., & Majumdar, G. (2021). In-situ generated carbon dot modified filter paper for heavy metals removal in water. *Environmental Nanotechnology, Monitoring & Management, 16*, 100582.

Bhoot, K., & Bhoot, M. T. (2019). Metal poisoning through soil. *IJRAR-International Journal of Research and Analytical Reviews (IJRAR), 6*(1), 506–509.

Bolan, N., Kunhikrishnan, A., Thangarajana, R., Kumpiene, J., Park, J., Makino, T., . . . & Scheckel, K. (2014). Remediation of heavy metal (-loid)s contaminated soils to mobilize or to immobilize? *Journal of Hazardous Materials, 26*, 141–166.

Briffa, J., Sinagra, E., & Blundell, R. (2020). Heavy metal pollution in the environment and their toxicological effects on humans. *Heliyon, 6*(9), e04691.

Burges, A., Fievet, V., Oustriere, N., Epelde, L., Garbisu, C., Becerril, J. M., & Mench, M. (2020). Long-term phytomanagement with compost and a sunflower—Tobacco rotation influences the structural microbial diversity of a Cu-contaminated soil. *Science of the Total Environment, 700*, 134529.

Cestone, B., Quartacci, M. F., & Navari-Izzo, F. (2010). Uptake and translocation of CuEDDS complexes by Brassica carinata. *Environmental Science & Technology, 44*(16), 6403–6408.

Chandrasekaran, A., Ravisankar, R., Harikrishnan, N., Satapathy, K. K., Prasad, M. V. R., & Kanagasabapathy, K. V. (2015). Multivariate statistical analysis of heavy metal concentration in soils of Yelagiri Hills, Tamilnadu, India—Spectroscopical approach. *Spectrochimica Acta Part A: Molecular and Biomolecular Spectroscopy, 137*, 589–600.

Chen, C. W., Chen, C. F., & Dong, C. D. (2012). Distribution and accumulation of mercury in sediments of Kaohsiung River Mouth, Taiwan. *Apcbee Procedia, 1*, 153–158.

Chen, J., Jiang, Y., Shi, H., Peng, Y., Fan, X., & Li, C. (2020). The molecular mechanisms of copper metabolism and its roles in human diseases. *Pflügers Archiv-European Journal of Physiology, 472*(10), 1415–1429.

Chibuike, G. U., & Obiora, S. C. (2014). Heavy metal polluted soils: Effect on plants and bioremediation methods. *Applied and Environmental Soil Science, 2014*.

Clements, W. H., Carlisle, D. M., Lazorchak, J. M., & Johnson, P. C. (2000). Heavy metals structure benthic communities in Colorado mountain streams. *Ecological Applications, 10*(2), 626–638.

Crini, G., & Lichtfouse, E. (2018). Wastewater treatment: An overview. *Green Adsorbents for Pollutant Removal*, 1–21.

de Souza Coracao, A. C., Dos Santos, F. S., Duarte, J. A. D., Lopes-Filho, E. A. P., De-Paula, J. C., Rocha, L. M., . . . & Teixeira, V. L. (2020). What do we know about the utilization of the Sargassum species as biosorbents of trace metals in Brazil? *Journal of Environmental Chemical Engineering, 8*(4), 103941.

Deepa, C. N., & Suresha, S. (2014). Biosorption of lead (II) from aqueous solution and industrial effluent by using leaves of *Araucaria cookii*: Application of response surface methodology. *IOSR Journal of Environmental Science, Toxicology and Food Technology, 8*(7), 67–79.

Dey, S., Haripavan, N., Basha, S. R., & Babu, G. V. (2021). Removal of ammonia and nitrates from contaminated water by using solid waste bio-adsorbents. *Current Research in Chemical Biology, 1*, 100005.

Dhaliwal, S. S., Singh, J., Taneja, P. K., & Mandal, A. (2020). Remediation techniques for removal of heavy metals from the soil contaminated through different sources: A review. *Environmental Science and Pollution Research, 27*(2), 1319–1333.

Dixit, R., Malaviya, D., Pandiyan, K., Singh, U. B., Sahu, A., Shukla, R., & Paul, D. (2015). Bioremediation of heavy metals from soil and aquatic environment: An overview of principles and criteria of fundamental processes. *Sustainability, 7*(2), 2189–2212.

Duruibe, J. O., Ogwuegbu, M. O. C., & Egwurugwu, J. N. (2007). Heavy metal pollution and human biotoxic effects. *International Journal of Physical Sciences, 2*(5), 112–118.

Fasani, E., Manara, A., Martini, F., Furini, A., & Dal, C. G. (2017). The potential of genetic engineering of plants for the remediation of soils contaminated with heavy metals. *Plant Cell& Environment, 41*, 1201–1232.

Fashola, M. O., Ngole-Jeme, V. M., & Babalola, O. O. (2020). Heavy metal immobilization potential of indigenous bacteria isolated from gold mine tailings. *International Journal of Environmental Research, 14*(1), 71–86.

Fatdillah, E. L. N., & Pauzi, N. I. M. (2018). Soil remediation technologies for heavy metals-A review. *International Journal of Engineering and Management Research (IJEMR), 8*(6), 171–175.

Florian, I. A., Lupan, I., Sur, L., Samasca, G., & Timiş, T. L. (2021). To be, or not to be . . . Guillain-Barré Syndrome. *Autoimmunity Reviews, 20*(12), 102983.

Fu, Z., & Xi, S. (2020). The effects of heavy metals on human metabolism. *Toxicology Mechanisms and Methods, 30*(3), 167–176.

Gadd, G. M. (2010). Metals, minerals and microbes: Geomicrobiology and bioremediation. *Microbiology, 156,* 609–643.

Gaur, N., Flora, G., Yadav, M., & Tiwari, A. (2014). A review with recent advancements on bioremediation-based abolition of heavy metals. *Environmental Science: Processes & Impacts, 16*(2), 180–193.

Gautam, P. K., Gautam, R. K., Banerjee, S., Chattopadhyaya, M. C., & Pandey, J. D. (2016). Heavy metals in the environment: Fate, transport, toxicity and remediation technologies. *Nova Science Publishers, 60,* 101–130.

Gauthier, P. T., Norwood, W. P., Prepas, E. E., & Pyle, G. G. (2014). Metal—PAH mixtures in the aquatic environment: A review of co-toxic mechanisms leading to more-than-additive outcomes. *Aquatic Toxicology, 154,* 253–269.

Ghag, S. B., Shekhawat, U. K. S., & Ganapathi, T. R. (2012). Petunia floral defensins with unique prodomains as novel candidates for development of Fusarium wilt resistance in transgenic banana plants. *PLoS One, 7*(6), e39557.

Ghani, A., & Ghani, A. (2011). Effect of chromium toxicity on growth, chlorophyll and some mineral nutrients of brassica juncea L. *Egyptian Academic Journal of Biological Sciences, H. Botany, 2*(1), 9–15.

Ghosh, U., Luthy, R. G., Cornelissen, G., Werner, D., & Menzie, C. A. (2011). *In-situ sorbent amendments: A new direction in contaminated sediment management* (pp. 1163–1168).

Gossuin, Y., & Vuong, Q. L. (2018). NMR relaxometry for adsorption studies: Proof of concept with copper adsorption on activated alumina. *Separation and Purification Technology, 202,* 138–143.

Han, B., Lv, Z., Han, X., Li, S., Han, B., Yang, Q., . . . & Zhang, Z. (2022). Harmful effects of inorganic mercury exposure on kidney cells: Mitochondrial dynamics disorder and excessive oxidative stress. *Biological Trace Element Research, 200*(4), 1591–1597.

He, J., Strezov, V., Kumar, R., Weldekidan, H., Jahan, S., Dastjerdi, B. H., & Kan, T. (2019). Pyrolysis of heavy metal contaminated Avicennia marina biomass from phytoremediation: Characterisation of biomass and pyrolysis products. *Journal of Cleaner Production, 234,* 1235–1245.

Hlihor, R. M., Figueiredo, H., Tavares, T., & Gavrilescu, M. (2017). Biosorption potential of dead and living *Arthrobacterviscosus* biomass in the removal of Cr (VI): Batch and column studies. *Process Safety and Environmental Protection, 108,* 44–56.

Hrynkiewicz, K., & Baum, C. (2014). Application of microorganisms in bioremediation of environment from heavy metals. In: *Environmental deterioration and human health.* Springer.

Huan, L., Haixia, Z., Longhua, W., Anna, L., Fang-Jie, Z., & Wenzhong, X. (2017). Heavy Metal ATPase 3 (HMA3) confers cadmium hypertolerance on the cadmium/zinc hyperaccumulator *Sedum plumbizincicola. New Phytologist., 15,* 687–698.

Huang, D., Hu, C., Zeng, G., Cheng, M., Xu, P., Gong, X., . . . & Xue, W. (2016). Combination of Fenton processes and biotreatment for wastewater treatment and soil remediation. *Science of the Total Environment, 574,* 1599–1610.

Igiri, B. E., Okoduwa, S. I., Idoko, G. O., Akabuogu, E. P., Adeyi, A. O., & Ejiogu, I. K. (2018). Toxicity and bioremediation of heavy metals contaminated ecosystem from tannery wastewater: A review. *Journal of Toxicology, 2018*(1), 2568038.

Iram, S., Shabbir, R., Zafar, H., & Javaid, M. (2015). Biosorption and bioaccumulation of copper and lead by heavy metal-resistant fungal isolates. *Arabian Journal for Science and Engineering, 40*(7), 1867–1873.

Jadia, C. D., & Fulekar, M. H. (2009). Phytoremediation of heavy metals: Recent techniques. *African Journal of Biotechnology, 8*(6).

Jain, S., & Arnepalli, D. N. (2019). Biominerlisation as a remediation technique: A critical review. *Geotechnical Characterisation and Geoenvironmental Engineering*, 155–162.

Jaishankar, M., Tseten, T., Anbalagan, N., Mathew, B. B., & Beeregowda, K. N. (2014). Toxicity, mechanism and health effects of some heavy metals. *Interdisciplinary Toxicology, 7*(2), 60.

Jaiswal, A., Verma, A., & Jaiswal, P. (2018). Detrimental effects of heavy metals in soil, plants, and aquatic ecosystems and in humans. *Journal of Environmental Pathology, Toxicology and Oncology, 37*(3).

Jankaite, A., & Vasarevičius, S. (2005). Remediation technologies for soils contaminated with heavy metals. *Journal of Environmental Engineering and Landscape Management, 13*(2), 109–113.

Jinadasa, N., Collins, D., Holford, P., Milham, P. J., & Conroy, J. P. (2016). Reactions to cadmium stress in a cadmium-tolerant variety of cabbage (Brassica oleracea L.): Is cadmium tolerance necessarily desirable in food crops? *Environment Science and Pollution Research, 23*, 5296–5306.

Johnston, N. R., & Strobel, S. A. (2020). Principles of fluoride toxicity and the cellular response: A review. *Archives of Toxicology, 94*(4), 1051–1069.

Joo, G., Lee, W., & Choi, Y. (2021). Heavy metal adsorption capacity of powdered Chlorella vulgaris biosorbent: Effect of chemical modification and growth media. *Environmental Science and Pollution Research, 28*(20), 25390–25399.

Kakar, R., Parab, A. R., Abdullah, A. A. A., & Yaqoob, S. B. (2022). Role of microbial community in microbial fuel cells. In: *Microbial fuel cells for environmental remediation* (pp. 139–166). Springer.

Kappes, H., & Haase, P. (2012). Slow, but steady: Dispersal of freshwater molluscs. *Aquatic Sciences, 74*(1), 1–14.

Kaur, R., Bhatti, S. S., Singh, S., Singh, J., & Singh, S. (2018). Phytoremediation of heavy metals using cotton plant: A field analysis. *Bulletin of Environmental Contamination and Toxicology, 101*, 637–643.

Krewski, D., Yokel, R. A., Nieboer, E., Borchelt, D., Cohen, J., Harry, J., & Rondeau, V. (2007). Human health risk assessment for aluminium, aluminium oxide, and aluminium hydroxide. *Journal of Toxicology and Environmental Health, Part B, 10*(S1), 1–269.

Lee, W. E., Gilbert, M., Murphy, S. T., & Grimes, R. W. (2013). Opportunities for advanced ceramics and composites in the nuclear sector. *Journal of the American Ceramic Society, 96*(7), 2005–2030.

Li, C., Yang, X., Xu, Y., Li, L., & Wang, Y. (2018). Cadmium detoxification induced by salt stress improves cadmium tolerance of multi-stress-tolerant *Pichia kudriavzevii*. *Environmental Pollution, 242*, 845–854.

Li, C., Zhou, K., Qin, W., Tian, C., Qi, M., Yan, X., & Han, W. (2019). A review on heavy metals contamination in soil: Effects, sources, and remediation techniques. *Soil and Sediment Contamination: An International Journal, 28*(4), 380–394.

Lianwen, L., Wei, L., Weiping, S., & Mingxin, G. (2018). Remediation techniques for heavy metal-contaminated soils: Principles and applicability. *Science of the Total Environment, 633*, 206–219.

Liu, L., Li, W., Song, W., & Guo, M. (2018). Remediation techniques for heavy metal-contaminated soils: Principles and applicability. *Science of the Total Environment, 633,* 206–219.

Liu, M., Dong, F., Zhang, W., Nie, X., Sun, S., Wei, H., & Zhang, G. (2016). Programmed gradient descent biosorption of strontium ions by *Saccaromyces cerevisiae* and ashing analysis: A decrement solution for nuclide and heavy metal disposal. *Journal of Hazardous Materials, 314,* 295–303.

Luo, Q. S., Zhang, X. H., & Wang, H. (2004). Mobilization of 2, 4-dichlorophenol in soils by non-uniform electrokinetics. *Acta Scientiae Circumstantiae, 24,* 1104–1109.

Mani, D., & Kumar, C. (2014). Biotechnological advances in bioremediation of heavy metals contaminated ecosystems: An overview with special reference to phytoremediation. *International Journal of Environmental Science and Technology, 11*(3), 843–872.

Martins, L. R., Lyra, F. H., Rugani, M. M., & Takahashi, J. A. (2016). Bioremediation of metallic ions by eight Penicillium species. *Journal of Environmental Engineering, 142*(9), C4015007.

Masindi, V., & Muedi, K. L. (2018). Environmental contamination by heavy metals. *Heavy Metals, 10,* 115–132.

Medfu Tarekegn, M., Zewdu Salilih, F., & Ishetu, A. I. (2020). Microbes used as a tool for bioremediation of heavy metal from the environment. *Cogent Food & Agriculture, 6*(1), 1783174.

Meena, R. S., Kumar, S., Datta, R., Lal, R., Vijayakumar, V., Brtnicky, M., . . . & Marfo, T. D. (2020). Impact of agrochemicals on soil microbiota and management: A review. *Land, 9*(2), 34.

Meitha, K., Pramesti, Y., & Suhandono, S. (2020). Reactive oxygen species and antioxidants in postharvest vegetables and fruits. *International Journal of Food Science, 2020*(1), 8817778.

Memon, A. R., & Schroder, P. (2009). Implications of metal accumulation mechanisms to phytoremediation. *Environment Science and Pollution Research, 16,* 162–175.

Mondal, N. C., Saxena, V. K., & Singh, V. S. (2005). Impact of pollution due to tanneries on groundwater regime. *Current Science,* 1988–1994.

Mondal, S., Pramanik, K., Ghosh, S. K., Pal, P., Mondal, T., Soren, T., & Maiti, T. K. (2021). Unraveling the role of plant growth-promoting rhizobacteria in the alleviation of arsenic phytotoxicity: A review. *Microbiological Research, 250,* 126809.

Morais, S., Costa, F. G., & Pereira, M. D. L. (2012). Heavy metals and human health. *Environmental Health—Emerging Issues and Practice, 10*(1), 227–245.

Mukhopadhyay, S., & Maiti, S. K. (2010). Phytoremediation of metal mine waste. *Applied Ecology and Environmental Research, 8,* 207–222.

Nachana'a Timothy, E. T. W. (2019). Environmental pollution by heavy metal: An overview. *Chemistry, 3*(2), 72–82.

Nath, S., Paul, P., Roy, R., Bhattacharjee, S., & Deb, B. (2019). Isolation and identification of metal-tolerant and antibiotic-resistant bacteria from soil samples of Cachar district of Assam, India. *SN Applied Sciences, 1*(7), 1–9.

Nekouei, F., Nekouei, S., Tyagi, I., & Gupta, V. K. (2015). Kinetic, thermodynamic and isotherm studies for acid blue 129 removal from liquids using copper oxide nanoparticle-modified activated carbon as a novel adsorbent. *Journal of Molecular Liquids, 201,* 124–133.

Okolo, N. V., Olowolafe, E. A., Akawu, I., & Okoduwa, S. I. R. (2016). Effects of industrial effluents on soil resources in Challawa industrial area, Kano, Nigeria. *Journal of Global Ecology and Environment, 5*(1), 1–10.

Ossai, I. C., Ahmed, A., Hassan, A., & Hamid, F. S. (2020). Remediation of soil and water contaminated with petroleum hydrocarbon: A review. *Environmental Technology & Innovation, 17,* 100526.

Pehoiu, G., Murarescu, O., Radulescu, C., Dulama, I. D., Teodorescu, S., Stirbescu, R. M., . . . & Stanescu, S. G. (2020). Heavy metals accumulation and translocation in native plants grown on tailing dumps and human health risk. *Plant and Soil, 456*(1), 405–424.

Perea-García, A., Puig, S., & Peñarrubia, L. (2022). The role of post-transcriptional modulators of metalloproteins in response to metal deficiencies. *Journal of Experimental Botany, 73*(6), 1735–1750.

Pourrut, B., Shahid, M., Dumat, C., Winterton, P., & Pinelli, E. (2011). Lead uptake, toxicity, and detoxification in plants. *Reviews of Environmental Contamination and Toxicology Volume, 213,* 113–136.

Qazilbash, A. A. (2004). Isolation and characterization of heavy metal tolerant biota from industrially polluted soils and their role in bioremediation [Doctoral dissertation]. Quaid-i-Azam University Islamabad.

Qin, G., Niu, Z., Yu, J., Li, Z., Ma, J., & Xiang, P. (2021). Soil heavy metal pollution and food safety in China: Effects, sources and removing technology. *Chemosphere, 267,* 129205.

Rajak, C., Singh, N., & Parashar, P. (2020). Metal toxicity and natural antidotes: Prevention is better than cure. *Environmental Science and Pollution Research, 27*(35), 43582–43598.

Rajapaksha, A. U., Vithanage, M., Ok, Y. S., & Oze, C. (2013). Cr (VI) formation related to Cr (III)-muscovite and birnessite interactions in ultramafic environments. *Environmental Science & Technology, 47*(17), 9722–9729.

Rajmohan, K. S., Chandrasekaran, R., & Varjani, S. (2020). A review on occurrence of pesticides in environment and current technologies for their remediation and management. *Indian Journal of Microbiology, 60*(2), 125–138.

Rasmussen, L. D., Sørensen, S. J., Turner, R. R., & Barkay, T. (2000). Application of a mer-lux biosensor for estimating bioavailable mercury in soil. *Soil Biology and Biochemistry, 32*(5), 639–646.

Rayu, S., Karpouzas, D. G., & Singh, B. K. (2012). Emerging technologies in bioremediation: Constraints and opportunities. *Biodegradation, 23*(6), 917–926.

Rehman, A. U., Nazir, S., Irshad, R., Tahir, K., urRehman, K., Islam, R. U., & Wahab, Z. (2021). Toxicity of heavy metals in plants and animals and their uptake by magnetic iron oxide nanoparticles. *Journal of Molecular Liquids, 321,* 114455.

Renu, K., Chakraborty, R., Myakala, H., Koti, R., Famurewa, A. C., Madhyastha, H., . . . & Gopalakrishnan, A. V. (2021). Molecular mechanism of heavy metals (Lead, Chromium, Arsenic, Mercury, Nickel and Cadmium)-induced hepatotoxicity—A review. *Chemosphere, 271,* 129735.

Rezania, S., Taib, S. M., Din, M. F. M., Dahalan, F. A., & Kamyab, H. (2016). Comprehensive review on phytotechnology: Heavy metals removal by diverse aquatic plants species from wastewater. *Journal of Hazardous Materials, 318,* 587–599.

Rezvani, M., Zaefarian, F., Miransari, M., & Nematzadeh, G. A. (2012). Uptake and translocation of cadmium and nutrients by Aeluropuslittoralis. *Archives of Agronomy and Soil Science, 58*(12), 1413–1425.

Rodríguez-Álvarez, M., Paz, S., Hardisson, A., González-Weller, D., Rubio, C., & Gutiérrez, Á. J. (2021). Assessment of toxic metals (Al, Cd, Pb) and trace elements (B, Ba, Co, Cr, Cu, Fe, Mn, Mo, Li, Zn, Ni, Sr, V) in the common kestrel (Falco tinnunculus) from the Canary Islands (Spain). *Biological Trace Element Research,* 1–11.

Sarker, B. C., Baten, M. A., Eqram, M., Haque, U., Das, A. K., Hossain, A., & Hasan, M. Z. (2015). Heavy metals concentration in textile and garments industries' wastewater of Bhaluka industrial area, Mymensingh, Bangladesh. *Current World Environment, 10*(1), 61.

Sarma, H., Islam, N. F., & Prasad, M. N. V. (2017). Plant-microbial association in petroleum and gas exploration sites in the state of Assam, north-east India—Significance for bioremediation. *Environmental Science and Pollution Research, 24*(9), 8744–8758.

Sarwar, N., Imran, M., Shaheen, M. R., Ishaque, W., Kamran, M. A., Matloob, A., . . . & Hussain, S. (2017). Phytoremediation strategies for soils contaminated with heavy metals: Modifications and future perspectives. *Chemosphere, 171*, 710–721.

Selim, S., AbdElgawad, H., Alsharari, S. S., Atif, M., Warrad, M., Hagagy, N., . . . & Abuelsoud, W. (2021). Soil enrichment with actinomycete mitigates the toxicity of arsenic oxide nanoparticles on wheat and maize growth and metabolism. *Physiologia Plantarum, 173*(3), 978–992.

Shah, S. B. (2021). Heavy metals in the marine environment—An overview. *Heavy Metals in Scleractinian Corals*, 1–26.

Shelar, M., Gawade, V., & Bhujbal, S. (2021). A review on heavy metal contamination in herbals. *Journal of Pharmaceutical Research International, 33*, 7–16.

Shifaw, E. (2018). Review of heavy metals pollution in China in agricultural and urban soils. *Journal of Health and Pollution, 8*(18).

Singh, J., & Kalamdhad, A. S. (2011). Effects of heavy metals on soil, plants, human health and aquatic life. *International Journal of Research in Chemistry and Environment, 1*(2), 15–21.

Singh, N., Kumar, A., Gupta, V. K., & Sharma, B. (2018). Biochemical and molecular bases of lead-induced toxicity in mammalian systems and possible mitigations. *Chemical Research in Toxicology, 31*(10), 1009–1021.

Tak, H. I., Ahmad, F., & Babalola, O. O. (2013). Advances in the application of plant growth-promoting rhizobacteria in phytoremediation of heavy metals. *Reviews of Environmental Contamination and Toxicology, 223*, 33–52.

Tandon, P. K., & Singh, S. B. (2016). Redox processes in water remediation. *Environmental Chemistry Letters, 14*(1), 15–25.

Thakare, M., Sarma, H., Datar, S., Roy, A., Pawar, P., Gupta, K., & Prasad, R. (2021). Understanding the holistic approach to plant-microbe remediation technologies for removing heavy metals and radionuclides from soil. *Current Research in Biotechnology, 3*, 84–98.

Tokunaga, S., & Hakuta, T. (2002). Acid washing and stabilization of an artificial arsenic-contaminated soil. *Chemosphere, 46*, 31–38.

Vardhan, K. H., Kumar, P. S., & Panda, R. C. (2019). A review on heavy metal pollution, toxicity and remedial measures: Current trends and future perspectives. *Journal of Molecular Liquids, 290*, 111197.

Vareda, J. P., Valente, A. J., & Durães, L. (2019). Assessment of heavy metal pollution from anthropogenic activities and remediation strategies: A review. *Journal of Environmental Management, 246*, 101–118.

Veit, H. M., Diehl, T. R., Salami, A. P., Rodrigues, J. D. S., Bernardes, A. M., & Tenório, J. A. S. (2005). Utilization of magnetic and electrostatic separation in the recycling of printed circuit boards scrap. *Waste Management, 25*(1), 67–74.

Wan, J., Zhang, C., Zeng, G., Huang, D., Hu, L., Huang, C., . . . & Wang, L. (2016). Synthesis and evaluation of a new class of stabilized nano-chlorapatite for Pb immobilization in sediment. *Journal of Hazardous Materials, 320*, 278–288.

Wang, X., Wang, G., Guo, T., Xing, Y., Mo, F., Wang, H., . . . & Zhang, F. (2021). Effects of plastic mulch and nitrogen fertilizer on the soil microbial community, enzymatic activity and yield performance in a dryland maize cropping system. *European Journal of Soil Science, 72*(1), 400–412.

Wuana, R. A., & Okieimen, F. E. (2011). Heavy metals in contaminated soils: A review of sources, chemistry, risks and best available strategies for remediation. *International Scholarly Research Notices, 2011.*

Xu, Q., Huang, X. F., & Cheng, J. J. (2006). Progress on electro-kinetic remediation and its combined methods for POPs from contaminated soils. *Environmental Science, 27,* 2363–2368.

Yadav, S., & Srivastava, J. (2014). Phytoremediation of cadmium toxicity by Brassica spp: A review. *International Journal of Biological Sciences, 3,* 47–52.

Yang, Z., Rui-lin, M., Wang-dong, N., & Hui, W. (2010). Selective leaching of base metals from copper smelter slag. *Hydrometallurgy, 103,* 25–29.

Yatoo, A. M., Ali, M., Baba, Z. A., & Hassan, B. (2021). Sustainable management of diseases and pests in crops by vermicompost and vermicompost tea: A review. *Agronomy for Sustainable Development, 41*(1), 1–26.

Yatoo, A. M., Ali, M., Zaheen, Z., Baba, Z. A., Ali, S., Rasool, S., . . . & Hamid, B. (2022). Assessment of pesticide toxicity on earthworms using multiple biomarkers: A review. *Environmental Chemistry Letters, 1,* 1–24.

Yeboah, I. B., Tuffour, H. O., Abubakari, A., Melenya, C., Bonsu, M., Quansah, C., & Adjei-Gyapong, T. (2019). Mobility and transport behavior of lead in agricultural soils. *Scientific African, 5,* e00117.

Yin, K., Lv, M., Wang, Q., Wu, Y., Liao, C., Zhang, W., & Chen, L. (2016). Simultaneous bioremediation and biodetection of mercury ion through surface display of carboxylesterase E2 from *Pseudomonas aeruginosa* PA1. *Water Research, 103,* 383–390.

Yin, K., Wang, Q., Lv, M., & Chen, L. (2019). Microorganism remediation strategies towards heavy metals. *Chemical Engineering Journal, 360,* 1553–1563.

Yusif, B. B., Bichi, K. A., Oyekunle, O. A., Girei, A. I., Garba, P. Y., & Garba, F. H. (2016). A review of tannery effluent treatment. *International Journal of Applied Mathematics and Mathematical Science, 2*(3), 29–43.

Zahid, A., Balke, K. D., Hassan, M. Q., & Flegr, M. (2006). Evaluation of aquifer environment under Hazaribagh leather processing zone of Dhaka city. *Environmental Geology, 50*(4), 495–504.

Zamri, M. F. M. A., Kamaruddin, M. A., Yusoff, M. S., Aziz, H. A., & Foo, K. Y. (2017). Semi-aerobic stabilized landfill leachate treatment by ion exchange resin: Isotherm and kinetic study. *Applied Water Science, 7*(2), 581–590.

Zhang, Q., & Wang, C. (2020). Natural and human factors affect the distribution of soil heavy metal pollution: A review. *Water, Air, & Soil Pollution, 231*(7), 1–13.

5 Biosurfactants and Their Potential in Remediation of Waste Disposal Sites

Roheela Ahmad, Tahir Ahmad Sheikh, Aamir Hassan Mir, Ali Mohd Yatoo, Shakeel Ahmad Mir, Ayman Javed Shah, Rifat Un Nisa, Javed Ahmad Bhat, Razia Gull, and Aanisa Manzoor Shah

5.1 INTRODUCTION

Due to the extensive and detrimental consequences that environmental pollution has on the ecosystem, socioeconomic issues, public health, and mental health, it is necessary to take early and appropriate action to restore and rehabilitate the environment (Ventriglio et al., 2021). For various industries or processes, biosurfactants (BSs; surface-active and biologically effective microbial amphiphilic compounds) are useful. Because each BS has a polar (hydrophilic) and nonpolar (hydrophobic) group in its structure, they are all amphiphilic substances. A hydrophilic group is made up of mono-, oligo-, or polysaccharides, peptides, or proteins, whereas a hydrophobic moiety is often made up of saturated, unsaturated, and hydroxylated fatty acids or fatty alcohols (Lang, 2002). The hydrophilic and hydrophobic proportion of elements represented by hydrophilic–lipophilic balance in surface-active substances is a distinguishing characteristic of BSs.

Living cells produce surface-active substances called BSs, which are made by different microbes. These substances differ from synthetic surfactants in a number of ways, including their mild manufacturing situations, versatility, compostable quality, and lower virulence. These substances serve crucial functions in the development and localization of respective bacteria and are primarily biosynthesized as secondary metabolites. They are also eco-friendly in the fields of waste treatment and bioremediation (Gayathiri et al., 2022). Glycolipids, fatty acids, lipopeptides, and polymers are different categories of BSs based on the biochemistry of their aquaphophic constituents.

The usefulness of BSs instead of synthetic/inorganic surfactants is a highly sought-after area of study because of their features that reduce surface tension, stabilize emulsions, promote foam, and degrade over time. Thus, BS-producing bacteria speed up bioremediation in hydrocarbon-contaminated locations. Yeasts and bacteria produce the majority of these. These organisms produce biochemicals with high surface activity such as glycolipid, phospholipid, and rhamnolipid. The creation of these compounds facilitates the transit of the hydrocarbon substrate into cells by emulsifying it. The mechanism of action used by BSs is revealed to involve swarming motility. The function of BSs and their spontaneous release are closely tied to hydrocarbon absorption. As a result, the bacteria that break down hydrocarbons produce the most. In certain rare instances, water-soluble substances like glucose and others appeared to also form BSs (Guerra-Santos et al., 1986; Cooper and Goldenberg, 1987). Additionally, these substances contain antibacterial capabilities that damage the membrane of food-competing microbes.

Research, development, and commercialization of biological agents have all significantly increased for BSs in recent years. Numerous amphiphilic metabolites are produced by microorganisms, many of which have distinctive structural characteristics. In addition to conventional surfactant classification techniques, a variety of criteria are used to classify microbial BSs, including structural similarity, width, moieties, water-hating character, capacity to change, and some other physicochemical properties. Microbial surfactants might have a few more benefits than lowering surface stress. Because of the multifunctionality of BSs, which in turn is dependent on the structural composition of each molecule, these BS molecules have a distinctive chance of being used in medication, farming, and other ecological solicitation, some of which are not located yet.

5.2 CATEGORIZATION AND PROPERTIES OF BSS

BSs are distinguished from chemically produced surfactants by their synthetic nature, molar mass, biophysical characteristics, action mechanism, and microbial genesis. Chemically produced surfactants are distinguished by their dissociation pattern in water (Table 5.1).

1. Low-molar-mass BSs such as glycolipids, phospholipids, and lipopeptides
2. High-molar-mass BSs that include proteins, lipopolysaccharides, and amphiphatic polysaccharides

Calvo et al. (2009) cited the fact that both types of BSs carry out different degrading actions; for example, BSs with higher mass stabilize oil–water combination, whereas low-molar-mass bioemulsifiers are good at reducing surface and interfacial tensions.

5.2.1 BS TYPES ACCORDING TO STRUCTURE

There are various types of BSs (Figure 5.1), which are classified as follows:

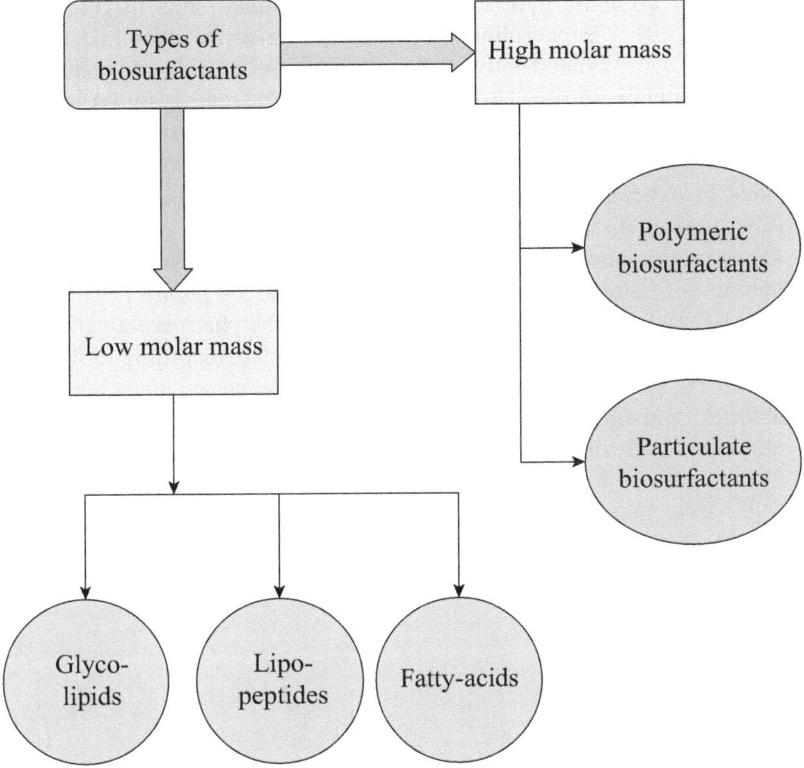

FIGURE 5.1   Types of biosurfactants

a. **Glycolipid (hydroxy aliphatic acids)**: A few examples of glycolipids include rhamnolipids, sophorolipids, trehalose lipids, and fructose lipids. Both glycolipids and lipopeptides (cycloheptapeptides containing amino acids connected to fatty acids) can be dissolved by polar and non-polar solutions (Perfumo et al., 2009; Singh et al., 2011). So far, seven rhamnolipid homologs have been discovered (Abalos et al., 2001). Rhamnolipids from *Pseudomonas aeruginosa* were reported to lower the surface tension and interfacial tension of n-hexadecane to 30 and 1 mN/m, respectively (Mulligan et al., 1989). Costa and co-workers in 2010 proclaimed the fact that rhamnolipids from *P. aeruginosa* L2–1 showed perfect (100%) blending against soybean oil and 69% against petroleum. Rhamnolipid is a crucial ingredient in the pharmaceutical sector due to its potent antibacterial effects against a variety of pathogens (Magalhães and Nitschke, 2013). *Pseudomonas* sp. that produces DYNA270s demonstrated the promise of decontamination of polypropylene coupons containing heavy oily sludges (Shen et al., 2011). Rhamnolipid is reported to boost the desorption efficiency of phenanthrene, demonstrating its use for the treatment of polycyclic aromatic hydrocarbon (PAH)—polluted soils in cold regions (Wu et al., 2017). Sophorolipids, a type of surface-active glycolipid, are composed of hydroxylated fatty acid surface-active glycolipids and a disaccharide sophorose unit. Both closed (lactonic sophorolipids) and open (acidic sophorolipids) types can be found in nature. In addition, novel bolaform sophorolipids and sophorosides (SSs), as well as SSs and glucosides, have been produced with promising metal coordination capabilities (Castelein et al., 2021; Price et al., 2012).

b. **Lipoproteins and lipopeptides**: The lipopeptides are cyclic peptides containing 8–17 amino acids and a lipid component. *Streptomyces* DPUA1566 may be used in bioremediation procedures in the pharmaceutical and cosmetic industries (Sanchita and Pritisnigdha, 2019). According to Ramani et al. (2010), *Pseudomonas gessardii* lipoprotein has strong metal ion-removal abilities, and *Bacillus atrophaeus* 5–2a lipopeptide is highly effective at removing crude oil (Zeraik and Nitschke, 2010). The lipoprotein generated

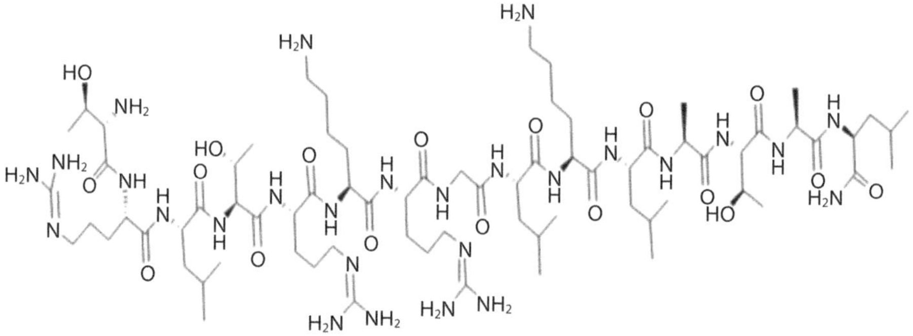

(A) Apolipoprotein

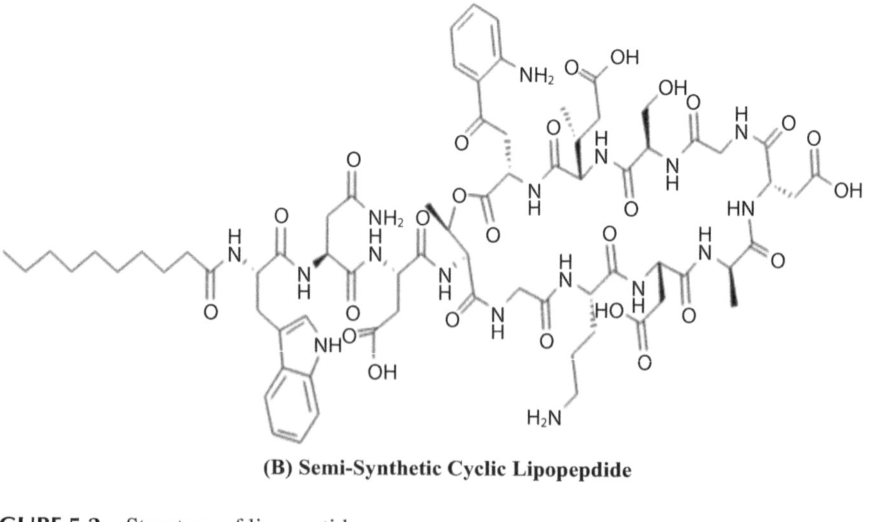

(B) Semi-Synthetic Cyclic Lipopepdide

FIGURE 5.2   Structure of lipopeptides

by *Bacillus subtilis* CMB32 significantly inhibits the anthracnose illness caused by *Colletotrichum gloeosporioides* (Kim et al., 2010). The structure of lipopeptides is depicted in Figure 5.2.

c. **Phospholipids, neutral lipids, and fatty acids**: Fatty acids and phospholipids are produced by a variety of microorganisms, including bacteria (such as *Thiobacillus thiooxidans*) and yeast, and they are in high demand right now because of their highly diverse and useful qualities as BSs (Rosas-Galvan et al., 2018).

d. **Polymeric BSs**: Biopolymeric complex mixture of polysaccharides, proteins, and lipopolysaccharides form the base components of exocellular polymeric surfactants that are produced from numerous bacterial species of various genera (Zokaei et al., 2018). The most efficient polymeric BSs are liposan, emulsan, and mannoprotein (Lang, 2002; Hatha et al., 2007). Pure emulsan has emulsifying actions at low doses. Emulsan made by *Acinetobacter calcoaceticus* PTCC1318 demonstrated properties for degrading crude oil (Amani and Kariminezhad,

TABLE 5.1
Biosurfactant classification

Biosurfactants		Molecular weight			
Group	Class	High	Low	Microorganism	References
Glycolipids	Rhamnolipids	×	√	*Pseudomonas aeruginosa, Pseudomonas* sp.	Sifour et al. (2007); Whang et al. (2008)
	Sophorolipids	×	√	*Torulopsis bombicola, Torulopsis apicola*	Pesce (2002)
	Trehalose lipids	×	√	*Nocardia* sp., *Arthrobacter* sp., *Corynebacterium* sp.	Franzetti et al. (2010)
Lipopeptides	Surfactin	×	√	*Bacillus subtilis*	Awashti et al. (1999)
	Lichenysin	×	√	*Bacillus licheniformis*	Thomas et al. (1993)
Fatty acids, phospholipids, and neutral lipids	Corynomycolic acid	×	√	*Corynebacterium lepus*	Gerson and Zajic (1978)
	Spiculisporic acid	×	√	*Penicillium spiculisporum*	Ishigami et al. (2000)
	Phosphatic dylethanolamine	×	√	*Rhodococcus erythropolis*	Appanna et al. (1995)
Polymeric biosurfactants	Emulsan	√	×	*Acinetobacter calcoaceticus* RAG-1	Zosim et al. (1982)
	Alasan	√	×	*Acinetobacter radioresistens* KA-53	Toren et al. (2001)
	Liposan	√	×	*Candida lipolytica*	Cirigliano and Carman (1984)
	Mannoprotein	√	×	*Saccharomyces cerevisiae*	Cameron et al. (1988)

2016). Liposan forms a stable emulsion with edible oils (Campos et al., 2013). Mannoprotein, which makes up the majority of *Saccharomyces cerevisiae*'s cell wall, works well as a bioemulsifier (Cameron et al., 1988). Formation of stable emulsions with hydrocarbons and other compounds by mannoproteins qualify their ability to be used as agents of remediation.

5.3 BSS AND HYDROCARBONS DEGRADATION

Hydrocarbons have significantly contaminated the environment as a result of their ubiquitous production and consumption. It is crucial to clean up polluted locations because of toxic, persistent, and detrimental effects on living things. Due to their hydrophobic nature, hydrocarbons have a restricted solubility in groundwater and have a tendency to partition (partitioning to the extent of 90–95%) in the soil medium. Because of the property of partitioning in the soil matrix, hydrocarbon pollutants are liable to reduced revival by commercial (physiochemical) treatments, restricted accessibility to oxidoreductive chemicals, and low accessibility to microorganisms whether utilized for on-site and off-site applications.

5.3.1 FUNCTION OF BSs IN DECONTAMINATION PROCESS

Utilizing BSs is a promising strategy that could increase the efficiency of bioremediation in hydrocarbon-contaminated areas. By two different methods, they can improve hydrocarbon bioremediation. The first method increases the bioavailability of the substrate for microorganisms, while the second method involves interacting with the cell surface to make it more hydrophobic, which makes it easier for hydrophobic substrates to attach to bacteria (Mulligan and Gibbs, 2004). BSs enhance the surface areas of insoluble substances, increasing their mobility and bioavailability and lowering surface and interfacial tensions in the process. As a result, BSs promote hydrocarbon removal and biodegradation. By mobilizing, solubilizing, or emulsifying hydrocarbons, the addition of BSs is anticipated to improve hydrocarbon biodegradation (Figure 5.3) (Nguyen et al., 2008; Nievas et al., 2008).

At the interface between two immiscible fluids or between a fluid and a solid, BSs build up. The repulsive forces between two dissimilar phases are decreased by reducing the tension of that exists between liquid–air surface and between liquid–liquid interfacial surface, which makes it easier for the two phases to mix and interact (Figure 5.4) (Soberon-Chavez and Maier, 2011).

The most powerful BSs, according to Soberon-Chavez and Maier (2011), may decrease the surface water tension from 72 to 30 mN m^{-1} and the tension between interfaces of a water molecule and n-hexadecane from 40 to 1 mN m^{-1}. The amount of BSs/bioemulsifiers produced is directly proportional to the concentration of surface-active chemicals, and the production of BSs continues till the critical micelle concentration (CMC) is attained. Above the CMC, BS molecules mix and form micelles, bilayers, and vesicles. Whang and co-workers in 2008 cited the fact that BSs create micelles to reduce surface and interfacial tension and increase the bioavailability/solubility of hydrophobic chemical substances.

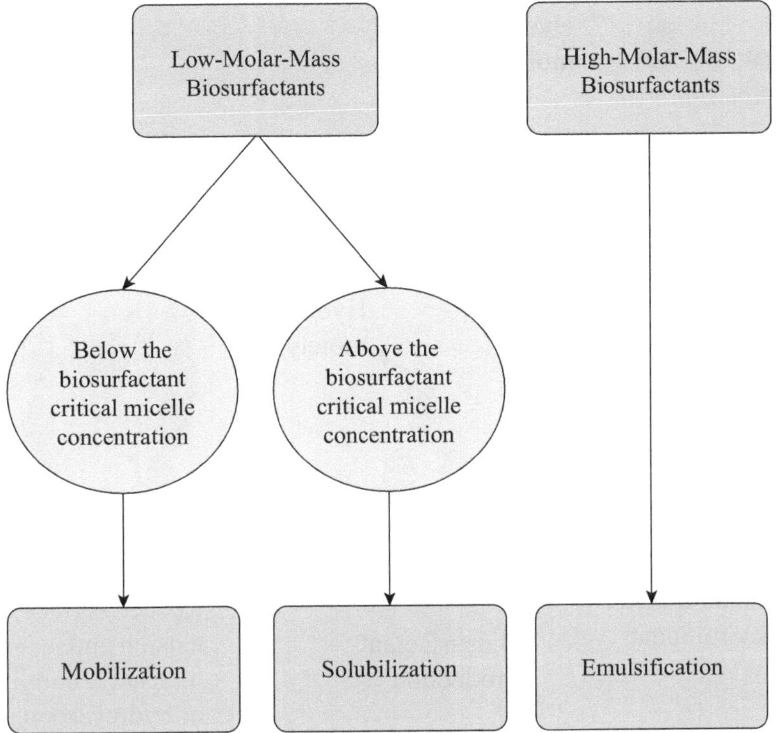

FIGURE 5.3 Mechanism of exclusion of hydrocarbons by BSs based on their molar mass and concentration (Urum and Pekdemir, 2004)

The CMC is frequently used to gauge surfactant effectiveness. Because low CMC BSs are effective, lowering the surface tension requires only a small amount of BS (Desai and Banat, 1997). In the creation of microemulsions, "which are stable liquid mixtures of oil and water (two liquid phase) and are composed of monolayers or aggregates of biosurfactants" micelle production is crucial (Nguyen et al., 2008). Franzetti and co-workers (2010) highlight the potential functions of BSs for hydrocarbon degradation via the alteration of surface cell hydrophobicity that enables them to adhere to micelles or emulsified oils. Bacteria and hydrocarbons can interact in three different ways:

- Utilization of hydrocarbons that are water soluble
- Direct interaction between big oil drops and cells
- Coming into touch with emulsified or pseudosolubilized oil

According to the authors, BSs can change the way hydrocarbons are accessed in response to different stages of microbe growth. They learned from their research that the strain BS 29 produced by *Gordonia* sp. releases cell-bound glycolipid BS

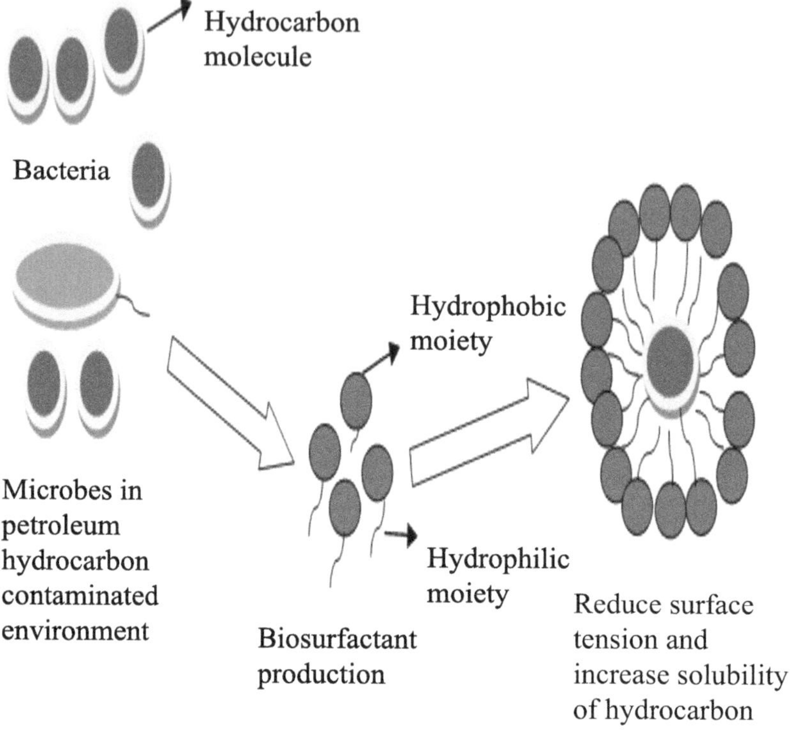

FIGURE 5.4   Emulsification process done by biosurfactants

and extracellular bioemulsifier when grown on hydrocarbons, and surface hydrophobicity changes were seen during the growth phase on hexadecane (Franzetti et al., 2009).

According to Cameotra and Singh (2009), internalization of the hydrocarbon into the cell occurs prior to the subsequent breakdown of the hydrocarbon. Hexadecane was disseminated into microdroplets by the action of the BS, which increased the hydrocarbon's accessibility to the bacterial cells. The hydrocarbon droplets coated with the BS were taken up, according to the electron microscopic investigations. It's interesting to note that "internalization" of "BS layered hydrocarbon droplets" was occurring through a process that resembled active pinocytosis. BS made by the *Pseudomonas* sp. LP1 strain has the ability to biodegrade crude oil and diesel. The crude and diesel oil's hydrocarbon components can be metabolized by the LP1 strain. According to reports, this strain of *Pseudomonas* sp. degrades crude oil by 92.34% and removes diesel oil by 95.29% (Obayori et al., 2009). According to findings presented by Kang et al. (2010), sophorolipid might be able to speed up the biodegradation of places contaminated with hydrocarbons that are only partially soluble in water and increase microbial consortium bioavailability for this process. They found that the addition of sophorolipid BS increased the degradation rate of hydrocarbons to the extent of 85% to 97%. In 2007, Das and Mukherjee looked at the efficacy of strains (*B. subtilis* DM-04, *P. aeruginosa* NM) in the remediation of sites highly contaminated with hydrocarbons from crude petroleum. For this, soil samples were treated with aqueous solutions of BSs produced by the appropriate bacterial strains. In addition to this, halite media with a certain content of either *P. aeruginosa* M and NM strains or *B. subtilis* DM-04 were utilized to inoculate the tested soil. After 120 days, the amount of total petroleum hydrocarbons (TPH) in the soil phase was assessed, and the level of biodegradation was determined by comparing it with non-inoculated (control) media. TPH levels drop to the extent of 84 to 21 and 39 g kg^{-1} of soil, by M and NM consortia of *P. aeruginosa* and *B. subtilis* in comparison with non-inoculated soil where the drop in the level of TPH was 83 g kg^{-1}.

The content of crude oil, the proportion of the structural classes of the hydrocarbons, and the bioavailability of the substrate are a few variables affecting the microbial breakdown of crude oil. The bioavailability of hydrocarbons, particularly the heavy fractions to be converted or for use in the bioremediation of hydrocarbon-contaminated soils, was one of the important components of microbial genetic engineering in the oil industries. The application of low mass transfer phases might have made it more difficult to remove PAHs from the soil that was polluted with them. PAHs are always detrimental to aquatic life and human health. Much of the research has been devoted to examining new chemicals that increase the bioavailability of contaminated hydrocarbon compounds by making them more soluble. The most effective and environmentally beneficial cleanup method for PAHs was thought to be bioremediation, which includes the microbial conversion of contaminants into beneficial metabolites. Zhuang et al. discovered and described bacteria that break down naphthalene-contaminated coastal sediments in 2002. The recovery of oil has frequently used microbial-enhanced oil recovery. Some bacteria release the oil sediments locked in reservoirs and rocks so they can use them in their metabolism to create a variety of metabolites (Karlapudi et al., 2018).

5.4   METAL REMEDIATION THROUGH THE USE OF BSS

Heavy metal pollution of soil habitats poses serious risks to people and other living things in the ecosystem. Heavy metals are exceedingly hazardous; thus, it has been discovered that even small amounts of them in the soil can have negative effects. Heavy metal–contaminated soil can now be cleaned up using a variety of approaches. These soils are cleaned up using either biological treatments or non-biological methods, like excavation and dumping contaminated soil in landfills (Asci et al., 2010). When removing metals from soil, biological approaches employ plants (phytoremediation) or microorganisms (bioremediation). It has been known for a long time that using microbes can help reduce metal contamination. Heavy metals cannot biodegrade and can only be transferred from one source to another. Microorganisms can also store metals by intracellular, metabolism-dependent (active) absorption and metabolism-independent (passive) uptake. By changing pH or by generating or releasing chemicals that alter the mobility of the metals, microorganisms can indirectly affect the mobility of metals (Ledin, 2000).

Metal-contaminated soil is treated using one of the mentioned methods: "soil washing" or "soil flushing." The first step is to remove contaminated soil from its original location, place it in a glass column, and clean it with a BS solution. For injecting and removing the BS solution from the soil during soil flushing, an in-situ technology, trenches and drainage pipes are used (Singh and Cameotra, 2004). It's intriguing to learn that soil metals can be removed using BSs.

5.4.1 Process of Mechanism of Metal Removal by BSs

The main factor that makes BSs useful for the remediation of heavy rock–polluted soil is their capacity to network complex formation with metals. The metal–BS complexes are desorbed from the soil matrix and dissolve in the soil solution as a result of the decrease in interfacial tension, and because of this bonding between the metal and BSs becomes stronger than the metal's connections to the soil colloids. The cationic BSs can compete for some negatively charged surfaces to substitute the negatively charged metal ions but not all of them (ion exchange). BS micelles also help in the elimination of metal ions from the soil surface via the bonding of polar head groups of micelles with metals that float in water (Figure 5.5).

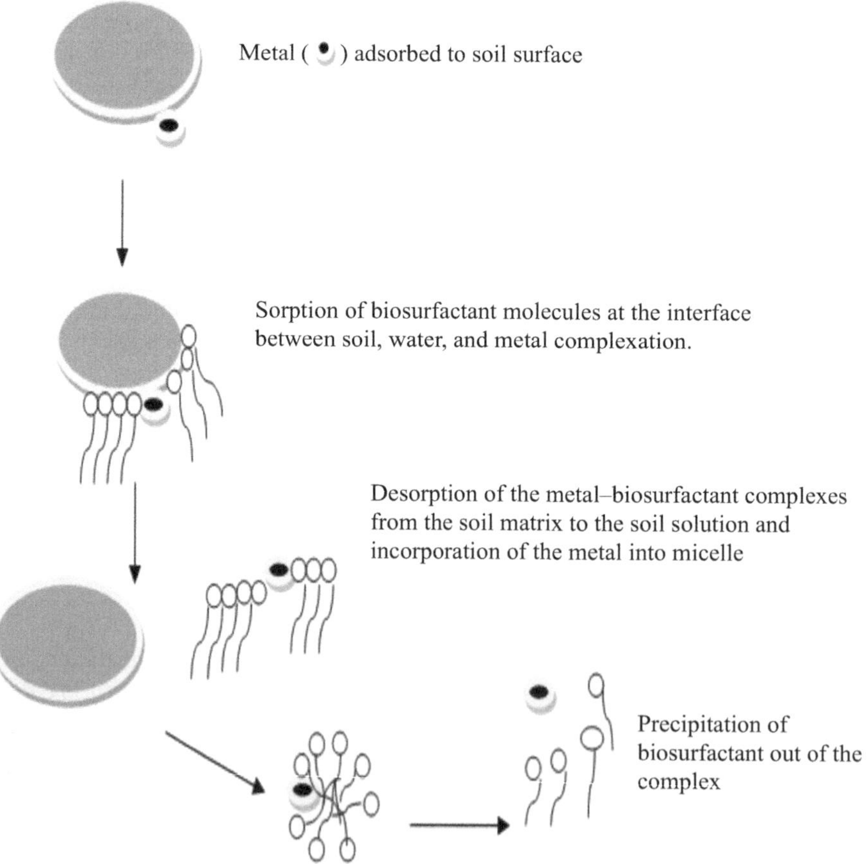

FIGURE 5.5   Metal removal mechanism via biosurfactants (Mulligan, 2005)

Juwarkar et al. (2008) discovered that when the soil was washed with 0.1% di-rhamnolipid BS produced by *P. aeruginosa*, BS2 selectively removed heavy metals from the soil in the order of Cd = Cr > Pb = Cu > Ni. Since many metals are largely found in the organic portion of the environment, adding inorganic materials like 1% NaOH to a rhamnolipid BS solution speeds up the removal of copper and nickel. A rhamnolipid BS can remove more metals as a result of the OH solubilizing this organic portion (Dahrazma and Mulligan, 2007). Heavy metal–contaminated soil can also be successfully remedied using the BS foam approach. Wang and Mulligan (2004) investigated the feasibility of extracting Cd and Ni using rhamnolipid foam from sandy soil. Utilizing rhamnolipid foam improved efficacy and permitted the exclusion of 73.2% and 68.1% of Cadmium (Cd) and Nickel (Ni), respectively, as opposed to its solution form that flushed only 61.7% and 51%, respectively. Gnanamani et al. (2010) looked into the BS-producing marine isolate *Bacillus* sp. MTCC 5514's bioremediation of chromium (VI). The remediation process for this strain required two steps: the BSs' trapping of Cr (III) and extracellular chrome reductase's conversion of Cr (VI) to Cr (III). The first mechanism changes the toxic state of chromium (Cr) to a less dangerous one, and the second protects bacterial cells against exposure to Cr (III). Both actions keep bacterial cells active all the time and help them to tolerate and be resilient in environments that are highly chromium rich in hexavalent and trivalent forms.

BSs' chemical makeup affects their capacity to remove heavy rock metals from the soil. Asci et al. (2010) used quartz as a soil component because it makes up the majority of the soil; to explore the metal ion recovery from this component utilizing rhamnolipid as a BS it was found that quartz and a 25 mM rhamnolipid recovered sorbed Cd and Zn to the extent of 91.6% and 87.2 %, respectively. By utilizing rhamnolipid BS, the mobilization of arsenic from mine tailings was also improved. The mobilization became rather stable when the rhamnolipid concentration was above 100 mg L^{-1} and increased with the BS concentration (Wang and Mulligan, 2009). According to Doong et al. (1998), the decontamination/remediation of heavy rock is linearly dependent on the concentration of surfactant both above and below the CMC.

5.5 RECENT DEVELOPMENTS

- **Compost's exposure to BS-producing bacteria:** Composting causes the stability of organic leftovers and the deterioration of the majority of ecologically friendly resources, primarily organic mixtures. Microbes are widely employed in the breakdown of organic matter, and, as a result, BSs have been widely used in biodegradation and bioremediation in recent decades. It is possible to leverage these advantages to raise compost quality as well. BSs being amphipathic reduce the surface tension between liquids and solids and hence boost the bioremediation of organic matter (De Giani et al., 2021). Additionally, BS-induced bacterial growth stimulation promoted the breakdown of the organic matter. Rhamnolipid and tween 80 were studied for their effects on bacterial variety, with rhamnolipid promoting microbial growth in composting. Synergisms between *Bacillus* sp. and *Streptomyces* sp. accelerated the breakdown of organic materials during the composting process. There is an increase in bacterial communities when BS-producing bacteria consortium is combined with cell suspension that contains BSs. This demonstrates that BSs had no detrimental effects on the growth of bacteria during composting and may perhaps have slightly stimulated it (Shi et al., 2006).

- **Vermicomposting of green waste and BS effects:** Due to the accelerated conversion of natural organic molecules into humic-like compounds caused by microbial inoculation and rhamnolipid inclusion, the additive treatments had higher quantities of humic acid (Zhang et al., 2008). By increasing the breakdown of organic carbon and the activity of nitrogen-fixing bacteria, rhamnolipid will increase nitrogen concentration, causing the combined total nitrogen concentrations to rise (Molina et al., 2013). Subramanian et al. (2010) conducted research on the use of rhamnolipid BS in vermicomposting and the results of their findings showed that the addition of rhamnolipid either alone or in combination with cellulolytic and lgnolytic fungus (*Paecilomyces fusisporus*) and a nitrogen-fixing free-living bacterium (*Azotobacter chrococcum*) to vermicompost green waste, where earthworm sp. (*Eisenia fetida*) is used for composting, significantly increased the growth rate of worms.

- **BSs in bionanotechnology:** Environmental remediation holds great promise for the production of nanoparticles via BS-mediated synthesis (Ni'matuzahroh et al., 2019). However, the BS-generated nanoparticles must be commercially feasible. Additionally, it must be energy-effective, toxicant removal-efficient, and environmentally friendly. Nanoparticle production may be stabilized and reduced by BS microorganisms (Garcia-Junco et al., 2003). Titanium, silver, and gold nanoparticles are among the materials that microorganisms help produce. Numerous researchers and scientists find this as the novel method for fabricating metallic nanoparticles, which opens up new avenues for investigation. Nanoparticles can be stable for around two months, thanks to *Brevibacterium casei* MSA19 BSs, which also prevent aggregate formation. This increases their utility as an environmentally beneficial component for diverse products. (Dadrasnia and Ismail, 2015). Before using nanoparticles in nanotechnologies, more research is required to stabilize them using BSs (Ni'matuzahroh et al., 2018).

5.6 CONCLUSION

There has been research on the use of BSs in bioremediation or environmental cleanup. Different techniques (physicochemical and biological) involving BSs can remove both organic and inorganic contaminants. They are extremely encouraging for usage in environmental biotechnologies because of their low toxicity and propensity to biodegrade. The high production costs of BSs continue to be a barrier to their commercial success. The extensive use of BS in the procedure is a key component of biological remediation technology. To make this procedure easier, fresh approaches need to be created, such as foams or micro-foams (colloidal gas aphrons) are two examples. The potential for in-situ BS synthesis by microorganisms is poorly understood. The majority of the research that was mentioned was carried out in a lab setting. More work is needed to assess the synthesis of BSs by in-situ microorganisms and their use in biological remediation systems. The architecture of BSs, their interactions with pollutants and soil, and the scale-up and efficient manufacturing of BSs all require more information. Commercially feasible biological and engineering solutions are necessary to reduce the cost of BS manufacturing.

As the BS market is continuously growing, they are used in a variety of specialized applications. Technical restrictions, namely, their costs and the drawbacks of combining technologies, are the industries' barriers. On the market, natural and renewable resources are becoming more and more popular. By employing agro-industrial waste of plant and animal origin, it may be possible to reduce manufacturing expenses and formulate them inexpensively and viable against synthetic BSs. A few issues that can be treated using bio-based surfactants include the removal of heavy rock metals, decontamination of polluted soil and water, skin conditions, food preservation, improved oil restoration, and plant disease eradication. Green manures used for vermicomposting frequently contain rhamnolipids, a form of commercially accessible BS.

REFERENCES

Abalos, A., Pinazo, A., Infante, M. R., Casals, M., García, F., & Manresa, A. (2001). Physicochemical and antimicrobial properties of new rhamnolipids produced by pseudomonas aeruginosa AT10 from soybean oil refinery wastes. *ACS Publications, 17,* 1367–1371.

Amani, H., & Kariminezhad, H. (2016). Study on emulsification of crude oil in water using emulsan biosurfactant for pipeline transportation. *Petroleum Science and Technology, 34,* 216–222.

Appanna, V. D., Finn, H., & Pierre, M., St. (1995). Exocellular phosphatidylethanolamine production and multiple-metal tolerance in Pseudomonas fluorescens. *FEMA Microbiology Letters, 131,* 53–56.

Asci, Y., Nurbas, M., & Acikel, Y. S. (2010). Investigation of sorption/desorption equilibria of heavy metal ions on/from quartz using rhamnolipid biosurfactant. *Journal of Environmental Management, 91,* 724–731.

Awashti, N., Kumar, A., Makkar, R., & Cameotra, S. (1999). Enhanced biodegradation of endosulfan, a chlorinated pesticide in presence of a biosurfactant. *Journal of Environmental Science and Health, Part B, 34,* 793–803.

Calvo, C., Manzanera, M., Silva-Castro, G. A., Uad, I., & Gonzalez-Lopez, J. (2009). Application of bioemulsifiers in soil oil bioremediation processes. Future Prospects. *Science of the Total Environment, 407,* 3634–3640.

Cameotra, S. S., & Singh, P. (2009). Synthesis of rhamnolipid biosurfactant and mode of hexadecane uptake by Pseudomonas species. *Microbial Cell Factories, 8,* 1–7.

Cameron, D. R., Cooper, D. G., & Neufeld, R. J. (1988). The mannoprotein of Saccharomyces cerevisiae is an effective bioemulsifier. *Applied and Environmental Microbiology, 54,* 1420–1425.

Campos, J. M., Stamford, T. L., Sarubbo, L. A., de Luna, J. M., Rufino, R. D., & Banat, I. M. (2013). Microbial biosurfactants as additives for food industries. *Biotechnology Progress, 29,* 1097–1108.

Castelein, M., Verbruggen, F., Van Renterghem, L., Spooren, J., Yurramendi, L., Du Laing, G., . . . & Williamson, A. J. (2021). Bioleaching of metals from secondary materials using glycolipid biosurfactants. *Minerals Engineering, 163,* 106665.

Cirigliano, M. C., & Carman, G. M. (1984). Purification and characterization of liposan, a bioemulsifier from Candida lipolytica. *Applied and Environmental Microbiology, 50,* 846–850.

Cooper, D. G., & Goldenberg, B. G. (1987). Surface active agents from two bacillus species. *Applied and Environmental Microbiology, 53,* 224–227.

Dadrasnia, A., & Ismail, S. (2015). Biosurfactant production by bacillus salmalaya for lubricating oil solubilization and biodegradation. *Journal of Environmental Research and Public Health, 12,* 9848–9863.

Dahrazma, B., & Mulligan, C. N. (2007). Investigation of the removal of heavy metals from sediments using rhamnolipid in a continuous flow configuration. *Chemosphere, 69,* 705–711.

Das, K., & Mukherjee, A. K. (2007). Crude petroleum-oil biodegradation efficiency of Bacillus subtilis and Pseudomonas aeruginosa strains isolated from a petroleum-oil contaminated soil from North-East India. *Bioresource Technology, 98*(7), 1339–1345.

De Giani, A., Zampolli, J., & Di Gennaro, P. (2021). Recent trends on biosurfactants with antimicrobial activity produced by bacteria associated with human health: Different perspectives on their properties, challenges, and potential applications. *Frontiers in Microbiology, 12,* 678.

Desai, J. D., & Banat, I. M. (1997). Microbial production of surfactants and their commercial potential. *Microbiology and Molecular Biology Reviews, 61,* 47–64.

Doong, R. A., Wu, Y. W., & Lei, W. G. (1998). Surfactant enhanced remediation of cadmium contaminated soils. *Water Science and Technology, 37,* 65–71.

Engwa, G. A., Ferdinand, P. U., Nwalo, F. N., & Unachukwu, M. N. (2019). Mechanism and health effects of heavy metal toxicity in humans. *Poisoning in the Modern World-New Tricks for an Old Dog, 10,* 70–90.

Franzetti, A., Caredda, P., Ruggeri, C., Colla La, P., Tamburini, E., Papacchini, M., & Bestetti, G. (2009). Potential applications of surface active compounds by Gordonia sp. strain BS29 in soil remediation technologies. *Chemosphere, 75,* 801–807.

Franzetti, A., Gandolfi, I., Bestetti, G., Smyth, T. J., & Banat, I. M. (2010). Production and applications of trehalose lipid biosurfactants. *European Journal of Lipid Science and Technology, 112,* 617–627.

Garcia-Junco, M., Gomez-Lahoz, C., Niqui-Arroyo, J. L., & Ortega-Calvo, J. J. (2003). Biodegradation and biosurfactant-enhanced partitioning of polycyclic aromatic hydrocarbons from nonaqueous-phase liquids. *Environmental Science and Technology, 37,* 2988–2996.

Gayathiri, E., Prakash, P., Karmegam, N., Varjani, S., Awasthi, M. K., & Ravindran, B. (2022). Biosurfactants: Potential and eco-friendly material for sustainable agriculture and environmental safety—A review. *Agronomy, 12,* 662. https://doi.org/10.3390/agronomy12030662

Georgiou, G., Lin, S. C., & Sharma, M. M. (1992). Surface–active compounds from microorganisms. *Bio/Technology, 10*(1), 60–65.

Gerson, O. F., & Zajic, J. E. (1978). Surfactant production from hydrocarbons by Corynebacterium lepus, sp. nov. and Pseudomonas asphaltenicus, sp. nov. *Developments in Industrial Microbiology, 19,* 577–599.

Gnanamani, A., Kavitha, V., Radhakrishnan, N., Rajakumar, G. S., Sekaran, G., & Mandal, A. B. (2010). Microbial products (biosurfactant and extracellular chromate reductase) of marine microorganism are the potential agents reduce the oxidative stress induced by toxic heavy metals. *Colloids and Surfaces B: Biointerfaces, 79,* 334–339.

Gorin, P. A. J., Spencer, J. F. T., & Tulloch, A. P. (1961). Hydroxy fatty acid glycosides of sophorose from Torulopsis magnoliae. *Canadian Journal of Chemistry, 39*(4), 846–855.

Guerra-Santos, L. H., Kappelli, O., & Fiechter, A. (1986). Dependence of pseudomonas aeruginosa continuous culture biosurfactant production on nutritional and environmental factors. *Applied Microbiology and Biotechnology, 24*, 443–448.

Hatha, sssA. A. M., Edward, G., & Rahman, K. S. M. P. (2007). Microbial biosurfactants-review. *Journal of Marine and Atmospheric Research, 3*, 1–17.

Ishigami, Y., Zhang, Y., & Ji, F. (2000). Spiculisporic acid: Functional development of biosurfactants. *Chimica Oggi-Chemistry Today, 18*, 32–34.

Juwarkar, A. A., Dubey, K. V., Nair, A., & Singh, S. K. (2008). Bioremediation of multimetal contaminated soil using biosurfactant—A novel approach. *Indian Journal of Microbiology, 48*, 142–146.

Kang, S. W., Kim, Y. B., Shin, J. D., & Kim, E. K. (2010). Enhanced biodegradation of hydrocarbons in soil by microbial biosurfactant, sophorolipid. *Applied Biochemistry and Biotechnology, 160*, 780–790.

Karlapudi, A. P., Venkateswarulu, T. C., Tammineedi, J., Kanumuri, L., Ravuru, B. K., Dirisala, V. R., & Kodali, V. P. (2018). Role of biosurfactants in bioremediation of oil pollution-a review. *Petroleum*. https://doi.org/10.1016/j.petlm.2018.03.007

Kim, P. I., Ryu, J., Kim, Y. H., & Chi, Y. T. (2010). Production of biosurfactant lipopeptides Iturin A, fengycin and surfactin A from Bacillus subtilis CMB32 for control of Colletotrichum gloeosporioides. *Journal of Microbiology and Biotechnology, 20*, 138–145.

Lang, S. (2002). Biological amphiphiles (microbial biosurfactants). *Current Opinion in Colloid & Interface Science, 7*(1–2), 12–20.

Ledin, M. (2000). Accumulation of metals by microorganisms—Processes and importance for soil systems. *Earth-Science Reviews, 51*(1–4), 1–31.

Pacwa-Płociniczak, M., Płaza, G. A., Piotrowska-Seget, Z., & Cameotra, S. S. (2011). Environmental applications of biosurfactants: Recent advances. *International Journal of Molecular Sciences, 12*(1), 633–654.

Magalhães, L., & Nitschke, M. (2013). Antimicrobial activity of rhamnolipids against Listeria monocytogenes and their synergistic interaction with nisin. *Food Control, 29*(1), 138–142.

Molina, M. J., Soriano, M. D., Ingelmo, F., & Llinares, J. (2013). Stabilisation of sewage sludge and vinasse bio-wastes by vermicomposting with rabbit manure using Eisenia fetida. *Bioresource Technology, 137*, 88–97.

Mulligan, C. N. (2005). Environmental applications for biosurfactants. *Environmental Pollution, 133*, 183–198.

Mulligan, C. N., & Gibbs, B. F. (2004). Types, production and applications of biosurfactants. *Proceedings of the National Academy of Sciences, 1*, 31–55.

Mulligan, C. N., Mahmourides, G., & Gibbs, B. F. (1989). The influence of phosphate metabolism on biosurfactant production by Pseudomonas aeruginosa. *Journal of Biotechnology, 12*, 199–209.

Nguyen, T. T., Youssef, N. H., McInerney, M. J., & Sabatini, D. A. (2008). Rhamnolipid biosurfactant mixtures for environmental remediation. *Water Research, 42*, 1735–1743.

Nievas, M. L., Commendatore, M. G., Estevas, J. L., & Bucalá, V. (2008). Biodegradation pattern of hydrocarbons from a fuel oil-type complex residue by an emulsifier-producing microbial consortium. *Journal of Hazardous Materials, 154*, 96–104.

Ni'matuzahroh, S. K., Ningrum, I. P., Pusfita, A. D., Marjayandari, L., Trikurniadewi, N., . . . & Yuliani, H. (2019). The potential of indigenous bacteria from oil sludge for biosurfactant production using hydrolysate of agricultural waste. *Biodiversitas, 20*, 1374–1379.

Ni'matuzahroh, Sari, S. K., Trikurniadewi, N., Pusfita, A. D., INingrum, P., Ibrahim, S. N. M. M., . . . & Surtiningsih, T. (2018). Utilization of rice straw hydrolysis product of penicillium sp. H9 as a substrate of biosurfactant production by LII61 hydrocarbonoclastic bacteria. *IOP Conference Series: Earth and Environmental Science, 217*, 012028.

Obayori, O. S., Ilori, M. O., Adebusoye, S. A., Oyetibo, G. O., Omotayo, A. E., & Amund, O. O. (2009). Degradation of hydrocarbons and biosurfactant production by Pseudomonas sp. strain LP1. *World Journal of Microbiology and Biotechnology, 25*, 1615–1623.

Perfumo, A., Smyth, T. J. P., Marchant, R., & Banat, I. M. (2009). Production and roles of biosurfactant and bioemulsifiers in accessing hydrophobic substrates. In: Kenneth, N. T. (Ed.), *Microbiology of hydrocarbons, oils, lipids and derived compounds* (pp. 1502–1512). Springer.

Pesce, L. (2002). A biotechnological method for the regeneration of hydrocarbons from dregs and muds, on the base of biosurfactants. World Patent 02/062, 495.

Price, N. P. J., Ray, K. J., Vermillion, K. E., Dunlap, C. A., & Kurtzman, C. P. (2012). Structural characterization of novel sophorolipid biosurfactants from a newly identified species of Candida yeast. *Carbohydrate Research, 348*, 33–41.

Ramani, K., Chockalingam, E., & Sekaran, G. (2010). Production of a novel extracellular acidic lipase from Pseudomonas gessardii using slaughterhouse waste as a substrate. *Journal of Industrial Microbiology and Biotechnology, 37*(5), 531–535.

Ramani, K., Jain, S. C., Mandal, A. B., & Sekaran, G. (2012). Microbial induced lipoprotein biosurfactant from slaughterhouse lipid waste and its application to the removal of metal ions from aqueous solution. *Colloids and Surfaces B: Biointerfaces, 97*, 254–263.

Rosas-Galvan, N. S., Martinez-Morales, F., Marquina-Bahena, S., Tinoco-Valencia, R., Serrano-Carreón, L., Bertrand, B., . . . & Trejo-Hernández, M. D. R. (2018). Improved production, purification, and characterization of biosurfactants produced bySerratia marcescensSM3 and its isogenic SMRG-5 strain. *Biotechnology and Applied Biochemistry, 65*, 690–700.

Rosenberg, E., & Ron, E. Z. (1999). High-and low-molecular-mass microbial surfactants. *Applied Microbiology and Biotechnology, 52*, 154–162.

Sanchita, K., & Pritisnigdha, P. (2019). Production and functional characterization of food compatible biosurfactants. *Applied Food Science Journal, 3*, 1–4.

Shen, H.-H., Lin, T.-W., Thomas, R. K., Taylor, D. J. F., & Penfold, J. (2011). Surfactin structures at interfaces and in solution: The effect of pH and cations. *The Journal of Physical Chemistry B, 115*, 4427–4435.

Shi, J. G., Zeng, G. M., & Yuan, X. Z. (2006). The stimulatory effects of surfactants on composting of waste rich in cellulose. *World Journal of Microbiology and Biotechnology, 22*, 1121–1127.

Sifour, M., Al-Jilawi, M. H., & Aziz, G. M. (2007). Emulsification properties of biosurfactant produced from Pseudomonas aeruginosa RB 28. *Pakistan Journal of Biological Sciences, 10*, 1331–1335.

Singh, B. R., Dwivedi, S., Al-Khedhairy, A. A., & Musarrat, J. (2011). Synthesis of stable cadmium sulfide nanoparticles using surfactin produced by Bacillus amyloliquifaciens strain KSU-109. *Colloids and Surfaces B: Biointerfaces, 85*, 207–213.

Singh, P., & Cameotra, S. S. (2004). Enhancement of metal bioremediation by use of microbial surfactants. *Biochemical and Biophysical Research Communications, 319*, 291–297.

Soberon-Chavez, G., & Maier, R. M. (2011). Biosurfactants: A general overview. In: Soberón-Chávez, G. (Ed.), *Biosurfactants* (pp. 1–11). Springer-Verlag.

Subramanian, S., Sivarajan, M., & Saravanapriya, S. (2010). Chemical changes during vermicomposting of sago industry solid wastes. *Journal of Hazardous Materials, 179*, 318–322.

Sumudumali, R. G. I., & Jayawardana, J. M. C. K. (2021). A review of biological monitoring of aquatic ecosystems approaches: With special reference to macroinvertebrates and pesticide pollution. *Environmental Management, 67*(2), 263–276.

Thomas, C. P., Duvall, M. L., Robertson, E. P., Barrett, K. B., & Bala, G. A. (1993). Surfactant-based EOR mediated by naturally occurring microorganisms. *SPE Reservoir Engineering, 8*(04), 285–291.

Toren, A., Navon-Venezia, S., Ron, E. Z., & Rosenberg, E. (2001). Emulsifying activity of purified alasan proteins from Acinetobacter radioresistens. *Applied and Environmental Microbiology, 67,* 1102–1106.

Urum, K., & Pekdemir, T. (2004). Evaluation of biosurfactants for crude oil contaminated soil washing. *Chemosphere, 57,* 1139–1150.

Ventriglio, A., Bellomo, A., di Gioia, I., Di Sabatino, D., Favale, D., De Berardis, D., & Cianconi, P. (2021). Environmental pollution and mental health: A narrative review of literature. *CNS Spectrums, 26*(1), 51–61.

Wang, S., & Mulligan, C. N. (2004). Rhamnolipid foam enhanced remediation of cadmium and nickel contaminated soil. *Water, Air, & Soil Pollution, 157,* 315–330.

Wang, S., & Mulligan, C. N. (2009). Arsenic mobilization from mine tailings in the presence of a biosurfactant. *Applied Geochemistry, 24,* 928–935.

Whang, L. M., Liu, P. W. G., Ma, C. C., & Cheng, S. S. (2008). Application of biosurfactant, rhamnolipid, and surfactin, for enhanced biodegradation of diesel-contaminated water and soil. *Journal of Hazardous Materials, 151,* 155–163.

Wu, Y.-S., Ngai, S.-C., Goh, B.-H., Chan, K.-G., Lee, L.-H., & Chuah, L.-H. (2017). Anticancer activities of surfactin and potential application of nanotechnology assisted surfactin delivery. *Frontiers in Pharmacology, 8,* 761.

Zeraik, A. E., & Nitschke, M. (2010). Biosurfactants as agents to reduce adhesion of pathogenic bacteria to polystyrene surfaces: Effect of temperature and hydrophobicity. *Current Microbiology, 61,* 554.

Zhang, Q., Weimin, C. A. I., & Juan, W. (2008). Stimulatory effects of biosurfactant produced by Pseudomonas aeruginosa BSZ-07 on rice straw decomposing. *Journal of Environmental Sciences, 20,* 975–980.

Zhuang, W. Q., Tay, J. H., Maszenan, A., & Tay, S. (2002). Bacillus naphthovorans sp. nov. from oil-contaminated tropical marine sediments and its role in naphthalene biodegradation. *Applied Microbiology and Biotechnology, 58,* 547–554.

Zokaei, E., Badoei-dalfrad, A., Ansari, M., Karami, Z., Eslaminejad, T., & Nematollahi-Mahani, S. N. (2018). Therapeutic potential of DNAzyme loaded on chitosan/cyclodextrin nanoparticle to recovery of chemosensitivity in the mcf-7 cell line. *Applied Biochemistry and Biotechnology, 183,* 126–136.

Zosim, Z., Gutnick, D., & Rosenberg, E. (1982). Properties of hydrocarbon-in-water emulsion stabilized by Acinetobacter RAG-1 emulsan. *Biotechnology and Bioengineering, 24*(2), 281–292.

6 Ecofriendly and Innovative Techniques for Managing Organic Pollutants

Malik Ahsaf Aziz, Shayesta Islam, Syed Hujjat ul Baligah, and Burhan Hamid

6.1 INTRODUCTION

Organic pollutants are constituted primarily of hydrocarbons of natural and human origin. Natural organic pollutants are those that emerge from biological sources including sugars, alkaloids, and terpenoids, while human-based pollutants are those that arise as a result of chemical reactions among different chemical species. Examples are dichloro-diphenyl trichloroethane (DDT), polychlorinated biphenyls (PCBs), chlordane, dieldrin, and so on (Evans et al., 2005; Zhang et al., 2003). The organic pollutants are carcinogenic, mutagenic, and teratogenic. These are generated either from point sources like treatment plants, industries, mining sites, or construction places or from non-point sources like agricultural runoff and urban runoff (Baranowska et al., 2005). Some organic pollutants are readily degraded in the environment, but most remain in the environment for a longer duration referred to as persistent organic pollutants. These persistent organic pollutants take centuries to fully deteriorate. Persistent organic pollutants have the potential to move to long distances and impede deterioration owing to their physicochemical characteristics, leading to accumulation in deeper areas; moreover, they have damaging effects on humans and the environment. Due to their lower solubility, they are resistant to chemical, biological, and photodegradation (Goktaş and MacLeod, 2016). But they can be degraded by different processes like biological degradation, atmospheric reactions, and abiotic alterations. These occur in a gaseous state and particle state in both air and water; owing to their gaseous state, emissions from one country can become the cause of contamination in another country. It has been reported that because of their semi-volatile characteristics, these are not restricted to one environment but can be readily absorbed by the particles in the atmosphere and ultimately reach to soil, water, and air environment (Adeola, 2004; Abelsohn et al., 2002; Qing et al., 2006) Moreover, because of their fat-loving character, persistent organic pollutants accumulate in the higher levels of food chain and cause various reproductive and behavioural disorders (Jones, 2021). The history of persistent organic pollutants started in 1962 with the publication of Silent Spring by Rachel Carson.

DOI: 10.1201/9781003359326-6

These originate from agricultural, industrial, electronic, and medical waste and are noxious even at low levels and highly fatal. The categories of persistent organic pollutants are organochlorine pesticides; hexachlorobenzene and other polychlorinated benzenes; polycyclic aromatic hydrocarbons (PAHs); polychlorinated naphthalenes; PCBs; polychlorinated dibenzo-*p*-dioxins; and dibenzofurans. DDT is the persistent organochlorine insecticide that was used to control malaria during the Second World War, but it has resulted in a negative impact on the environment, particularly birds. It has been reported by Carson that the consumption of DDT by birds causes thinning of egg shells, which ultimately leads to early hatching and death (Doug, 2012). Her book finally led to the ban on DDTs in 1972. During the Stockholm Convention on persistent

TABLE 6.1

List of chemicals decided to phase out by member countries in various amendments of the Stockholm Convention

S. no	Name of the organic pollutants	Category
According to the 2001 Amendment of the Stockholm Convention		
01	Polychlorinated biphenyls	Industrial waste/secondary product
02	Polychlorinated dibenzo-p-dioxins (PCDDs)	Secondary product
03	Polychlorinated dibenzofurans PCDF	Secondary product
04	Chlordane	Pesticide
05	Mirex	Pesticide
06	Endrin	Pesticide
07	Aldrin	Pesticide
08	Dieldrin	Pesticide
09	Hexachlorobenzene (HCB)	Pesticide
10	Heptachlor	Pesticide
11	Toxaphene	Pesticide
12	Dichloro-diphenyl trichloroethane	Pesticide
According to the 2009 Amendment of the Stockholm Convention		
1	Lindane	Pesticide
02	Chlordecone	Pesticide
03	Pentachloro benzene	Pesticide/secondary product
04	Alpha-hexachlorocyclohexane (HCH)	Pesticide/secondary product
05	Beta-HCH	Pesticide/secondary product
06	Perfluorooctanoic acid (PFOA) and perfluorooctane sulfonate (PFOS)	Industrial
07	Hexabromobiphenyl	Industrial
08	Hexa-brominated diphenyl ethers and hepta-brominated diphenyl ethers	Industrial
09	Tetra-brominated diphenyl ethers and penta-brominated diphenyl ethers	Industrial
According to the 2009 Amendment of the Stockholm Convention		
01	Endosulfan	Pesticide

organic pollutants (international treaty), signed on 22 May 2001 and effective from 17 May 2004, the member countries decided to phase out the manufacture, utilization, and liberation of 12 key persistent organic pollutants popularly known as "the dirty dozen". Later on, ten more chemical substances were added to the group of persistent organic pollutants in the 2009 and 2011 Amendments, presented in Table 6.1 (Eljarrat and Barcelo, 2003).

The familiar categories of persistent organic pollutants include hydrophobic and lipophilic compounds. They get strongly attached to solid substances, mostly organic matter in all ecosystems. Furthermore, these penetrate lipids more readily than water inside the cell and get deposited in fatty tissues. As a consequence of deposition in fatty tissues, they transfer to the higher levels of the food chain. They drift into the gaseous phase under natural conditions of temperature, from soil, plants, and water, as well as rove to longer distances owing to impedance to break down (Ashraf, 2017). Pesticides are the prominent persistent organic pollutants, especially organochlorines. It has been reported that pesticides have resulted in the loss of biodiversity in agricultural ecosystems (European Commission, 2018a). In Germany, during the past few decades, a significant loss of approximately 70% in the insect population, a 50% reduction in the bird population in Europe, and a significant impact on pollinators has been reported (Donald et al., 2001; Potts et al., 2010; Hallmann et al., 2017). Moreover, it is regarded as the second significant factor in the global decline of the insect population (Sánchez-Bayo and Wyckhuys, 2019). Other factors include habitat destruction, transformation of agricultural land, climate change, and introduced species.

6.2 ORGANIC POLLUTANTS IN WATER

Pollutants in water are of varying types with organic pollutants performing a critical function. Based on their deterioration extent by microorganisms, they are classified into two categories.

Easily degradable: Examples include polysaccharides and methanol, which have a simple design and are water-loving and thus can easily be degraded by microorganisms. But few chemicals like acetone become noxious at higher levels in the water.

Persistent organic pollutants: Examples include PAHs, PCBs, and DDT. They deteriorate at a very low speed. Few persistent organic pollutants like pesticides have been in use on a large scale over decades. The level and toxicity caused by these persistent organic pollutants are less than dissolvable organic contaminants, but they accumulate in sediments for a longer time and enter in food chain. Some of the noxious substances in water are:

a) **Organic matter**: Organic matter is defined as the remnant of living organisms. It is of two types, viz humus and non-humus. The organic matter in water influences its physical and chemical properties as well as other purification processes.

b) **Formaldehyde**: The origin of this compound in water is various industrial effluents like organic synthesis, chemical industry, synthetic fibre, wood processing, and the paint industry. It has the tendency to integrate with a variety of materials because of its greater reducibility. In humans it damages the CNS and causes skin and respiratory tract cancer (Soysal and Demiral, 2007; Muzi et al., 2004).

c) **Phenols:** The source of phenols in water is cooking and refining plants, paper manufacturing, and insulation substance manufacturing. They cause cancer in humans and are of significant interest even at lower levels. In aquatic ecosystems, they diminish the growth potential of organisms (Roger et al., 2003).

d) **PCBs:** These are utilized as coolants in various electric equipment like transformers and capacitors. These are cancer-causing agents and accumulate in lipid tissues. In addition to this, they cause brain damage and disorders in internal organs (Ruder et al., 2006).

e) **PAHs:** These organisms have a recalcitrant nature and consist of two or more benzene rings. Incomplete combustion of organic matter, pyrolysis of organic matter, and agriculture refuse are the prominent manmade sources of PAHs. These are found in oil deposits and enter into aquatic environments through leaking, from the air environment or from soil. Because of their extremely carcinogenic nature, they pose a threat to the health of organisms (WHO, 2021).

f) **Organ phosphorus pesticides:** These are substantially utilized to limit weeds and disease-causing agents but are noxious to non-target organisms like arthropods, fish, birds, human beings, animals, and plants. These remain in the environment for a longer duration. Water-containing organ phosphorus pesticides cause adverse environmental issues. Few organophosphate pesticides have acute toxicity to humans and livestock (Mulla et al., 2019).

g) **Petroleum hydrocarbons:** These enter into the environment from various industrial activities like oil manufacturing, refining, and transportation. These are noxious to the water-inhibiting organisms, as they create an oil layer that drops the swap of oxygen between the atmosphere and water bodies; further, they adversely affect the land-bearing organisms (Karthick et al., 2019).

h) **Atrazine:** This is a commonly used herbicide, employed to enhance the production of commercial crops. The source of atrazine in water is mainly the industries synthesizing this chemical and the agricultural fields where it is sprayed. At high temperatures, it vaporizes and liberates noxious gases like CO and NO_x, which causes respiratory irritations. Owing to its cost-effective nature and success in controlling weeds, it is used more commonly than other herbicides, leading to reproductive disorders in both aquatic plants and animals, which ultimately affects community structure (Graymore et al., 2001).

6.3　ORGANIC POLLUTANTS IN SOIL

The subsistence of civilization and the continual of agriculture depend on soil and water resources. But both of these resources have been severely deteriorated owing to different natural and human activities. Among all the soil-polluting substances, organic pollutants are the most deleterious pollutants for humans and the environment. Usually, these pollutants are found in low concentrations and persist in the environment for longer periods of time (Bajaj and Singh, 2015). A list of different organic pollutants in the soil and their source is given in Table 6.2.

TABLE 6.2
Different organic pollutants and their source in both soil and water (Tanwira et al., 2021)

Organic pollutant	Source
Polychlorinated biphenyls (PCBs)	Electronic equipments, microscope oils, fire of incinerators, short circuits, transformers, capacitors, oils of motors, and hydraulics
Petroleum hydrocarbon	Oil spills, oil wells, oil storage, and oil refineries
Agrochemicals	Insecticides
Polycyclic aromatic hydrocarbons (PAHs)	Incomplete combustion of coal, tobacco, garbage, oil, and wood
Explosives (RDX, TNT)	Cladding, mining, and ordinance factories' use in wars and fireworks

6.4 IMPACT OF PERSISTENT ORGANIC POLLUTANTS ON ENVIRONMENT

Persistent organic pollutants are a potential source of harm worldwide owing to their tendency to reach significant distances mediated by air and water, through the phenomenon of evaporation and redeposition from the origin. The most important characteristic of persistent organic pollutants is to combine with other gases in the atmosphere due to their semi-volatile nature. After combining with other gases, they are subjected to other phenomena like bioaccumulation, deterioration, and sedimentation. The global distillation theory envisages that gaseous pollutants will be transferred from warmer areas, viz tropical or temperate areas to colder regions, thereby impacting vapour pressure and Henry's constant, leading to condensation and accumulation of persistent organic pollutants in the land, plants, and other places, from where they can enter into the food chains (Fernández and Grimalt, 2003). Persistent organic pollutants arise as a consequence of human activities and ultimately enter into the food chain. The ways of transport of persistent organic pollutants from the source to the food chain are depicted in Figure 6.1.

6.5 IMPACT OF PERSISTENT ORGANIC POLLUTANTS ON HUMANS

The remaining persistent organic pollutants have been reported in the fatty tissues of many people from various countries like Europe, Asia, Africa, and North America. The endurance and toxicity of pesticides help wipe out target species but cause deleterious effects on humans as well. For example, it reduces the semen characteristics, causes malformation in reproductive functions of males, variation in sex ratio, endometriosis, precocious puberty, and early menarche (Jorgensen et al., 2001). Escalation in the cases of breast cancer and testicular cancer is of significant concern (Damstra, 2002). The repelling agents like organometallic compounds that adhere to proteins like organomercurials, fat-soluble pollutants viz dioxins, PCBs, polybrominated diphenyl ether, chlorinated pesticides, and persistent non-lipophilic

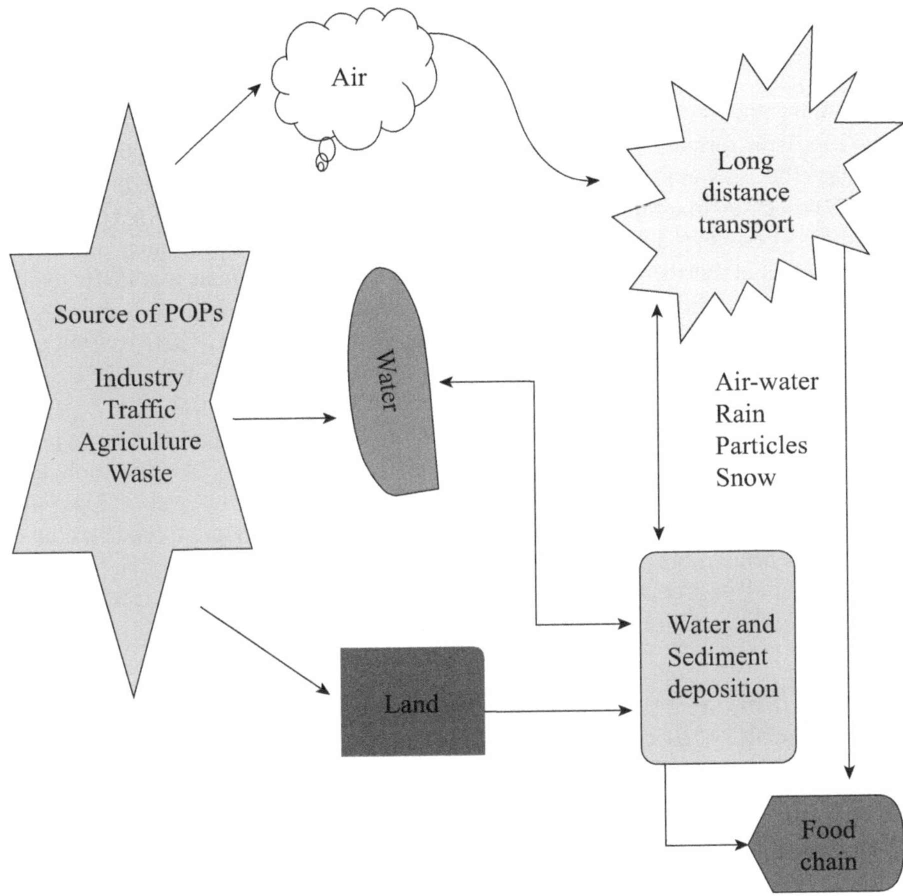

FIGURE 6.1   Transport of organic pollutants from the source to the food chain

compounds such as perfluorinated compounds have become a matter of trouble to humans as well. Toxicity is not necessarily related to the endurance of the chemical. It has been reported that many volatile chlorinated compounds have escalated the possibility of type-2 diabetes (Lee et al., 2007) and that too at very low levels; furthermore, it does not seem to pursue a linear dose—response curve (Lee et al., 2010). Type 2 diabetes is an insulin receptor disease, and the mechanism behind the cause is not known; however, it is most likely the result of gene induction (Lee et al., 2007).

6.6   IMPACT OF ORGANIC POLLUTANTS ON PLANTS

The development of plants is affected by a number of factors like nutrition, genetic makeup, hormones, and the environment. It has been revealed that subjection of plants to contaminants enhances or diminishes the development, seed germination, expansion of pollen tubes, and photosynthesis. The root is an indispensable part of plants and helps in the absorption of not only water and minerals but also contaminants. Organic pollutants have a harmful impact on the metabolic process of plants, viz photosynthesis (Kreslavski et al., 2017; Tomar and Jajoo, 2013a; Sharma et al., 2017), protein formation, and lipid, nucleic acid, (Paskovz et al., 2006), and hormone synthesis (Vanova et al., 2011; Kummerova et al., 2010) by impeding the biological processes like mitosis, germination, enzymatic activities, and generation of hormones. Pesticides impede the germination process in plants, which ultimately affects the yield. Furthermore, the submission of organic pollutants like PAHs to UV radiation causes photo-oxidation; the photo-oxidation products have a more noxious impact on plants (Grenvald et al., 2013; Tomar and Jajoo, 2015). Similar to abiotic stresses, organic pollutants stimulate the concentration of reactive oxygen species (ROS) (Liao and Chen, 2007; Ye et al., 2003; Zhang et al., 2007), which in turn stimulates damage to genetic material and lipid peroxidation. Research has revealed that pesticides diminish the concentration of pigments like chlorophyll in grapes and carotenoids in tobacco (Garcia et al., 2003; Saladin et al., 2003a) and photosynthetic pigments in maize due to the application of pyriproxyfen (Coskun et al., 2015). Furthermore, the application of aldicarb, phorate fensulfothion, and so on in chickpeas (Tiyagi et al., 2004) and tricyclazole in tomatoes resulted in a reduction of pigment concentration (Shanmugapriya et al., 2013).

6.7 ECOFRIENDLY TECHNOLOGIES FOR REMOVAL OF ORGANIC POLLUTANTS IN WATER

The damaging impact of persistent organic pollutants on all domains of the environment and the inefficiency of traditional techniques for the elimination of persistent organic pollutants necessitates the demand for advanced ecofriendly technologies for eliminating persistent organic pollutants (Kang, 2014). The various nanotechnologies used are described in the following subsections.

6.7.1 NANOCATALYSIS

Nanocatalysis is the utilization of semiconducting nanomaterials for transforming pollutants into environmentally friendly substances. Among nanomaterials metals and their oxides have attained considerable importance. Fenton-based catalysts, electrocatalysts, photocatalysts, and doped multifunctional nanocatalysts are utilized for the efficient deterioration of organic pollutants in water (Doll and Frimmel, 2005; Li et al., 2014). Photocatalysis escalates the degradation of persistent organic pollutants by employing oxidants with the liberation of ROS; furthermore, it includes the activation of persistent organic pollutants using promoter and under light conditions and depends on the production of HO, O, O, and OH radicals, which deteriorate approximately all organic molecules (Hadei et al., 2021; Kurade et al., 2021). The universally used semiconductors for the deterioration of persistent organic pollutants are TIO and zinc oxide. Hydrophobic persistent organic pollutants can be effectively eliminated from water with the help of heterogeneous photocatalysts. The determining factors for choosing photocatalyst are the surface, pore volume, and structure that can be modulated to amplify the deterioration efficiency (Valizadeh et al., 2021; Choi et al., 2004; Bano et al., 2021).

6.7.2 NANOADSORPTION

Nanoadsorbents are utilized for the efficient adsorption of numerous organic pollutants owing to their extraordinary outer area, sorption points, adjustable pore size, and low distance between particles (Tara et al., 2020; Yadav et al., 2020). Easy modulation to design them for a particular pollutant is the biggest benefit of these nanoadsorbents, and reports reveal their success in remediating various persistent organic pollutants like biphenyls, detergents, dyes, hydrocarbon, phenols, pharmaceuticals, and pesticides (Harja and Ciobanu, 2020). This is a simple and secure technology for eliminating persistent organic pollutants from water courses. The linkages involved in this process are electrostatic force, hydrogen bonding, van der Waals force, electron exchange reactions, and so on, and the substances used are nanomaterials like clay, zeolite, alumina, metal and their oxides, activated carbon, carbon nanomaterials, nanocomposite, nanosheets, nanotubes, chitosan-based polymers, and graphene nanomaterials (Sadegh and Ali, 2021). Efficient separation of persistent organic pollutants involves the utilization of magnetic nanoparticles, particularly iron oxides. Similarly, the adsorption efficiency of activated carbon is aided by small porous structures (Kurniawan et al., 2012; Subbaiah et al., 2021). Carbon-based nanomaterials interact with pollutants through various mechanisms, including hydrophilicity, hydrogen bonding, and covalent and electrostatic linkages. These interactions are facilitated by the nanomaterials' elasticity and the presence of specific adsorption sites (Gul et al., 2021; Hussain et al., 2022). The carbon nanotubes have been manipulated by enhancing the porosity to increase the adsorption efficiency of organic pollutants (Hu et al., 2022). The advantages of nanoadsorption are given in Figure 6.2.

6.7.3 NANOFILTRATION

The extraordinary successful techniques for handling waste water are membrane based like microfiltration, reverse osmosis (RO), ultra-filtration, and nanofiltration (NF), which are operated by pressure and regarded as an optional technique for the remediation of organic micropollutants from water courses (Mustereț and Teodosiu, 2007). The separation effectiveness of this technique is better than traditional methods, but this is a very economical process (Luis et al., 2011). Among all the filtration techniques, the efficient techniques for remediating trace amounts of organic pollutants are NF and RO, with NF being the more energy-saving technique in comparison to RO (Vazquez-Nunez et al., 2020; Ng et al., 2021; Lau and Ismail, 2009). The high remediation efficacy of NF for micropollutants is governed by a small size of pores, outstanding efficiency, and user-friendliness (Oatley-Radcliffe et al., 2017). The materials employed in the manufacture of NF membranes are polypropylene, polyvinyl fluoride, polyacrylonitrile, and cellulose acetate (Mulyanti and Susanto, 2018; Yoon and Lueptow, 2005; Agenson et al., 2003; Paul and Jons, 2016). It has been reported that nanofibres are more efficient for separating pesticides from

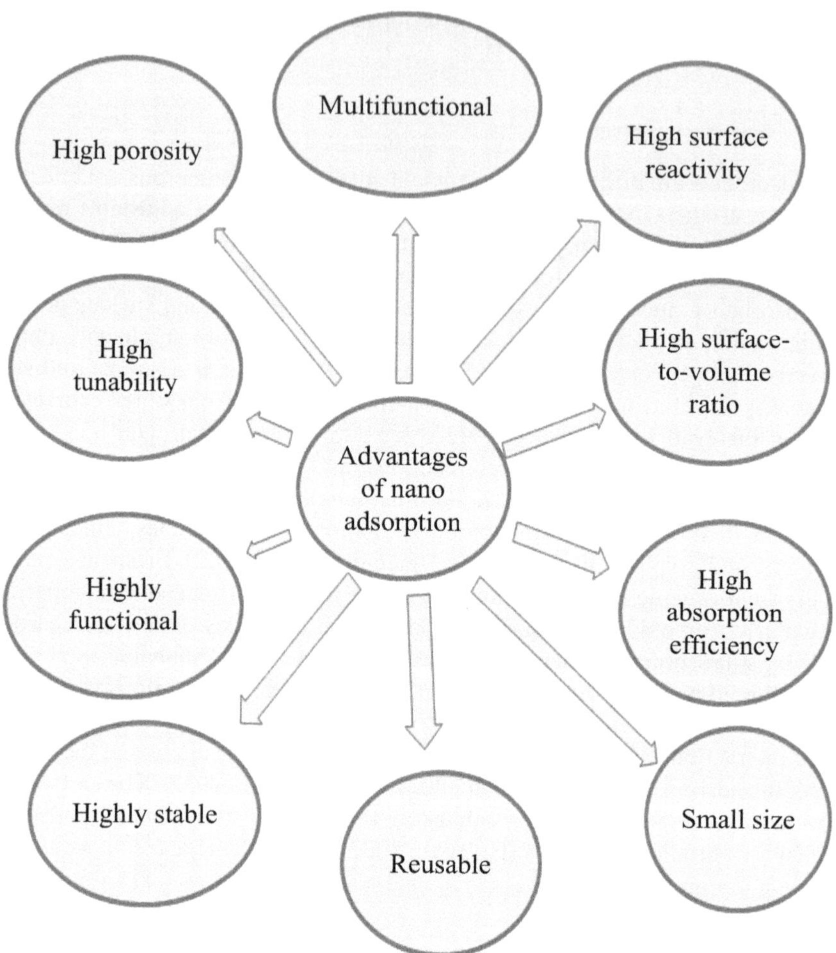

FIGURE 6.2   Advantages of nanoadsorption

water through molecular propagation mechanism and have firm adsorption structures owing to their loose bundles in comparison to nanotubes and nanoparticles;
in addition to this, the photocatalytic character can be enhanced by utilizing semiconducting material in manufacturing (Koe et al., 2020). ZnO–cellulose acetate
and TiO–graphene nanofibres, owing to their powerful photocatalytic activity, have
remediation potential for dye compounds (Molinari et al., 2000). Single or a blend
of filtration techniques have been employed for excellent separation of a number of
organic and inorganic contaminants; in addition to this integration of filtration, biological and chemical techniques are successful in the removal of persistent organic
pollutants from water (Argurio et al., 2018). The success of this technique relies on
membrane type, membrane modules, membrane composition, and interaction with
contaminants (Zhang et al., 2016).

6.8 ECOFRIENDLY TECHNOLOGIES FOR REMOVAL OF ORGANIC POLLUTANTS FROM SOIL

Organic compounds like pesticides sprinkled on plants or seeds are flushed out by rainfall and settle in the soil (Sitaramaraju et al., 2014). In addition to this, these pesticides enter into the soil either accidentally or by rinsing application equipment. Usually, the pesticides are organic and do not deteriorate easily, and thus they remain in the soil for a longer duration (Baez et al., 2015; Jablonowski et al., 2012; Navarro et al., 2007; Turgut et al., 2010). Thus, the need of the hour is to remediate the soil to reduce the detrimental consequences on the environment. The ecofriendly remediation techniques are as follows.

6.8.1 PHYTOREMEDIATION

Phytoremediation is an ecofriendly, effective, and viable technique for solving the problem of contaminated soils. The plants are grown at contaminated sites to remove pollutants, without impeding the normal development of plants. The various types of phytoremediation that have been reported to be successful in removing organic pollutants from the soil are phytostabilization, rhizodegradation, rhizofiltration, phytodegradation, and phytovolatilization (Tanglahu et al., 2011). The removal of organic pollutants by plants depends on various factors, viz type of plant species, characteristics of the medium, the chemistry of rhizosphere, amendments, characteristics of contaminants, and environmental factors.

Phytoextraction: This process involves the transfer of pollutants from plants roots to the above-ground parts (USEPA, 2000). The plants centralize and precipitate noxious elements from the polluted areas to the above-ground parts, viz shoot, leaves, and so on. Hyperaccumulators are commonly used for phytoextraction, because of their potential to accumulate 100-fold more pollutants in comparison to non-accumulating (Raskin and Ensley, 2000). The metals removed through this process are nickel, zinc, and copper. The benefits of this process are cost-effectiveness, permanent elimination of contaminants, and the amount of waste to be rejected reduced by about 95% (USEPA, 2000). Some of the hyperaccumulators reported in southern China, including Guangxi, Yunnan, Hunan, Zhejiang, Fujian, Guizhou, and Sichuan Provinces, are given in Table 6.3.

Phytovolatilization: In this mechanism, plants take contaminants from the soil and transform them into volatile forms, and liberate them into the atmosphere (USEPA, 2000). The volatile substance is released into the atmosphere through leaves (EPA, 2000). This is used for remediation in the soil, sediments, and water. The primary pollutant remediated through this process is Hg; it has been reported to exist with volatile organic substances like trichloroethene, and inorganic volatile chemicals, viz selenium and arsenic (EPA, 2000). The merits are that the pollutant, viz mercuric ion, may get transformed to less toxic elemental Hg, but the demerit is it may return back from the atmosphere through precipitation and get accumulated in water bodies, which will ultimately lead to the generation of methylmercury by anaerobic bacteria (USEPA, 2000).

Phytostabilization: This mechanism is employed for the remediation of soil, sediments, and so on and is part of in-situ inactivation (USEPA, 2000). In this mechanism

TABLE 6.3

Some of the reported hyperaccumulators and the metals accumulated

Hyperaccumulator	Family
Wang et al. (2006) reported *Pteris cretica* and *Pteris oshimensis* as As accumulator	Pteridaceae
Wang et al. (2007) reported *Pteris fauriei, P. oshimensis,* and *P. cretica* as As accumulator	Pteridaceae
Liu et al. (2016) reported *Centella asiatica* as Cd accumulator	Apiaceae
Liu et al. (2010) reported *Phytolacca americana* as Cd accumulator	Phytolaccaceae
Tang et al. (2009) reported *Picris divaricata* as Cd accumulator and *P. divaricata* as Zn accumulator	Asteraceae
Hu et al. (2007) reported *Potentilla griffithi* as Cd accumulator	Rosaceae
Li et al. (2010) reported *Sedum plumbizincicola* as Cd and Zn accumulator	Crassulaceae
Hou et al. (2012) reported *Isache globos* and *Pogonatherum crinitum* as Pb accumulator	Gramineae
Liu et al. (2014) reported *Celosia argentea* as Mn accumulator	Amaranthaceae

plants don't take up the contaminants but are immobilized in the root zone through adsorption by roots. In other words, it reduces the mobility of contaminants and availability in the food chain. The process involved in phytostabilization include sorption, precipitation, complexation, or metal valence reduction. The metals remediated through phytostabilization are lead (Pb), arsenic (As), cadmium (Cd), chromium (Cr), copper (Cu), and zinc (Zn) (Zhang et al., 2009).

Phytodegradation: This mechanism transforms complex forms of contaminants to simpler substances or integrates these molecules in the tissues of plants; it is also known as phyto transformation (Trap et al., 2005). The advantage of phytodegradation over phytoextraction and phytovolatilization is the remediation of organic contaminants like chlorinated solvents, herbicides, and munitions, whereas remediation by phytoextraction and phytovolatilization occurs only when the dissolubility and hydrphobicity are within acceptable levels (EPA, 2000).

Rhizodegradation: This is also known as phytostimulation and involves the disintegration of contaminants within the root zone aided by the microorganisms inhibiting the root zone. It has been reported that the rhizosphere has a 100 times larger population of microorganisms than the outside root zone (USEPA, 2000), which may be due to the secretion of exudates like sugars, amino acids, enzymes, and other compounds that enhance bacterial growth. Furthermore, roots act as a growth substrate for microbes and means of oxygen exchange with the environment. It has been successfully used for remediating numerous organic pollutants, viz petroleum hydrocarbons, PAHs, chlorinated solvents, pesticides, PCBs, benzene, toluene, ethylbenzene, and xylenes (EPA, 2000).

6.8.2 Microremediation

Remediation of contaminated sites using microbes is considered an outstanding technique. It has been reported that a great variety of microorganisms take part in the bioremediation process (Bhatt et al., 2019). The microbes produce a number of enzymes, which have the capability to degrade contaminants to produce fewer toxic substances. The degradation potential of a combination of microbes for contaminants has proved to be much greater than a single microbe (Mishra et al., 2021). The management of organic pollutant-contaminated soils aiding microbes is an ecofriendly, cost-effective, and efficient technique (Yair et al., 2008). Microbes digest organic pollutants through aerobic and anaerobic processes, but the anaerobic process is the most desirable technique, particularly for chlorinated compounds. However, few organic pollutants like trichloroethylene upon anaerobic degradation lead to the production of more toxic substances like dichloroethylene and vinyl chlorides (Chen and Chang, 2020). Thus, aerobic degradation can sometimes be considered a better option for the remediation of organic pollutants, owing to the presence of specific enzymes with the tendency to deteriorate various categories of organic pollutants. The oxygenase enzyme of the aerobes performs a critical role in the deterioration of pollutants. For example, *Pseudomonas* sp. has the potential to degrade trichloroethylene and its products, viz dichloroethylene, and volatile compounds into non-noxious substances like carbon dioxide and chloride ions due to the presence of oxygenase enzyme (Yoshikawa et al., 2017). However, effective deterioration relies on understanding the transfer of pollutants into the cells of microbes and their absorption. It

---

TABLE 6.4

Different microbial strains and their glycol conjugates used in bioremediation

Microstrains and their remediation	Glycol conjugates
Zhou et al. (2020) reported the remediation of organic pollutants with the help of *Acinetobacter* sp. *Y1*	Methyl hexadcanoate and methyl octadecanoate
Aguila-Torres et al. (2020) reported the remediation of cypermethrin with the help of *Pseudomonas* and *Rhodococcus*	Biosurfactants
Deng et al. (2020) reported the remediation of crude oil using *Achromobacter* sp. *A-8*	Biosurfactants
Gupta et al. (2020) reported remediation of methyl parathion, ethyl parathion, and trifuralin using *Acinetobacter baumannii BJ5*	Glycolipid biosurfactant
Wattanaphon et al. (2008) reported the remediation of hexachlorocyclohexane (HCH) using *Burkholderia cenocepacia BSP3*	Glycolipid
Sharma et al. (2009) reported the remediation of HCH using *Pseudomonas aeruginosa* WH-1	Biosurfactants
Singh et al. (2009) reported the remediation of Chlorpyrifos utilizing *Pseudomonas* sp.	Rhamnolipids
Onbasli and Aslim (2009) reported the remediation of 2,4-D, benzene, toluene, xylene, and gasoline using *Pseudomonas aeruginosa* B1, *Pseudomonas fuorescens* B5, *Pseudomonas stutzeri* B11, and *Pseudomonas putida* B15	Exopolysaccharides (EPSs)
Zhou et al. (2011) reported the remediation of phenol using *Penicillium simplicissimum*	Tea saponin and rhamnolipid

has been reported that glycol lipids and glycol conjugates perform a critical role in the transportation of organic pollutants across microbial membranes (Liston et al., 2017). These are placed either on the interior side or the exterior side and function as emulsifiers and are called "biosurfactants" (Chong and Li, 2017). This gave birth to "microbial glycobiotechnology", which involves a wide range of methods, with the main aim of decontaminating different types of pollutants. A list of different microbial strains and their glycol conjugates used in bioremediation is given in Table 6.4.

6.9 CONCLUSION

There are numerous ways to remediate polluted sites of soil, water, and air environment. To eliminate pollutants from the soil, phytoremediation and microremediation are widely used techniques. These are ecofriendly, cost-effective, and successful techniques. Plants promote the elimination of pollutants through different mechanisms, viz uptake through roots and accumulation in aerial parts, stabilization, degradation in the root zone to non-toxic substances, and so on. Microbes secrete a variety of glycolipids and glycol conjugates, which aid in the remediation of organic contaminants. On the other hand, a number of filtration techniques are utilized for the remediation of pollutants in water. These remediation techniques are very essential to prevent damage to the environment and to keep the resources available to future generations.

REFERENCES

Abelsohn, A., Gibson, B. L., Sanborn, M. D., & Weir, E. (2002). Identifying and managing adverse environmental health effects: Persistent organic pollutants. *Canadian Medical Association Journal, 166*, 1549–1554.

Adeola, F. O. (2004). Boon or bane: The environmental and health impacts of persistent organic pollutants (Persistent organic pollutants). *Human Ecology Review, 11*, 27–35.

Agenson, K. O., Oh, J. I., & Urase, T. (2003). Retention of a wide variety of organic pollutants by different nanofiltration/reverse osmosis membranes: Controlling parameters of process. *Journal of Membrane Science, 225*, 91–103.

Aguila-Torres, P., Maldonado, J., Gaete, A., Figueroa, J., González, A., Miranda, R., . . . & González, M. (2020). Biochemical and genomic characterization of the cypermethrin-degrading and biosurfactant-producing bacterial strains isolated from marine sediments of the Chilean Northern Patagonia. *Marine Drugs, 18*, 252.

Argurio, P., Fontananova, E., Molinari, R., & Drioli, E. (2018). Photocatalytic membranes in photocatalytic membrane reactors. *Processes, 6*, 162.

Ashraf, M. A. (2017). Persistent Organic Pollutants (POPs): A global issue, a global challenge. *Environmental Science and Pollution Research, 24*(5), 4223–4227.

Baez, M. E., Espinoza, J., Silva, R., & Fuentes, E. (2015). Sorption-desorption behavior of pesticides and their degradation products in volcanic and non-volcanic soils: Interpretation of interactions through two-way principal component analysis. *Environment Science and Pollution Research, 22*(11), 8576–8585.

Bajaj, S., & Singh, K. D. (2015). Biodegradation of persistent organic pollutants in soil, water and pristine sites by cold-adapted microorganisms: Mini review. *International Biodeterioration and Biodegradation, 100*, 98–105.

Bano, K., Kaushal, S., & Singh, P. P. (2021). A review on photocatalytic degradation of hazardous pesticides using heterojunctions. *Polyhedron, 209*, 115–465.

Baranowska, I., Barchańska, H., & Pyrsz, A. (2005). Distribution of pesticides and heavy metals in trophic chain. *Chemosphere, 60*, 1590–1599.

Bhatt, P., Huang, Y., Zhan, H., & Chen, S. (2019). Insight into microbial applications for the biodegradation of pyrethroid insecticides. *Frontiers in Microbiology, 10*, 1778.

Chen, T., & Chang, S. (2020). Potential microbial indicators for better bioremediation of an aquifer contaminated with vinyl chloride or 1,1-dichloroethene. *Water Air & Soil Pollution, 231*, 239.

Choi, E., Cho, I. H., & Park, J. (2004). The effect of operational parameters on the photocatalytic degradation of pesticide. *Journal of Environmental Science* and *Health Part B, 39*, 53–64.

Chong, H., & Li, Q. (2017). Microbial production of rhamnolipids: Opportunities, challenges and strategies. *Microbial Cell Factories, 16*, 137.

Coskun, Y., Kilic, S., & Duran, R. E. (2015). The effects of the insecticide pyriproxyfen on germination, development and growth responses of maize seedlings. *Fresenius Environmental Bulletin, 24*, 278–284.

Damstra, T. (2002). Potential effects of certain persistent organic pollutants and endocrine disrupting chemicals on the health of children. *Journal of Toxicology: Clinical Toxicology, 40*(4), 457–465.

Deng, Z., Jiang, Y., Chen, K., Li, J., Zheng, C., Gao, F., & Liu, X. (2020). One biosurfactant producing bacteria Achromobacter sp: A-8 and its potential use in microbial enhanced oil recovery and bioremediation. *Frontiers in Microbiology, 11*, 247.

Doll, T. E., & Frimmel, F. H. (2005). Removal of selected persistent organic pollutants by heterogeneous photocatalysis in water. *Catalysis Today, 101*, 195–202.

Donald, P. F., Green, R. E., & Heath, M. F. (2001). Agricultural intensification and the collapse of Europe's farmland bird populations. *Proceedings of the Royal Society of London. B, 268*, 25–29.

Doug, S. R. (2012). *How important was Rachel Carson's Silent Spring in the recovery of bald eagles and other bird species?* Scientific American.

Eljarrat, E., & Barcelo, D. (2003). Priority lists for persistent organic pollutants and emerging contaminants based on their relative toxic potency in environmental samples. *Trends in Analytical Chemistry, 1*, 655–665.

EPA. (2000). A citizen's guide to phytoremediation. EPA 542-F-98–011. United States Environmental Protection Agency, p. 6. http//www.bugsatwork.com/XYCLONYX/EPA_GUIDES/PHYTO.PDF

European Commission. (2018a). Science for environment policy: Flying insects in West German nature reserves suffer decline of more than 76% (1973–2000). *European Commission DG Environment News Alert Service.* http://ec.europa.eu/environment/integration/research/newsalert/pdf/flying_insects_west_german_nature_reserves_suffer_decline_more_than_76pc_1973_2000_511nal_en.pdf. Accessed on 27 March 2019.

Evans, M. S., Muir, D., Lockhart, W. L., Stern, G., Ryan, M., & Roach, P. (2005). Persistent organic pollutants and metals in the freshwater biota of the Canadian Subarctic and Arctic: An overview. *Science of the Total Environment, 351–352*, 94–147.

Fernández, P., & Grimalt, J. O. (2003). On the global distribution of persistent organic pollutants. *Chimia (Aarau), 57*(9), 514–521.

Garcia, P. C., Rivero, R. M., Ruiz, J. M., & Romero, L. (2003). The role of fungicides in the physiology of higher plants: Implications for defense responses. *Botanical Review, 69*, 162–172.

Goktaş, R. K., & MacLeod, M. (2016). Remoteness from sources of persistent organic pollutants in the multi -media global environment. *Environmental Pollution, 217*, 33–41.

Graymore, M., Stagnitti, F., & Allinson, G. (2001). Impacts of atrazine in aquatic ecosystems. *Environment International, 26*(7–8), 483–495.

Grenvald, J. C., Nielsen, T. G., & Hjorth, M. (2013). Effects of pyrene exposure and temperature on early development of two co-existing Arctic copepods. *Ecotoxicology, 22*, 184–198.

Gul, A., Khaligh, N. G., & Julkapli, N. M. (2021). Surface modification of carbon-based nano-adsorbents for the advanced wastewater treatment. *Journal of Molecular Structure, 1235*, 130148.

Gupta, B., Puri, S., Thakur, I. S., & Kaur, J. (2020). Enhanced pyrene degradation by a biosurfactant producing Acinetobacter baumannii BJ5: Growth kinetics, toxicity and substrate inhibition studies. *Environmental Technology Innovation, 19*, 100–804.

Hadei, M., Mesdaghinia, A., Nabizadeh, R., Mahvi, A. H., Rabbani, S., & Naddafi, K. (2021). A comprehensiv e systematic review of photocatalytic degradation of pesticides using nano TiO2. *Environmental Science and Pollution Research, 28*, 13055–13071.

Hallmann, C. A., Sorg, M., Jongejans, E., Siepel, H., Hofland, N., & Schwan, H. (2017). More than 75 percent decline over 27 years in total flying insect biomass in protected areas. *PLoS One, 12*, e0185809.

Harja, M., & Ciobanu, G. (2020). Eco-friendly nano-adsorbents for pollutant removal from wastewaters. In: Kharissova, O., Martínez, L., & Kharisov, B. (Eds.), *Handbook of nanomaterials and nanocomposites for energy and environmental applications* (pp. 1–22). Springer.

Hou, X. L., Chang, Q. S., Liu, G. F., Liu, A. Q., & Cai, L. P. (2012). Two 1223 lead-hyperaccumulators: *Pogonatherum crinitum* and *Lsache globosa*. *Chinese Journal of Environmental Engineering, 6*(3), 989–994.

Hu, P. J., Zhou, X. Y., Qiu, R. L., Tang, Y. T., & Ying, R. R. (2007). Cadmium tolerance 1212 and accumulation features of Zn-hyperaccmulator *Potentilla griffithi* var. 1213 velutina. *Journal of Agro-Environment Science, 26*(6), 2221–2224, 1214.

Hu, X., You, S., Li, F., & Liu, Y. (2022). Recent advances in antimony removal using carbon-based nanomaterials: A review. *Frontiers of Environmental Science & Engineering, 16*, 48.

Hussain, N., Bilal, M., & Iqbal, H. M. (2022). Carbon-based nanomaterials with multipurpose attributes for water treatment: Greening the 21st-century nanostructure materials deployment. *Biomaterials and Polymers Horizon, 1*, 48–58.

Jablonowski, N. D., Linden, A., Koppchen, S., Thiele, B., Hofmann, D., Mittelstaedt, W., . . . & Burauel, P. (2012). Long-term persistence of various 14C-labeled pesticides in soils. *Environmental Pollution, 168*, 29–36.

Jones, K. C. (2021). Persistent organic pollutants (Persistent organic pollutants) and related chemicals in the global environment: Some personal reflections. *Environmental Science and Technology.* https://doi.org/10.1021/acs.est.0c08093

Jorgensen, N., Andersen, A. G., Eustache, F., Irvine, D. S., Suominen, J., & Petersen, J. H. (2001). Regional differences in semen quality in Europe. *Human Reproduction, 16*(5), 1012–1019.

Kang, J. W. (2014). Removing environmental organic pollutants with bioremediation and phytoremediation. *Biotechnology Letters, 36*, 1129–1139.

Karthick, A., Roy, B., & Chattopadhyay, P. (2019). A review on the application of chemical surfactant and surfactant foam for remediation of petroleum oil contaminated soil. *Journal of Environmental Management, 243*, 187–205.

Koe, W. S., Lee, J. W., Chong, W. C., Pang, Y. L., & Sim, L. C. (2020). An overview of photocatalytic degradation: Photocatalysts, mechanisms, and development of photocatalytic membrane. *Environmental Science and Pollution Research, 27*, 2522–2565.

Kreslavski, V. D., Brestic, M., & Zharmukhamedov, S. K. (2017). Mechanisms of inhibitory effects of polycyclic aromatic hydrocarbons in photosynthetic primary processes in pea leaves and thylakoid preparations. *Plant Biology.* https://doi.org/10.1111/plb.12598

Kummerova, M., Vanova, L., & Fiserova, H. (2010). Understanding the effect of organic pollutant fluoranthene on pea in vitro using cytokinins, ethylene, ethane and carbon dioxide as indicators. *Plant Growth Regulation, 61*, 161–174.

Kurade, M. B., Ha, Y. H., Xiong, J. Q., Govindwar, S. P., Jang, M., & Jeon, B. H. (2021). Phytoremediation as a green biotechnology tool for emerging environmental pollution: A step forward towards sustainable rehabilitation of the environment. *Chemical Engineering Journal, 415*, 129040.

Kurniawan, T. A., Sillanpaa, M. E., & Sillanpaa, M. (2012). Nanoadsorbents for remediation of aquatic enviro nment: Local and practical solutions for global water pollution problems. *Critical Reviews in Environmental Science and Technology, 42*, 1233–1295.

Lau, W. J., & Ismail, A. F. (2009). Polymeric nanofiltration membranes for textile dye wastewater treatment: Preparation, performance evaluation, transport modelling, and fouling control—A review. *Desalination, 245*, 321–348.

Lee, D. H., Lee, I. K., Porta, M., Steffes, M., & Jacobs, D. R. (2007). Relationship between serum concentrations of persistent organic pollutants and the prevalence of metabolic syndrome among nondiabetic adults: Results from the National Health and Nutrition Examination Survey 1999–2002. *Diabetologia, 50*(9), 1841–1851.

Lee, D. H., Steffes, M. W., Sjodin, A., Jones, R. S., Needham, L. L., & Jacobs, D. R. (2010). Low dose of some persistent organic pollutants predicts type 2 diabetes: A nested case-control study. *Environmental Health Perspectives, 118*(9), 1235–1242.

Li, K., Yan, L., Zeng, Z., Luo, S., Luo, X., Liu, X., & Guo, Y. (2014). Fabrication of H3PW12O40-doped carbon nitride nanotubes by one-step hydrothermal treatment strategy and their efficient visible-light p hotocatalytic activity toward representative aqueous persistent organic pollutants degradation. *Applied Catalysis B: Environmental, 156*, 141–152.

Li, S. L., Li, N., Xu, L. S., Tan, W. N., Zhou, S. B., Wu, L. H., & Luo, Y. M. (2010). Characters of Zn and Cd accumulation and distribution in leaves of Sedum plumbizincicola at different ages. *Soils, 42*(3), 446–452.

Liao, Y., & Chen, G. (2007). Physiological adaptability of three mangrove species to salt stress. *Acta Ecologica Sinica, 27*, 2208–2214.

Liston, S. D., Mann, E., & Whitfeld, C. (2017). Glycolipid substrates for ABC transporters required for the assembly of bacterial cell-envelope and cellsurface glycoconjugates. *Biochimica et Biophysica Acta—Molecular and Cell Biology of Lipids, 1862*, 1394–1403.

Liu, J., Shang, W. W., Zhang, X. H., Zhu, Y. N., & Yu, K. (2014). Mn accumulation and tolerance in *Celosia argentea* Linn: A new Mn-hyperaccumulating plant species. *Journal of Hazardous Materials, 267*(1), 136–141.

Liu, K. H., Zhou, Z. M., Yu, F. M., Chen, M. L., Chen, C. S., Zhu, J., & Jiang, Y. (2016). A newly found cadmium hyperaccumulator—*Centella asiatica* Linn. *Fresenius Environmental Bulletin, 25*(9), 2668–2675.

Liu, X. Q., Peng, K. J., Wang, A. G., Lian, C. L., & Shen, Z. G. (2010). Cadmium accumulation and distribution in populations of *Phytolacca americana* L. and the role of transpiration. *Chemosphere, 78*(9), 1136–1141.

Luis, P., Saquib, M., Vinckier, C., & Van der Bruggen, B. (2011). Effect of membrane filtration on ozonation efficiency for removal of atrazine from surface water. *Industrial & Engineering Chemistry Research, 50*, 8686–8692.

Mishra, S., Lin, Z., Pang, S., Zhang, W., Bhatt, P., & Chen, S. (2021). Recent advanced technologies for the characterization of xenobiotic-degrading microorganisms and microbial communities. *Frontiers in Bioengineering and Biotechnology, 9*, 632059.

Molinari, R., Mungari, M., Drioli, E., Di Paola, A., Loddo, V., Palmisano, L., & Schiavello, M. (2000). Study on a photocatalytic membrane reactor for water purification. *Catalysis Today, 55*, 71–78.

Mulla, S. I., Ameen, F., Talwar, M. P., Eqani, S. A. M. A. S., Bharagava, R. M., Saxena, G., . . . & Ninnekar, H. Z. (2019). Organophosphate pesticides: Impact on environment, toxicity, and their degradation. *Bioremediation of Industrial Waste for Environmental Safety*, 265–290.

Mulyanti, R., & Susanto, H. (2018). Wastewater treatment by nanofiltration membranes. In: *IOP conference series: Earth and environmental science* (Vol. 142, pp. 12–17). IOP Publishing.

Mustereț, C. P., & Teodosiu, C. (2007). Removal of persistent organic pollutants from textile wastewater by membrane processes. *Environmental Engineering and Management, 6*, 175–187.

Muzi, G., Dell'omo, M., Murgia, N., & Abritti, G. (2004). Chemical pollution of indoor air and its effect on health. *Giornale Italiano di Medicina del Lavoro ed Ergonomia, 26*(4), 364–369.

Navarro, S., Vela, N., & Navarro, G. (2007). Review: An overview on the environmental behavior of pesticide residues in soils. *Spanish Journal of Agricultural Research, 5*(3), 357–375.

Ng, L. Y., Chua, H. S., & Ng, C. Y. (2021). Incorporation of graphene oxide-based nanocomposite in the polymeric membrane for water and wastewater treatment: A review on recent development. *Journal of Environmental Chemical Engineering, 9*, 105–994.

Oatley-Radcliffe, D. L., Walters, M., Ainscough, T. J., Williams, P. M., Mohammad, A. W., & Hilal, N. (2017). Nano filtration membranes and processes: A review of research trends over the past decade. *Journal of Water Process Engineering, 19*, 164–171.

Onbasli, D., & Aslim, B. (2009). Effects of some organic pollutants on the exopolysaccharides (EPSs) produced by some Pseudomonas spp. strains. *Journal of Hazardous Materials, 168*, 64–67.

Paskovz, V., Hilscherova, K., Feldmannova, M., & Blaha, L. (2006). Toxic effects and oxidative stress in higher plants exposed to polycyclic aromatic hydrocarbons and their N-heterocyclic derivatives. *Environmental Toxicology and Chemistry, 25*, 3238–3245.

Paul, M., & Jons, S. D. (2016). Chemistry and fabrication of polymeric nanofiltration membranes: A review. *Polymer, 103*, 417–456.

Potts, S. G., Biesmeijer, J. C., Kremen, C., Neumann, P., Schweiger, O., & Kunin, W. E. (2010). Global pollinator declines: Trends, impacts and drivers. *Trends in Ecology & Evolution, 25*, 345–353.

Qing Li, Q., Loganath, A., Seng Chong, Y., Tan, J., & Philip Obbard, J. (2006). Persistent organic pollutants and adverse health effects in humans. *Journal of Toxicology and Environmental Health Part A, 69*, 1987–2005.

Raskin, I., & Ensley, B. D. (2000). Recent developments for in situ treatment of metal contaminated soils. In: *Phytoremediation of toxic metals: Using plants to clean up the environment*. John Wiley & Sons Inc. http//clu-n.org/techfocus

Roger, L., Breton, R., Teed, S., & Moore, D. R. J. (2003). An ecological risk assessment of phenol in the aquatic environment. *Human and Ecological Risk Assessment, 9*(2), 549–568.

Ruder, A. M., Hein, M. J., Nilsen, N., Waters, M. A., Laber, P., Davis-King, K., . . . & Whelan, E. (2006). Mortality among workers exposed to Polychlorinated Biphenyls (PCBs) in an electrical capacitor manufacturing plant in Indiana. *Environment Health Perspective, 114*(1), 18–23.

Sadegh, H., & Ali, G. A. (2021). Potential applications of nanomaterials in wastewater treatment: Nano adsorbents performance. In *Research anthology on synthesis, characterization, and applications of nanomaterials* (pp. 1230–1240). IGI Global.

Saladin, G., Magné, C., & Clément, C. (2003a). Effects of fludioxonil and pyrimethanil, two fungicides used against Botrytis cinerea, on carbohydrate physiology in Vitis vinifera L. *Pest Manage Science, 59*, 1083–1092.

Sánchez-Bayo, F., & Wyckhuys, K. A. (2019). Worldwide decline of the entomofauna: A review of its drivers. *Biological Conservation, 232*, 8–27.

Shanmugapriya, A. K., Sivakumar, T., & Panneerselvam, R. (2013). Difenoconazole and Tricyclazole induced changes in photosynthetic pigments of Lycopersicon esculentum L. *International Journal of Agriculture and Food Science, 3*, 72–75.

Sharma, C., Mathur, S., Tomar, R. S., & Jajoo, A. (2017). Investigating role of Triton X-100 in ameliorating deleterious effects of anthracene in wheat plants. *Photosynthetica, 55*, 1–8.

Sharma, S., Singh, P., Raj, M., Chadha, B. S., & Saini, H. S. (2009). Aqueous phase partitioning of hexachlorocyclohexane (HCH) isomers by biosurfactant produced by Pseudomonas aeruginosa WH-2. *Journal of Hazardous Materials, 171*, 1178–1182.

Singh, P. B., Sharma, S., Saini, H. S., & Chadha, B. S. (2009). Biosurfactant production by Pseudomonas sp. and its role in aqueous phase partitioning and biodegradation of chlorpyrifos. *Letters in Applied Microbiology, 49*, 378–383.

Sitaramaraju, S., Prasad, N. V. V. S. D., Chengareddy, V., & Narayana, E. (2014). Impact of pesticides used for crop production on the environment. *Journal of Chemical and Pharmaceutical Sciences, 3*, 75–79.

Soysal, A., & Demiral, Y. (2007). Indoor air pollution. *TAF Preventive Medicine Bulletin, 6*(3), 221–226.

Subbaiah, M. P., Kalimuthu, P., Jung, J., & Jeon, B. H. (2021). Recent advances in effective capture of inorganic mercury from aqueous solutions through sulfurized 2D-material-based adsorbents. *Journal of Materials Chemistry, 9*, 18086–18101.

Tang, Y. T., Qiu, R. L., Zeng, X. W., Fang, X. H., Yu, F. M., Zhou, X. Y., & Wu, Y. D. (2009). Zn and Cd hyperaccumulating characteristics of Picris divaricata Vant. *International Journal of Environment* and *Pollution, 38*(3–4), 26–38.

Tanglahu, B. V., Abdullah, S. R. S., Basri, H., Idrish, M., & Anuar, N. (2011). A review on heavy metals (As, Pb, and Hg) uptake by plants through phytoremediation. *International Journal of Chemical Engineering*, 1–31.

Tanwira, K., Amna, Javeda, M. T., Shahid, M., Akram, M. S., & Alia, Q. (2021). Antioxidant defense systems in bioremediation of organic pollutants. In: *Handbook of bioremediation.* https://doi.org/10.1016/B978-0-12-819382-2.00032-6

Tara, N., Siddiqui, S. I., Rathi, G., Chaudhry, S. A., & Asiri, A. M. (2020). Nano-engineered adsorbent for the removal of dyes from water: A review. *Current Analytical Chemistry, 16*, 14–40.

Tiyagi, S. A., Ajaz, S., & Azam, M. F. (2004). Effect of some pesticides on plant growth, root nodulation and chlorophyll content of chickpea. *Archives of Agronomy and Soil Science, 50*, 529–533.

Tomar, R. S., & Jajoo, A. (2013a). A quick investigation of the detrimental effects of environmental pollutant polycyclic aromatic hydrocarbon fluoranthene on the photosynthetic processes. *Ecotoxicology, 22*, 1313–1318.

Tomar, R. S., & Jajoo, A. (2015). Photomodified fluoranthene exerts more harmful effects as compared to intact fluoranthene by inhibiting growth and photosynthetic processes. *Ecotoxicology and Environmental Safety, 122*, 31–36.

Trap, S., Kohler, A., Larsen, L. C., Zambrano, K. C., & Karlson, U. (2005). Phytotoxicity of fresh and weathered diesel and gasoline to willow and poplar trees. *Journal of Soils and Sediments, 1,* 71–76.

Turgut, C., Erdogan, O., Ates, D., Gokbulut, C., & Cutright, T. J. (2010). Persistence and behavior of pesticides in cotton production in Turkish soils. *Environmental Monitoring and Assessment, 162*(1), 201–208.

United States Environmental Protection Agency (USEPA). (2000). Introduction to phytoremediation. EPA 600/R-99/107, U.S. Environmental Protection Agency, Office of Research and Development, Cincinnati, OH.

Valizadeh, S., Lee, S. S., Baek, K., Choi, Y. J., Jeon, B. H., Rhee, G. H., & Park, Y. K. (2021). Bioremediation strategies with biochar for Polychlorinated Biphenyls (PCBs)-contaminated soils: A review. *Environmental Research, 200,* 111–757.

Vanova, L., Kummerova, M., & Votrubova, O. (2011). Fluoranthene-induced production of ethylene and formation of lysigenous intercellular spaces in pea plants cultivated in vitro. *Acta Physiologiae Plantarum, 33,* 1037–1042.

Vazquez-Nunez, E., Molina-Guerrero, C. E., Pena-Castro, J. M., Fernandez-Luqueno, F., & de la Rosa-Á lvarez, M. (2020). Use of nanotechnology for the bioremediation of contaminants: A review. *Processes, 8,* 826.

Wang, H. B., Wong, M. H., Lan, C. Y., Baker, A. J. M., Qin, Y. R., Shu, W. S., . . . & Ye, Z. H. (2007). Uptake and accumulation of arsenic by 11 Pteris taxa 1198 from southern China. *Environmental Pollution, 145*(1), 225–233.

Wang, H. B., Ye, Z. H., Shu, W. S., Li, W. C., Wong, M. H., & Lan, C. Y. (2006). Arsenic 1194 uptake and accumulation in fern species growing at arsenic-contaminated sites 1195 of southern China: Field surveys. *International Journal of Phytoremediation, 8*(1), 1–11.

Wattanaphon, H. T., Kerdsin, A., Thammacharoen, C., Sangvanich, P., & Vangnai, A. S. (2008). A biosurfactant from Burkholderia cenocepacia BSP3 and its enhancement of pesticide solubilization. *Journal of Applied Microbiology, 23,* 1365–2672.

WHO. (2021). Human health effects of polycyclic aromatic hydrocarbons as ambient air pollutants—Report of the working group on polycyclic aromatic hydrocarbons of the joint task force on the health aspects of air pollution. Copenhagen: WHO Regional Office for Europe; 2021. Licence: CC BY-NC-SA 3.0 IGO

Yadav, V. K., Ali, D., Khan, S. H., Gnanamoorthy, G., Choudhary, N., Yadav, K. K., & Manhrdas, S. (2020). Synthesis and characterization of amorphous iron oxide nanoparticles by the sono chemical method and their application for the remediation of heavy metals from wastewater. *Nanomaterials, 10,* 1551.

Yadav, V. K., Choudhary, N., Khan, S. H., Malik, P., Inwati, G. K., Suriyaprabha, R., & Ravi, R. K. (2020). Synthesis and characterisation of nano-biosorbents and their applications for waste water treatment. In: *Handbook of research on emerging developments and environmental impacts of ecological chemistry* (pp. 252–290). IGI Global.

Yair, S., Ofer, B., Arik, E., Shai, S., Yossi, R., Tzvika, D., & Amir, K. (2008). Organophosphate degrading microorganisms and enzymes as biocatalysts in environmental and personal decontamination applications. *Critical Reviews in Biotechnology, 28,* 265–275.

Ye, Y., Tam, N. F. Y., Wong, Y. S., & Lu, C. Y. (2003). Growth and physiological responses of two mangrove species (Bruguiera gymnorrhiza and Kandelia candel) to water logging. *Environmental and Experimental Botany, 49,* 209–221.

Yoon, Y., & Lueptow, R. M. (2005). Removal of organic contaminants by RO and NF membranes. *Journal of Membrane Science, 261,* 76–86.

Yoshikawa, M., Zhang, M., & Toyota, K. (2017). Biodegradation of volatile organic compounds and their effects on biodegradability under co-existing conditions. *Microbes and Environments, 32,* 188–200.

Zhang, H., Zheng, L. C., & Yi, X. Y. (2009). Remediation of soil co-contaminated with pyrene and cadmium by growing maize (Zea mays L.). *International Journal of Environmental Science and Technology, 6*, 249–258.

Zhang, W., Ding, L., Luo, J., Jaffrin, M. Y., & Tang, B. (2016). Membrane fouling in Photocatalytic Membrane Reactors (PMRs) for water and wastewater treatment: A critical review. *Chemical Engineering Journal, 302*, 446–458.

Zhang, Z. L., Hong, H. S., Zhou, J. L., Huang, J., & Yu, G. (2003). Fate and assessment of persistent organic pollutants in water and sediment from Minjiang River Estuary, Southeast China. *Chemosphere, 52*, 1423–1430.

Zhang, Z. L., Wei, N., Wu, Q. X., & Ping, M. L. (2007). Antioxidant response of Cucumis sativus L. to fungicide carbendazim. *Pesticide Biochemistry and Physiology, 89*, 54–59.

Zhou, H., Huang, X., Liang, Y., Li, Y., Xie, Q., Zhang, C., & You, S. (2020). Enhanced bioremediation of hydraulic fracturing fowback and produced water using an indigenous biosurfactant-producing bacteria Acinetobacter sp, Y2. *Chemical Engineering Journal, 397*, 125–348.

Zhou, M. F., Yuan, X. Z., Zhong, H., Liu, Z. F., Li, H., Jiang, L. L., & Zeng, G. M. (2011). Effect of biosurfactants on Laccase production and phenol biodegradation in solid-state fermentation. *Applied Biochemistry and Biotechnology, 164*, 104–114.

7 Bioassessment and Bioremediation of Heavy Metals and Pesticides from Wastewater Using Efficient Microbes

An Insight into the Mechanism and Factors Involved

*Aanisa Manzoor Shah, Tahir Ahmad Sheikh,
Inayat Mustafa Khan, Tajamul Islam Shah, Shabir
Ahmad Bangroo, Yasir Hanif, Ali Mohd Yatoo,
Aabiroo Rashid, Roheela Ahmad, and Razia Gull*

7.1 INTRODUCTION

Environmental degradation on account of anthropogenic activities poses a hefty threat to the sustainable, sparkling and green environment. Irrational expansion of the global population and industrialization has directed the production of an enormous amount of environmental contaminants at the expense of environmental and human health (Asim and Rao, 2021). The current trend of water shortage is scary in view of the projection put forward by the World Bank and World Health Organization that by 2025 not less than three billion populations globally will be devoid of access to clean water (Bankole et al., 2019). In addition, to meet the food demand of the escalating population in 2050, it has been estimated that agricultural production would have to increase by 70% (Velasco-Muñoz et al., 2019; Fischer et al., 2007). This may lead to a 53% increase in global water consumption (López-Serrano et al., 2020). Water being an indispensable element of life, spines the maintenance of an adequate food supply and a productive environment for all living organisms. Consequently, there is a call for the transformation of traditional system to a closed-loop wastewater management circle that warrants sustainability to sustain crop production with this

DOI: 10.1201/9781003359326-7

limited resource base (Jhansi and Mishra, 2013). The water reuse concept would aid in alleviating the water stresses and overexploitation of groundwater sources as water demands endure to outweigh conventional freshwater sources (Morris et al., 2021).

The deterioration and degeneration of terrestrial and aquatic ecosystems have been primarily attributed to noxious pollutants encompassing dyes, heavy metals, pharmaceuticals, personal care stuff, fertilizer and pesticides, organic bacteria, viruses, micro-cystins and antibiotics discharged as a consequence of anthropogenic activities (Yatoo et al., 2022; Egbosiuba et al., 2020). The term "heavy metals" encircles the metal elements possessing a density greater than 5 g cm^{-3}, comprising a sum of physiologically influential (Zn, Cu, Fe, Mn), extremely toxic (Cr (VI) As, Pb, Hg, Cd, Sb) and less toxic (Mo, Co, Au, Cr (III), Ag) elements (Odobasic et al., 2019). Some heavy metals play an imperative part in the growth and development of animals as well as plants, as long as they reside in the range of permissible limits; otherwise, they tend to be amply detrimental, owing to their enduring persistence, non-degradability and subsequent biomagnification across trophic levels (Khalef et al., 2022). The inevitable exposure of humans to heavy metals poses numerous acute and chronic toxic imprints and cast off detrimental cumulative effects (Costa, 2019; Gazwi et al., 2020). The employment of pesticides to declining yield losses through the elimination of weeds, pests and pathogens nurtures grave health as well as environmental concerns in debt of long-lasting persistence (Masia et al., 2013, Silva et al., 2019), bioaccumulation (Goutner et al., 2012; Silva et al., 2022) or toxicity to target and non-target species (Yatoo et al., 2022; Colin et al., 2019; Kim et al., 2017; Ullah et al., 2018). In this regard, monitoring and assessment of such noxious pollutants from water streams demands a hasty address and, therefore, pays promptly to two of the sustainable development goals framed by United Nations (UN SDG), viz. the "clean water and sanitation" and "life below water".

The era of wastewater treatment has obliged the employment of numerous techniques encasing physical, chemical and biological tracks for remediating heavy metals and pesticides (Raklami et al., 2022). Conventional approaches encasing chemical precipitation, membrane filtration, ion exchange, coagulation, solvent extraction, complexation and so on have been exploited to eradicate pesticides and heavy metals from wastewater (Hussain et al., 2021). However, these approaches are not devoid of certain disadvantages such as high energy demands, low efficiency, precipitation of venomous elements, cost ineptness and generation of substantial quantities of toxic sludge apart from being impractical with low metal concentrations (Hussain et al., 2021). In view of these aspects, there is a call for novel remediation approaches that foster environmental health while being economically feasible too. Microbial remediation is the answer to these quests and encompasses the employment of microorganisms comprising fungi, algae, bacteria and so on for the removal, immobilization, transformation or detoxification of environmental contaminants (Cepoi et al., 2021). The biological approaches play on the mechanisms exploited by exogenous or endogenous microorganisms to manage pollutant stress encompassing enzymatic detoxification and volatilization, intracellular or extracellular sequestration, production of metal chelators and precipitation (Chang et al., 2021). These microbial remediation strategies possess an edge over physico-chemical approaches in that they are simple, cost-effective, reliable and non-destructive; have extensive coverage; and are environmentally friendly (Liu et al., 2020). Nonetheless, there exist certain bottlenecks in their widespread application. Yet, such constraints could be outshined by exploring novel strains with advanced molecular toolkits, engineering the existing ones with the specified genes of interest and through the exploitation of meta-organism concept (combination of microbial and phytoremediation strategies) as if either one of them plunges, the other one perks it up!

7.2 SOURCES OF HEAVY METALS AND PESTICIDES

The advent of global industrialization exposes terrestrial and aquatic ecospheres to a wide array of noxious pollutants (Tarfeen et al., 2022). These contaminants carve their negative imprints on the environment as well as living organisms (Aluko et al., 2021). In addition, these contaminants are acknowledged to biomagnify across the food chain, manifesting a lethal impact even at slightly higher concentrations than required for normal metabolic processes (Sayyed et al., 2019). Heavy metal stems as a consequence of natural processes encircling disintegration of rocks, soil formation, volcanic eruptions and anthropogenic activities comprising industrialization, agriculture, mining and domestic effluents and improper waste disposal (Alengebawy et al., 2021). The predominant industries that pay to heavy metal pollution comprise textiles, fertilizers, tanneries, metallurgy, galvanizing factories, varnishes, pharmaceuticals and pesticide-manufacturing units (Figure 7.1) (Chabukdhara and Nema, 2012). Metallurgical industries directly pour out contaminants in the course of metal extraction and processing in addition to the production of indirect pollution too (Wuana and Okieimen, 2011). Highly polluted effluents from tanning and textile industries make their way to water sources. Enormous volumes of left-over rocks with a modest content of heavy metals are created during mining procedures, which eventually accumulate in living tissues as leaching transports them to the ground and water areas (Li and Yu, 2015). Agricultural-based conducts further augment the heavy metal pollution on account of inorganic chemical consumption such as natural phosphates have been marked to enclose associated heavy metals as impurities. In parallel, the application of pesticides adds to this as plentiful inorganic pesticides include Pb(II), As (V/III), Hg(II) and Cu(II) as dynamic ingredients. Irrigation in conjunction with municipal and industrial wastewater over extended periods or at high-frequency upshots in the accumulation of pollutants in soil (Tarfeen et al., 2022).

The escalating population across the globe demands an exponential increase in food production with a limited resource base. For such accomplishments, the extensive exploitation of agro-chemicals plays its part, but at the expenditure of environmental health (Yatoo et al., 2021). The bold label of pesticides shelters a wide array of herbicides, insecticides, fungicides, defoliants, nematicides and rodenticides (Tsaboula et al., 2019). Sniffing the present scenario marks pesticides as indispensable toolkits for farmers to combat food security issues, predominately in developing nations. An expenditure of 2 million tonnes of pesticides has been documented (Yatoo et al., 2022); the picture is worse in developing nations, contributing around one-quarter of global pesticide consumption. Industries account for hefty volumes of pesticide residues, which contaminate the terrestrial and

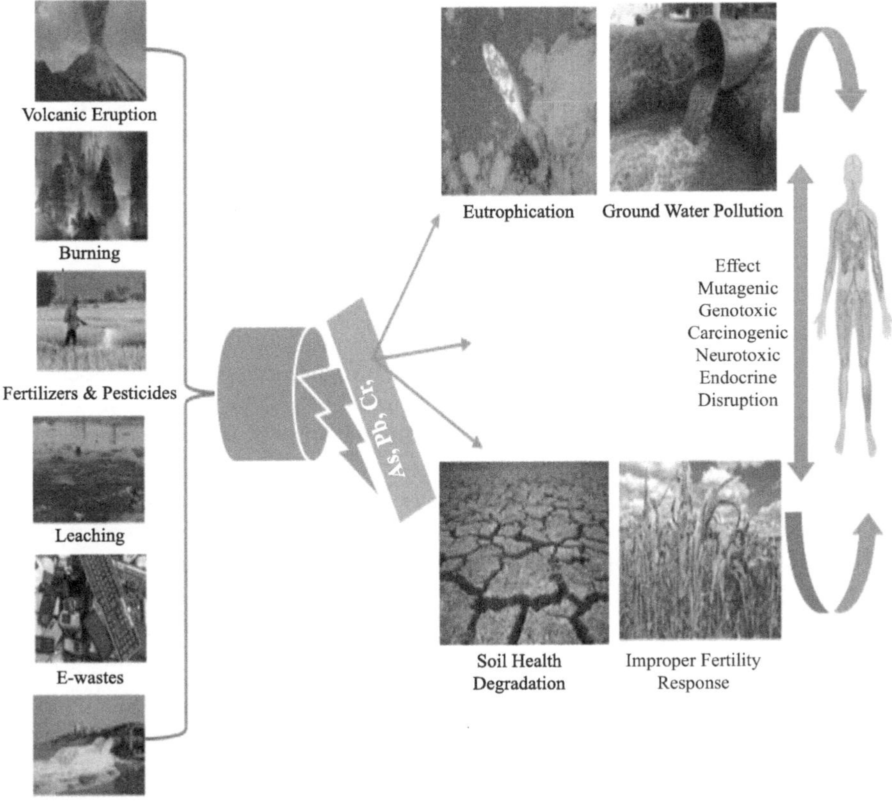

FIGURE 7.1   Sources of environmental contaminants

aquatic ecosystems and impose their ill imprints on living beings through food intake (Chopra et al., 2011). However, the pesticide mobility in soil is governed by their solubility in water and rate of adsorption on soil colloidal complex. Soil organic matter content on account of binding sites regulates pesticide retention in soil and, thereby, reduces the bioavailability and degradation of these pesticides (Becerra-Castro et al., 2013).

7.3   BIOASSESSMENT OF HEAVY METALS AND PESTICIDES

The deleterious imprints of heavy metal and pesticide pollution call for real-time monitoring and diagnosis of contaminated environments, so that preventive actions can be adopted ahead of time and further deterioration of soil quality can be controlled, as well as appropriate remediation techniques can be used on time. Monitoring the physico-chemical qualities of soil takes time, therefore, the magnitude of soil pollution can be determined by analysing the soil's microbiological and biochemical properties following contamination. In addition, the creation of early warning systems that are quick, dependable and sensitive to a diverse array of substances is terribly desired. These systems need to exhibit the ability to monitor, identify and analyse the imprints of toxicants. Microbial interventions serve as a blooming bud in the detection of such noxious pollutants. In addition, the microbial consortia integrated with electro-chemical sensors aid in toxicity assessment of these environmental contaminants. The employment of biosensors has facilitated the detection of pesticides and heavy metals in wastewater as well as contaminated soil (Ahmad and Kumar, 2020).

7.4   BIOASSAY TESTS

In the current world, a range of different bioassay tests have been developed as a result of the requirement for observing contaminated ecosystems to assess and accomplish the risk management of environmental impairment. Bioassays intend to evaluate the toxic effects of different compounds on biota through the exposure of test organisms to polluted waters or sediment sections. These toxicity tests are accomplished with the employment of a wide array of microorganisms, algae, fish and invertebrates (Chavan et al., 2016). Therefore, to investigate and assess the possible impacts of toxicants, bioassay techniques rely on the response of testing organisms (Ekelund and Hader, 2018). Physiological response is the primary research parameter when utilizing microalgae in bioassays. Setting the toxicity criteria for the bioassays is extremely important since physiological factors encasing growth, quantum yield, photosynthesis, motility, chlorophyll content and fluorescence react to contaminants contrarily. The non-invasive measurement of in vivo fluorescence, which makes it easier to track activity changes that impact PS II in microalgae, is a quick procedure to follow. Fluorescence constitutes one of three courses in which energy acquired from the absorbed light of chlorophyll molecules is released back as red fluorescence (Fai et al., 2007). PAM fluorescence has been employed to measure photosynthetic yield and rate of electron transport while assessing the imprints of cadmium and nickel on photosynthetic aspects of *E. gracile*. A quick approach for monitoring the environment that relies on photosynthetic activity is the employment of biosensors, which accurately reflect the physiological effects pertaining to active substances.

7.5   TOXICITY TESTS

The ill imprints of pollutants, by and large, on the environment cannot be quantified in the dearth of estimating their toxicity. A distinct test organism is subjected to a series of dilutions of a potentially harmful substance during a controlled laboratory experiment. Kress (2019) advanced two biosensors to assess the impact of isothiocyanate, phenolic and aldehyde terpenes on microorganisms. One of the biosensors manifested the combination of characteristics exhibited by commercial biosensors by transforming E. coli HB101 with the luxCDABE gene acquired from *Vibrio fischeri*, while the other biosensor comprised the transformation product of *Acinetobacter baylyi* ADP1_recA_lux and luxCDABE gene acquired from *Photorhabdus luminescens*. The intensity of the light produced by the respective transgenic bacteria in the company of toxicants, which harms DNA, is strongly associated with the level of recA expression.

7.6 BIO-INDICATION

Changes in the population- or community-level organization of the biota can be used to gauge the health of any ecosystem. An overall assessment of the environmental quality is accomplished by contrasting the population or community structures in places that are thought to be under the stress of pollution with those that are generally clean in the same geographic area. Bio-indication represents the comparison of diversity indices or species composition at different locations to attain the estimation of pollution loadings prevailing in that location (Bellinger and Sigee, 2010). Biotic indices are alterations at the population level that are employed to measure the pollution load in a region, whereas species diversity indices are typically used to describe variations at the community level. The abundance, absence or presence of particular organisms is used to calculate biotic pollution indices, which provide information about the kind and amount of pollution present in diverse ecosystems. Predictions of SO, levels are frequently acquired by marking the lichen flora in a given site since, lichens are typically regarded to be predominantly responsive to atmospheric SO levels (Ristić et al., 2017).

The absence of a species can be linked to a variety of other ecological phenomena, such as competitive exclusion, predation pressures and other unfavourable physical conditions in addition to the pollutant loadings existing in the environment, biotic indices illustrating pollution that just reflect the absence or presence of a species in an area are challenging. Species diversity declines are a common indicator of changes in community organization. While a pollution-stressed ecosystem may only be able to host a part of the species found in a healthy environment, a sound, unpolluted environment will maintain a large population of creatures representing a diversity of species. This is the fundamental tenet of experiments using bio-indication to gauge species diversity. It is theorized that pollutants are responsible for this decline in species diversity by exerting substantial selection pressures that favours resistant species while concurrently selecting against contaminant-prone species. Fundamentally, comparative studies are involved in bio-indication courses at the community level. Therefore, the scope is strictly restricted in view of the fact that species are limited to specific geographical locations and comparative assessments can only be directed if the same biological species prevail at all desired locations. As a result, bio-indication surveys are no longer as popular, and ecotoxicologists are now concentrating more on quantitative techniques like biomonitoring.

7.7 BIOMONITORING

Biomonitoring is the process of exploiting organisms in an environment to detect and measure toxicants (Chaphekar, 1991). To identify and evaluate contamination in any ecosphere, this method makes use of organisms that exhibit sensitivity as well as specificity to the contaminants in question. Biomonitoring exploits the capability of microorganisms to accrue pollutants in their tissues via bio-concentration-manifesting the acquisition from the existing circumference and biomagnification, representing the uptake via trophic levels. Biomonitoring can be accomplished through the introduction of external microbes or through the assessment of endemic organisms, reflecting active and passive biomonitoring, respectively. Biomonitoring possesses an edge over the other assessment techniques in that bio-accumulated sub-lethal contents of pollutants occurring in the microbial tissues signify the total quantity of contaminants accrued by microbes over an integrated epoch. In this way, pounded or irregular release upshots that routine chemical monitoring is expected to miss can be detected and integrated via biomonitoring.

Lichens represent a bold label of microorganisms employed for biosensing of environmental contaminants. Some of the species enveloped by this label entail *Xanthoria parietina*, *Parmelia sulcata*, *Hypogymnia physodes* and *Parmelia caperata* (Paoli et al., 2012). The imprints of toxic metal accumulation documented on lichens encompass discoloured or pinkish necrotic spots apart from growth retardation (Tovar-Sánchez et al., 2019). The bacteria multiply quickly, are simple to spot and analyse and react swiftly to changes in the environment like pH, temperature or heavy metal toxicity. These illustrations serve bacteria as worthy candidates for biosensing of environmental contaminants (Bouskill et al., 2007). In this manner, heavy metal-contaminated water and soil have been monitored using bioluminescent bacteria like *Photobacterium phosphoreum* and *Aliivibrio fischeri* (An et al., 2012). *Vogesella indigofera* is a bacterium that normally produces indigoidine, which gives it a blue hue. When this bacterium grows in the presence of Cr (VI), however, it produces less pigment, and the amount of pigment it produces depends on the amount of Cr (VI) present (Gu and Cheung, 2001). *Serratia marcescens*, a gram-negative bacterium produces a red pigment acknowledged as prodigiosin. However, the bacterium growing in sub-inhibitory concentrations of toxic heavy metals such as Cr, Pb and Cd, drastically reduce the pigment production, reflecting it as a potential bio-indicator of heavy metal pollution (Cristani et al., 2012). The application of bio-markers for risk assessment subsequent to the pesticide application illustrates that a reduction in the thickness of the eggshell offers a solution to bio-monitor the contaminants as different extents of eggshell abrasions prevail in the presence of DDE, DDT or dicofol in the environment. Inhibitors of acetylcholinesterase organophosphate serve as biomarkers of microorganisms following the use of carbamate pesticides and can be detected conveniently and more reliably in comparison to chemical analysis (Wołejko et al., 2022).

7.8 MICROBIAL REMEDIATION OF HEAVY METALS

Bioassessment of these noxious pollutants necessitates a follow-up with remediative techniques for the restoration of degraded ecosystems. Bioremediation embraces the probability of degrading or rendering contaminants harmless while exploiting natural biological bustle in the ecosystem (Medfu Tarekegn et al., 2020). Microbial remediation with the employment of different organisms such as bacteria, fungi, algae and yeast plays its part in remediation approaches (Pande et al., 2022). Microorganisms alter the ionic assert of metallic ions that subsequently alter the bioavailability, solubility and their transport in soil and aquatic ecosystems (Ayangbenro and Babalola,

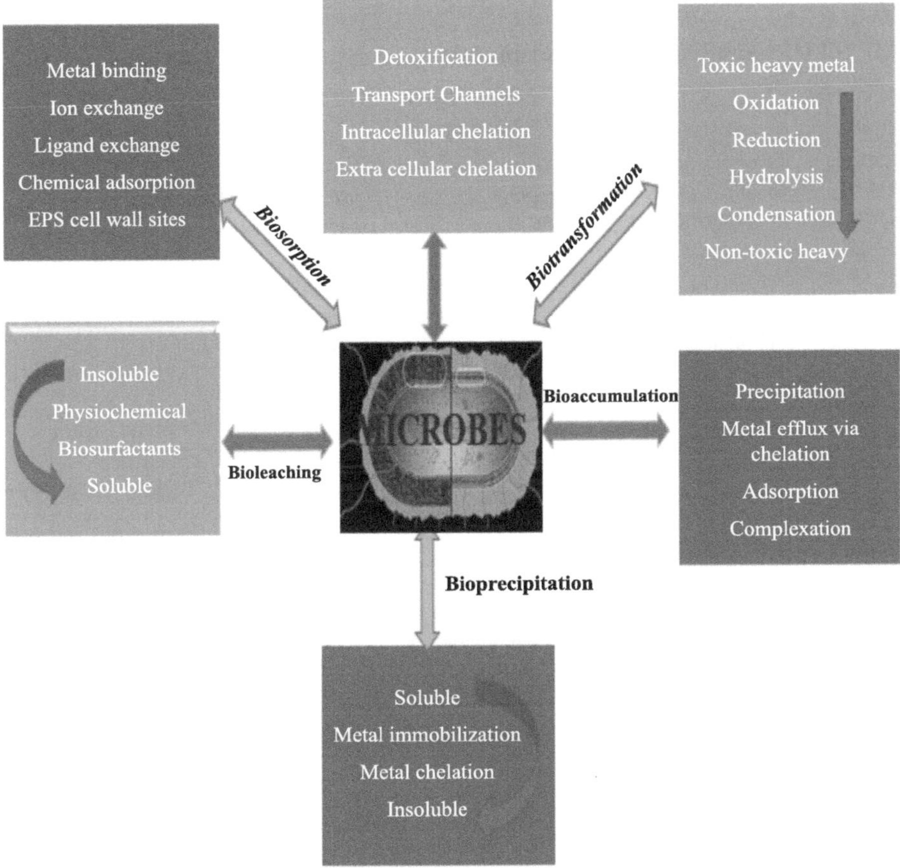

FIGURE 7.2 Microbial remediation of heavy metals

2017). The chelation, oxidation–reduction, alteration of metallic complexes and biomethylation processes that follow the immobilization or mobilization of heavy metals help in microbial remediation (Pratush et al., 2018). It is possible to isolate the bacteria that break down heavy metals from aerobic and anaerobic environments. Nonetheless, aerobic microorganisms are much more receptive to bioremediation than anaerobic germs (Azubuike et al., 2016). Microorganisms exploit mechanisms such as bioaccumulation, biosorption, bioleaching and biotransformation to survive metal-contaminated environments (Table 7.1; Figure 7.2).

7.9 BIOSORPTION

Biosorption connotes an interactive process including the non-specific binding of metal ions to proteins and polysaccharides prevalent on the cell surface. It exhibits the property that both living and dead microorganisms can contribute binding

TABLE 7.1

Microbial remediation of heavy metals

Microbial group	Microorganism	Heavy metal contamination	Resistance mechanism	References
Bacteria	*Pseudomonas aeruginosa*	Cadmium	Biosorption	Chellaiah (2018)
	Bacillus subtilis X3	Lead	Bioimmobilization	Qiao et al. (2019)
	P. aeruginosa and *Bacillus cereus*	Cadmiun and lead	Bioaugmentation	Nath et al. (2018)
	Bacillus sp. *KL1*	Nickel	Biosorption	Taran et al. (2019)
	Desulfovibrio desulfuricans	Cadmium, zinc and copper	Extracellular Sequestration	Yue et al. (2015)
	P. aeruginosa	Palladium, zinc and copper	Bioprecipitation	Li et al. (2019)
	Escherichia coli	Cadmium and mercury	Active export	Lerebours et al. (2016)
	Bacillus firmus	Mercury	Enzymatic detoxification	Noroozi et al. (2017)
Fungi	*Penicillium chrysogenum A15*	Lead		Povedano-Priego et al. (2017)
	Aspergillus niger	Lead, copper and chromium		Dursun et al. (2003)
	Botrytis cinereal	Lead		Akar et al. (2005)
	Rhizopus oryzae	Copper		Fu et al. (2012)
Algae	*Asparagopsis armata*	Lead, cadmium and nickel	Biosorption	Romera et al. (2007)
	Oedogonium hatei	Chromium		Gupta and Rastogi (2009)
	Spirulina platensis	Copper		Al-Homaidan et al. (2014)
	Cystoseira barbata	Cadmium and nickel		Yalcin et al. (2012)

sites, which can bind heavy metals even in extremely diluted solutions (Vasudevan et al., 2001).

The metal adsorption by bacteria is accounted to the presence of peptidoglycan layer in them. Teichoic acid, amino acids and meso-diaminopimelic acid exist in multi-layer peptidoglycan in gram-positive bacteria, whereas gram-negative exhibit only one layer (Nanda et al., 2019). This layer comprises phospholipids, glycoproteins, enzymes and lipopolysaccharides, which perform as ligands and furnish active metal-binding sites (Gupta et al., 2015). The cell wall contains a number of acidic groups, including teichoic acid, which are sources of carboxyl groups essential for metal uptake. Consequently, gram-positive bacteria are capable of absorbing more metal ions in comparison to gram-negative bacteria. Chemical modifications of extracellular polysaccharides as well as biological activity encircle methylation, sulphonylation, acetylation, phosphorylation and carboxymethylation (Verma and

Kuila, 2019). In addition, biofilm formation by bacteria illustrates metal adsorption. The ability of *Staphylococcus aureus* biofilms to bio-precipitate U(IV) was acknowledged, and the remediation of U(IV) was further aided by the addition of acid phosphatase (Tayang and Songachan, 2021).

Fungi have been exploited at length for the accomplishment of heavy metal adsorption, on account of their proficient metal-uptake capacity and excellent biosorption features. These characteristics are marked in debt of fungal surface composition encircling proteins, polysaccharides and lipids such as glucans, chitins and mannins. However, the biosorption capacities are strongly regulated by pH and accessibility of metal ions. *Saccharomyces cerevisiae* has proven its capacity to eliminate poisonous metals from filthy wastewater through the execution of bio-sorption (Fu et al., 2012). Algae have been recognized for notable heavy metal absorption propensities, owing to the polymeric substances, polysaccharides (cellulose and alginate), lipids and reactive groups encasing hydroxyl, thiol, carboxyl, phosphate, amino, sulphonate and imidazole in the algal cell wall (Priatni et al., 2017). Deprotonated sulphate, laminarin and monomeric alcohols are other compounds found in algae that appeal to both positive as well as negative metal ion species. For metal ions, the biosorption abilities of various microalgal strains including *Chlorella vulgaris*, *Spirulina platensis*, *Sargassam* sp. and *Oscillatoria* sp. have been documented (Leong and Chang, 2020; Riskuwa-Shehu et al., 2019).

7.10 BIOTRANSFORMATION

Biotransformation connotes a process of altering the structural properties of a chemical complex, resulting in the development of new polar molecules (Asha and Vidyavathi, 2009; Pervaiz et al., 2013). Specifically, the harmful metals and organic molecules are changed by this interaction of metal and bacteria into a state that is relatively less harmful. The emergence of this contrivance in the microbes enables them to adapt to environmental fluctuations. Microbial cells in debt of a high surface-to-volume ratio, a rapid rate of growth, rapid metabolic activities and convenient upkeep of sterility offer the most candidates for biotransformation (Tayang and Songachan, 2021). The acquisition of these transformations is realized via the creation of new carbon bonds, the introduction of reactive groups, isomerization, oxidation–reduction, methylation, hydrolysis, demethylation and condensation (Pande et al., 2022).

Microbial transformation is extensively employed for transforming a diverse array of pollutants encasing hydrocarbons, heavy metals, pharmaceuticals and so on (Karigar and Rao, 2011). *Micrococcus* sp. and *Acinetobacter* sp. have been recognized for the conversion of As (III) into less soluble as well as harmless As(III) through the execution of the oxidation process (Nagvenkar and Ramaiah, 2010). Numerous iron and sulphur-reducing bacteria may convert Cr(VI) to Cr (III), including *Acidithiobacillus* and *Desulfovibrio* (Singh et al., 2012).

7.11 BIOLEACHING

The process of bioleaching crafts microorganisms encircling bacteria and fungi which aid in the solubilization of metal sulphides and oxides from secondary wastes and ore deposits (Jafari et al., 2019). These solubilized metals are subsequently purified using appropriate techniques such as ion exchange, adsorption, membrane separation and selective extraction. Bioleaching manifests a lucrative and ecologically sustainable approach, for it consumes less energy and emits no hazardous emissions (Asghari et al., 2013). The process can be accomplished through contact and non-contact procedures. A non-random physical interaction prevails between mineral sulphide and the bacterial cell in the contact mechanism. However, in the non-contact approach, no physical contact exists between the mineral surface and the bacterial cell (Anjum et al., 2012). The generation of lixiviant by bacteria aids in the chemical oxidization of sulphide minerals. However, this reaction prevails only in acidic environments having less than 5.0 pH (Vestola et al., 2010). Bioleaching involves a diverse spectrum of microorganisms, with acidophiles playing a predominant role. On summarizing the mechanism accomplished by the microbes, three basic phases are identified: (i) microbial oxidation–reduction process in solution; (ii) acid production via organic and inorganic pathways; and (iii) extraction of metals from the matrix. Bioleaching of arsenic with individual as well as intermixed cultures of *Acidithiobacillus thiooxidans* and *Acidithiobacillus ferrooxidans* has been observed where the mixed culture of consortia possessed an edge over the individual employment of them (Zhang and Gu, 2007). The propensity of fungi in mobilizing and bioleaching metals from left-over ores and industrial wastes is predominantly accounted for the production of diverse complexion agents and diverse organic acids in an environment encompassing oxalic acid, citric acid and gluconic acid apart from their biochemical capabilities and comparatively heightened resistance to agitating parameters like pH and temperature. These include a wide array of species such as *Cladosporium* sp., *Aspergillus* sp., *Rhizopus* sp., *Alternaria* sp., *Penicillium* sp. and *Mucor* sp. (Medfu Tarekegn et al., 2020).

7.12 BIOMINERALIZATION

Biomineralization of metal ions confers a usual course regarding mineral formation. Mineral production largely relies on the occurrence of amplified variable and extreme reactive surfaces encircling cell walls and extra organic layers such as S-layer and EPS with varied levels of hydration, content and structure. Additionally, the prevalence of organic ligands (hydroxyl, amine, phosphoryl, carboxyl and sulphur) stem in the deprotonation and rise of net negative surface charge of microbes in response to pH increase (Gavrilescu, 2004). Potential hazardous metals exhibiting positive charges precipitate unevenly into more stable and condensed mineral products (Zhang et al., 2019). Mechanisms comprising carbonate precipitation, phosphate precipitation, oxalate precipitation and so on can all result in the immobilization or complexation of metals (Yang et al., 2016). Zhang et al. (2019) marked the potential of *Bacillus* sp. to discharge inorganic phosphate and sequester toxic heavy metal ions through the formation of insoluble metal phosphate complexes. *Bacillus subtilis* FZUL-33 assisted in the lead biomineralization on account of the PO_4^{3-} formation released in the course of acephate degradation (Lin et al., 2016). In parallel, the organophosphate degradation by *Pseudomonas putida* marks the production of phosphate and carbonated minerals, which in turn account for the acceleration of Cd precipitation (Li et al., 2016).

7.13 BIOACCUMULATION

Bioaccumulation results when the rate of absorption allied to a specific contaminant exceeds the rate of squandering it and represents a metabolically active process governed by an import-storage system. The system aids in the transportation of heavy metal ions across lipid bilayer into the cytoplasm or intracellular spaces with the employment of transporter proteins. Metal-binding substances like proteins and peptide ligands sequester the metal ions, and the heavy metals that cling to these substances can be found in particulate, insoluble and by-product forms (Mishra and Malik, 2013). Using transmission electron microscopy, Rani and Goel (2009) portrayed that *P. putida* 62 BN exhibited periplasmic and intracellular accumulation of the metals. The accumulation of about 20% nickel (II) by budding cells of *Bacillus cereus* M116 has been marked by Naskar et al. (2020). Furthermore, the accumulation of cadmium and lead by *Monodictys pelagic* and *Aspergillus niger* was documented by Sher and Rehman (2019).

7.14 MICROBIAL REMEDIATION OF PESTICIDES

Microbial remediation is a useful strategy for fixing pesticide-contaminated environments. The microbial remediation process converts harmful chemicals/substances into low-level toxic substances. The key benefits of using a microbiome for the remediation of pesticides are its ease of culture, increased microbial biomass and rapid mutation. Microbial decomposition can be swift and complete under the appropriate circumstances (temperature, appropriate humidity, air movement, adequate pH). *Bacillus* sp., *Mycobacterium* sp., *Pseudomonas* sp., *Pandoraea* sp., *Klebsiella* sp. and *Phanerochaete chrysosporium* are the most often utilized bacterial species for pesticide bioremediation (Rani and Dhania, 2014; Senthil Kumar, 2018). Table 7.2 illustrates the species of microorganisms used to remediate pesticides.

Biodegradation in natural environments entails the exchange of substrates and compounds among a strongly coordinated microbial population, a phenomenon known as metabolic cooperation (Ramos et al., 2008). The ability of microorganisms to interact with substances, both chemically as well as physically, results in structural alterations or complete breakdown of the specific analyte. Bacteria, actinomycetes and fungi are the primary converters and chemical degraders among microbial diversity. Pesticides and allied xenobiotics are generally biotransformed by fungi by the inception of minor structural adjustments to the molecule, making it less harmless (Figure 7.3). The bio-transformed pesticide, therefore, is discharged into the ecosystem, allowing it to be further degraded by microbes. This capacity is generated in response to the synthesis of extracellular enzymes that work on a diverse range of contaminants. These extracellular enzymes encompass oxidases, laccases, manganese peroxidase and lignin peroxidase are engaged in the breakdown of lignin. cytochrome P450, glutathione S-transferases (GSTs) and esterases represent the three major enzyme groups entangled in degradation processes. Further, esterases (also hydrolases), mixed-function oxidases and hydrolases are primarily involved in pesticide degradation in the first metabolic stage following the GST system in its second stage

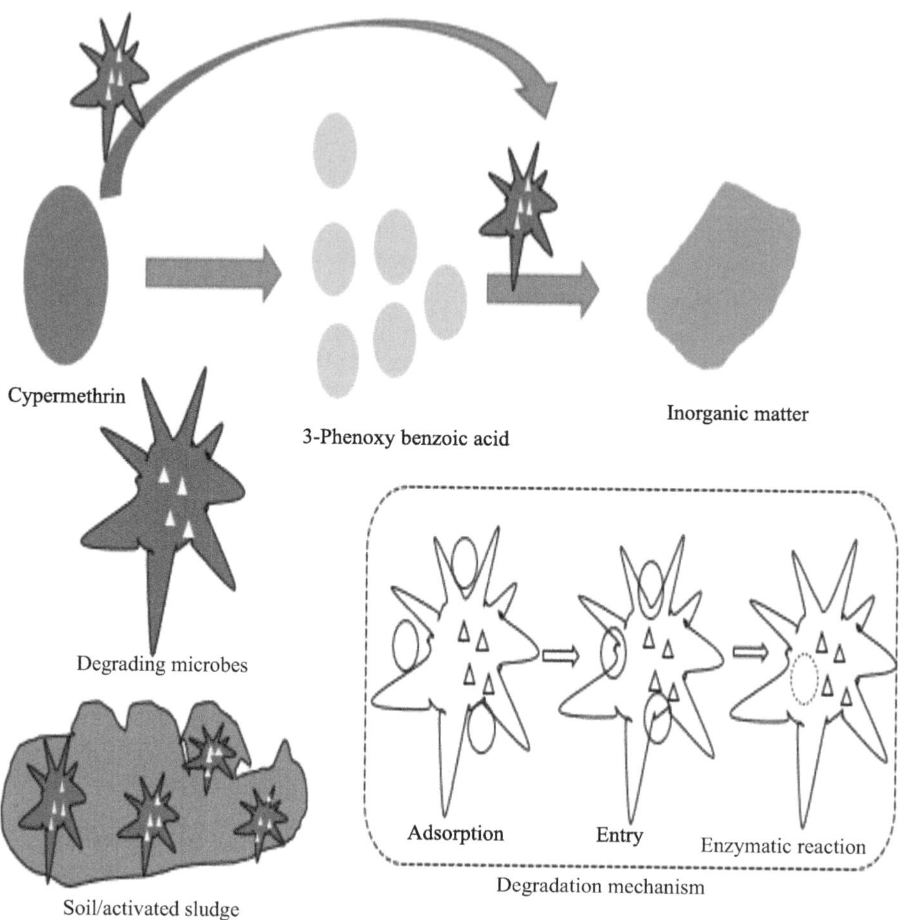

FIGURE 7.3 Microbial remediation of pesticides

(Tovar-Sánchez et al., 2019). A diverse array of enzymes catalyses metabolic processes circumferencing oxidation, dehalogenation, hydrolysis, inclusion of a hydroxyl group to a benzene ring, reduction, replacing sulphur with oxygen, side chain metabolism and ring cleavage. The biodegradation process relies on the ability of microbes to purify or modify the contaminant molecules, which depends on both accessibility and bioavailability.

7.14.1 Bacterial Remediation of Pesticides

Bacterial remediation serves as the pertinent bioremediation approach owing to quick growth, financial feasibility in hand with ease of operation. It can proceed via aerobic and anaerobic settings. Diuron, a potent algicide or herbicide found in water and soil gets metabolized by *Anthrobacter sulfonivorans*, *Variovorax soli* and *Advenella* sp. of bacterial consortia (Villaverde et al., 2017). Briceno et al. (2016) investigated

TABLE 7.2
Microbial remediation of pesticides

Type of microorganism	Potential microbial species	Target pesticides	References
Bacteria	*Chlorella vulgaris*	Diazinon	Kurade et al. (2016)
	Pseudomonas aeruginosa	Chlorpyrifos	Kharabsheh et al. (2017)
	Streptomyces consortium	Lindane	Saez et al. (2014)
	Alcaligenes faecalis	Endosulfan	Kong et al. (2013)
	Ochrobactrum sp.	Dichloro-diphenyl trichloroethane (DDT)	Pan et al. (2017)
Fungi	*Aspergillus flavus*	Malathion	Derbalah et al. (2020)
	Trichoderma sp.	Carbofuran	Saikia and Gopal (2004)
	Trichoderma harzianum, T.viride	Pirimicarb	Eapen et al. (2007)
	Phlebia lindtneri, Phlebia brevispora	DDT	Xiao et al. (2011)
Algae	*Dunaliella* sp., *Cylindrotheca* sp.	Naphthalene, DDT	Biswas et al. (2015)
	Chlamydomonas mexicana	Atrazine	Kabra et al. (2014)
	Spirulina platensis	Chlorpyrifos	Thengodkar and Sivakami (2010)
	Gracilaria verrucosa	2,4-D	Ata et al. (2012)

the elimination of diazinon using sole and mixed cultures of *Streptomyces* sp. The explorers discovered that the mixed culture removed more of the diazinon pesticide in contrast to the sole culture.

7.14.2 FUNGAL REMEDIATION OF PESTICIDES

Fungal remediation of different pesticides is executed through the discharge of an assemblage of extracellular enzymes such as polyphenol oxidases, laccases and lignin peroxidases that portray a significant function in the bioremediation pathway. Intracellular enzymes including methyl-transferases, reductases and cytochrome oxygenases too are engaged in the restoration of contaminants. Pesticides such as clothianidin are bio-transformed through the action of white rot fungi like *Phanerochaete sordida*, which aids in the transformation of pesticides into non-toxic compound *N*-(2-chlorothiazol-5-methyl)-*N*-methyl urea (Mori et al., 2017). Extracellular enzymes including laccase, manganese peroxidase and lignin peroxidase are present in white rot fungi, implying that it is an excellent pesticide degrader. The knack of fungi to exploit organic compounds as a substrate for development and the long mycelial network render them the ideal approach for bioremediation.

7.14.3 Cyanobacterial Remediation of Pesticides

Algae transform organic molecules into new compounds having a high economic benefit. The species of microalgae is strongly advisable for habitats affected by the lindane pesticide, which is extremely detrimental to both the ecology and mankind. According to Thengodkar and Sivakami (2010), pesticide breakdown involving microbial enzymes such as alkaline phosphatase, released by *Spirulina platensis* exhibits the capacity of hydrolysing the pesticides such as chlorpyrifos to its prime non-toxic component 3, 5, 6-trichloro-2-pyridinol. Kabra et al. (2014) examined the breakdown of atrazine on account of the microalgal species *Chlamydomonas mexicana*, and the carbohydrate levels increase, demonstrating that *C. mexicana* can remove pesticides from contaminated streams.

7.15 FACTORS AFFECTING MICROBIAL BIOREMEDIATION

To reap maximum benefits for microbial remediation of pollutants, it becomes crucial to take into account the factors influencing the bioremediation process. The intricacy of the process amplifies in debt of venomous and hydrophobic nature of these environmental contaminants. The complexities are even more varied owing to the prevalent environmental circumstances as well as community structure and diversity of microorganisms. The proficiency of the microbial bioremediation process is governed by different physical, chemical and biological aspects encircling temperature, pH, nutrient status, dissolved oxygen content, prevalence of electron acceptors and donors, contaminant load and so on (Mohan et al., 2006). The response of bioremediation approaches relies on a multitude of aspects to optimally play the exclusion of poisonous contaminants.

7.15.1 Concentration of the Contaminant

The concentration of contaminant engraves its imprints on microbial activity in that if the contaminant concentration is low, the induction of bacterial degradation enzymes is inhibited. However, toxic effects only become apparent at extremely high pollutant concentrations (Adams et al., 2015). The synergistic affiliations between several elements of a contaminant increase the degradation rates of catabolic enzymes.

7.15.2 Nutrient Availability

Microorganisms possess a drive for nutrient acquisition entailing carbon, phosphorus, nitrogen, potassium and calcium to ascertain their growth. Additionally, the comparative concentrations of the nutrients readily available have an effect on how the pollutant degrades. The content of nitrogen, phosphorus and potassium in excess has been acknowledged to pose a negative bearing on biodegradation (Van Hamme et al., 2003). Furthermore, the accessibility of organic in hand with inorganic contaminants to microorganisms, commonly reflected as bioavailability, affects the rate of biodegradation (Mishra et al., 2021).

7.15.3 pH

The pH is indispensable for microbial biosorption because various bacteria frequently have varied optimal pH levels. Primarily, the pH exhibits a predominant effect on the activity of enzymes in microorganisms, which in itself has an impression on how quickly heavy metals are metabolized by bacteria. The microorganism's surface charge is impacted by pH, which in turn impacts how well it can bind to heavy metal ions (Galiulin and Galiulina, 2008). Additionally, pH possesses its imprint on the mobility and hydration of several metal ions in the soil. The rate of heavy metal remediation by microbes increases in response to the pH increase throughout a constrained range, but it starts to decline after the pH climbs to a particular limit (Wierzba, 2015). Specifically, the removal rate increased for Pb^{2+} and Zn^{2+} for pH values between 2.0 and 5.5. At pH 5.5, the adsorption capacity is seven times higher (70 mg/g) than at pH 2.0 (10 mg/g). The rate of removal, however, drops to the identical level as that of pH 2.0 for pH values greater than 5.5.

7.15.4 Ambient Temperature

The heavy metal adsorption rate is altered by the ambient temperature influences through influencing the development and proliferation of microbes (Fang et al., 2011). The ideal temperature for different microbes varies often. *Sulfolobus solfatataricus* and *Acidianus brierleyi* exhibit extreme thermophilic nature, while *Thiobacillus acidophilus*, *Thiobacillus ferrooxidans* and *Thiobacillus tepidarius* represent mesophilic characteristics. *Bacillus jeotgali* and *Bacillus licheniformis* have been recognized for absorption of heavy metals such as Cd^{2+}, Cr^{2+} and Zn^{2+}, where the optimum temperature for the same microorganism employed for different heavy metals was different (Zouboulis et al., 2004 and Rodríguez-Tirado et al., 2012). However, the ideal temperatures are often between 25°C and 35°C (Gan et al., 2012).

7.15.5 Oxygen Availability

The process of bioremediation is either aerobic or anaerobic depending on the availability of oxygen. The function of monooxygenase and dioxygenase enzymes in the oxidation of aromatic rings depends on the transitory initial stages of aerobic metabolism of PAH oxygen (Sihag et al., 2014). Ferrous iron, sulphates and nitrates in their substituted forms are required as electron acceptors during the anaerobic oxidation of aromatic compounds. However, the process produces significant phosphorus levels and ferrous ions, which on their own contaminates the ecosystem too.

7.16 ADVANCES IN MICROBIAL BIOREMEDIATION

The aspects covered previously have a lot of promise and offer a lot of benefits, but there are still some obstacles to be overcome. Approaches like nanotechnology and genetic engineering constitute two contemporary biotechnological tools that can be exploited to boost the effectiveness of traditional bioremediation methods (Giri et al., 2021). Nano-bioremediation illustrates a novel approach encircling an affiliation of nanoparticles (NPs) and microbes to remediate noxious pollutants. Nanoparticles can be employed directly or via adsorption to eliminate contaminants through the promotion of microbial growth, the formation of biosurfactants and enhanced enzymatic activities (Bhatt et al., 2022). The application of artificial microbial consortia, plasmid mediation and enzymes amplifies the potential for pollutant remediation (Sellami et al., 2022). The enzymatic complex exuded by basidiomycetes is effective at breaking down tenacious pesticides like pentachlorophenol and polychlorinated biphenyls as well as polyphenolic substances like lignin (PCBs) (Bentil, 2021). Recent advances in system and synthetic biology enable the creation of more effective microbes possessing elevated degrading capacities (Tran et al., 2021). *Arabidopsis thaliana*, genetically modified with an organophosphate hydrolase gene procured from *Pseudomonas* sp. marked an amplified compliance to chlorpyrifos signifying the prospective of these transgenics for remediation (Asraful Islam et al., 2021). In parallel, the successful construction of a recombinant bacterium employing promoter engineering exhibited a robust capability of mineralizing the persistent residues of para-nitrophenol (Huo et al., 2022).

7.17 CONCLUSION

Heavy metal and pesticide residue accrual in terrestrial and aquatic ecospheres has magnified on account of human interferences and amplified the persistence of toxic remnants at the expense of human and environmental health. Therefore, a call for robust monitoring and assessment techniques is buzzing, which must be followed by proper remediation approaches. The process of remediating environmental contaminants in soil is enduring and laborious. However, the technology is relentlessly evolving, from a lone remediation approach exhibiting solo effect to an amalgam of benefits. High-cost techniques possessing hefty side effects are replaced with green, low-cost and eco-friendly approaches, where novel and more proficient restoration tools are incessantly synthesized from conventional resources. In contrast to the physical and chemical approaches of remediation, microbial remediation reflects an edge for being carried out in situ, saving treatment expenditure and having no ill imprints on the environment apart from reaping the benefits financially too. However, the bottlenecks encasing specificity and high time consumption need to be addressed at a finer scale using advanced approaches such as omics, gene editing tools or nano-remediation. Consequently, significant aspects such as time, environment and economy must be taken into consideration while employing any remediation technique for an efficient remediation plan.

REFERENCES

Adams, G. O., Fufeyin, P. T., Okoro, S. E., & Ehinomen, I. (2015). Bioremediation, biostimulation and bioaugmention: A review. *International Journal of Environmental Bioremediation & Biodegradation*, *3*(1), 28–39.

Ahmad, R. G., & Kumar, V. (2020). Microorganism based biosensors to detect soil pollutants. *Plant Archives, 20*(2), 2509–2516.

Akar, T., Tunali, S., & Kiran, I. (2005). Botrytis cinerea as a new fungal biosorbent for removal of Pb (II) from aqueous solutions. *Biochemical Engineering Journal, 25*(3), 227–235.

Alengebawy, A., Abdelkhalek, S. T., Qureshi, S. R., & Wang, M. Q. (2021). Heavy metals and pesticides toxicity in agricultural soil and plants: Ecological risks and human health implications. *Toxics, 9*(3), 42.

Al-Homaidan, A. A., Al-Houri, H. J., Al-Hazzani, A. A., Elgaaly, G., & Moubayed, N. M. (2014). Biosorption of copper ions from aqueous solutions by Spirulina platensis biomass. *Arabian Journal of Chemistry, 7*(1), 57–62.

Aluko, O. A., Opoku, E. E. O., & Ibrahim, M. (2021). Investigating the environmental effect of globalization: Insights from selected industrialized countries. *Journal of Environmental Management, 281*, 111892.

An, J., Jeong, S., Moon, H. S., Jho, E. H., & Nam, K. (2012). Prediction of Cd and Pb toxicity to Vibrio fischeri using biotic ligand-based models in soil. *Journal of Hazardous Materials, 203*, 69–76.

Anjum, F., Shahid, M., & Akcil, A. (2012). Biohydrometallurgy techniques of low grade ores: A review on black shale. *Hydrometallurgy, 117*, 1–12.

Asghari, I., Mousavi, S. M., Amiri, F., & Tavassoli, S. (2013). Bioleaching of spent refinery catalysts: A review. *Journal of Industrial and Engineering Chemistry, 19*(4), 1069–1081.

Asha, S., & Vidyavathi, M. (2009). Cunninghamella–A microbial model for drug metabolism studies–A review. *Biotechnology Advances, 27*(1), 16–29.

Asim, M., & Rao, K. (2021). Assessment of heavy metal pollution in Yamuna River, Delhi-NCR, using heavy metal pollution index and GIS. *Environmental Monitoring and Assessment, 193*(2), 1–16.

Asraful Islam, S. M., Yeasmin, S., & Saiful Islam, M. (2021). Organophosphorus pesticide tolerance of transgenic Arabidopsis thaliana by bacterial oph B gene encode organophosphorus hydrolase. *Journal of Environmental Science and Health, Part B, 56*(12), 1051–1056.

Ata, A., Nalcaci, O. O., & Ovez, B. (2012). Macro algae Gracilaria verrucosa as a biosorbent: A study of sorption mechanisms. *Algal Research, 1*(2), 194–204.

Ayangbenro, A. S., & Babalola, O. O. (2017). A new strategy for heavy metal polluted environments: A review of microbial biosorbents. *International Journal of Environmental Research and Public Health, 14*(1), 94.

Azubuike, C. C., Chikere, C. B., & Okpokwasili, G. C. (2016). Bioremediation techniques—Classification based on site of application: Principles, advantages, limitations and prospects. *World Journal of Microbiology and Biotechnology, 32*(11), 1–18.

Bankole, M. T., Abdulkareem, A. S., Mohammed, I. A., Ochigbo, S. S., Tijani, J. O., Abubakre, O. K., & Roos, W. D. (2019). Selected heavy metals removal from electroplating wastewater by purified and polyhydroxylbutyrate functionalized carbon nanotubes adsorbents. *Scientific Reports, 9*(1), 1–19.

Becerra-Castro, C., Prieto-Fernández, Á., Kidd, P. S., Weyens, N., Rodríguez-Garrido, B., Touceda-González, M., . . . & Vangronsveld, J. (2013). Improving performance of Cytisus striatus on substrates contaminated with hexachlorocyclohexane (HCH) isomers using bacterial inoculants: Developing a phytoremediation strategy. *Plant and Soil, 362*, 247–260.

Bellinger, E. G., & Sigee, D. C. (2010). Introduction to freshwater algae. *Freshwater algae: Identification and use as bioindicators*, 1–40.

Bentil, J. A. (2021). Biocatalytic potential of basidiomycetes: Relevance, challenges and research interventions in industrial processes. *Scientific African, 11*, e00717.

Bhatt, P., Pandey, S. C., Joshi, S., Chaudhary, P., Pathak, V. M., Huang, Y., . . . & Chen, S. (2022). Nanobioremediation: A sustainable approach for the removal of toxic pollutants from the environment. *Journal of Hazardous Materials, 427*, 128033.

Biswas, K. (2015). Biological agents of bioremediation: A concise review. *Frontiers in Environmental Microbiology, 1*(3).

Bouskill, N. J., Barnhart, E. P., Galloway, T. S., Handy, R. D., & Ford, T. E. (2007). Quantification of changing Pseudomonas aeruginosa sodA, htpX and mt gene abundance in response to trace metal toxicity: A potential in situ biomarker of environmental health. *FEMS Microbiology Ecology, 60*(2), 276–286.

Briceno, G., Schalchli, H., Rubilar, O., Tortella, G. R., Mutis, A., Benimeli, C. S., . . . & Diez, M. C. (2016). Increased diazinon hydrolysis to 2-isopropyl-6-methyl-4-pyrimidinol in liquid medium by a specific Streptomyces mixed culture. *Chemosphere, 156*, 195–203.

Cepoi, L., Zinicovscaia, I., Valuta, A., Codreanu, L., Rudi, L., Chiriac, T., . . . & Peshkova, A. (2021). Bioremediation capacity of edaphic cyanobacteria Nostoc linckia for chromium in association with other heavy-metals-contaminated soils. *Environments, 9*(1), 1.

Chabukdhara, M., & Nema, A. K. (2012). Assessment of heavy metal contamination in Hindon River sediments: A chemometric and geochemical approach. *Chemosphere, 87*(8), 945–953.

Chang, J., Si, G., Dong, J., Yang, Q., Shi, Y., Chen, Y., . . . & Chen, J. (2021). Transcriptomic analyses reveal the pathways associated with the volatilization and resistance of mercury (II) in the fungus Lecythophora sp. DC-F1. *Science of the Total Environment, 752*, 142172.

Chaphekar, S. B. (1991). An overview on bio-indicators. *Journal of Environmental Biology, 12*, 163–168.

Chavan, M., Thacker, N. P., & Tarar, J. L. (2016). Toxicity evaluation of pesticide industry wastewater through fish bioassay. *International Journal of Applied Sciences, 3*(3), 331–339.

Chellaiah, E. R. (2018). Cadmium (heavy metals) bioremediation by Pseudomonas aeruginosa: A minireview. *Applied Water Science, 8*(6), 1–10.

Chopra, A. K., Sharma, M. K., & Chamoli, S. (2011). Bioaccumulation of organochlorine pesticides in aquatic system—An overview. *Environmental Monitoring and Assessment, 173*, 905–916.

Colin, T., Meikle, W. G., Wu, X., & Barron, A. B. (2019). Traces of a neonicotinoid induce precocious foraging and reduce foraging performance in honey bees. *Environmental Science & Technology, 53*(14), 8252–8261.

Costa, M. (2019). Review of arsenic toxicity, speciation and polyadenylation of canonical histones. *Toxicology and Applied Pharmacology, 375*, 1–4.

Cristani, M., Naccari, C., Nostro, A., Pizzimenti, A., Trombetta, D., & Pizzimenti, F. (2012). Possible use of Serratia marcescens in toxic metal biosorption (removal). *Environmental Science and Pollution Research, 19*(1), 161–168.

Derbalah, A., Khattab, I., & Saad Allah, M. (2020). Isolation and molecular identification of Aspergillus flavus and the study of its potential for malathion biodegradation in water. *World Journal of Microbiology and Biotechnology, 36*(7), 1–11.

Dursun, A. Y., Uslu, G., Cuci, Y., & Aksu, Z. (2003). Bioaccumulation of copper (II), lead (II) and chromium (VI) by growing Aspergillus niger. *Process Biochemistry, 38*(12), 1647–1651.

Eapen, S., Singh, S., & D'souza, S. F. (2007). Advances in development of transgenic plants for remediation of xenobiotic pollutants. *Biotechnology Advances, 25*(5), 442–451.

Egbosiuba, T. C., Abdulkareem, A. S., Kovo, A. S., Afolabi, E. A., Tijani, J. O., Auta, M., & Roos, W. D. (2020). Ultrasonic enhanced adsorption of methylene blue onto the optimized surface area of activated carbon: Adsorption isotherm, kinetics and thermodynamics. *Chemical Engineering Research and Design, 153*, 315–336.

Ekelund, N. G., & Hader, D. P. (2018). Environmental monitoring using bioassays. In *Bioassays* (pp. 419–437). Elsevier.

Fai, P. B., Grant, A., & Reid, B. (2007). Chlorophyll a fluorescence as a biomarker for rapid toxicity assessment. *Environmental Toxicology and Chemistry: An International Journal, 26*(7), 1520–1531.

Fang, L., Zhou, C., Cai, P., Chen, W., Rong, X., Dai, K., . . . & Huang, Q. (2011). Binding characteristics of copper and cadmium by cyanobacterium Spirulina platensis. *Journal of Hazardous Materials, 190*(1–3), 810–815.

Fischer, G., Tubiello, F. N., Van Velthuizen, H., & Wiberg, D. A. (2007). Climate change impacts on irrigation water requirements: Effects of mitigation, 1990–2080. *Technological Forecasting and Social Change, 74*(7), 1083–1107.

Fu, Y. Q., Li, S., Zhu, H. Y., Jiang, R., & Yin, L. F. (2012). Biosorption of copper (II) from aqueous solution by mycelial pellets of Rhizopus oryzae. *African Journal of Biotechnology, 11*(6), 1403–1411.

Galiulin, R. V., & Galiulina, R. A. (2008). Removing heavy metals from soil with plants. *Herald of the Russian Academy of Sciences, 78*(2), 141–143.

Gan, W., He, Y., Zhang, X., Shan, Y., Zheng, L., & Lin, Y. (2012). Speciation analysis of heavy metals in soils polluted by electroplating and effect of washing to the removal of the pollutants. *Journal of Ecology and Rural Environment, 28*(1), 82–87.

Gavrilescu, M. (2004). Removal of heavy metals from the environment by biosorption. *Engineering in Life Sciences, 4*(3), 219–232.

Gazwi, H. S., Yassien, E. E., & Hassan, H. M. (2020). Mitigation of lead neurotoxicity by the ethanolic extract of Laurus leaf in rats. *Ecotoxicology and Environmental Safety, 192*, 110297.

Giri, B. S., Geed, S., Vikrant, K., Lee, S. S., Kim, K. H., Kailasa, S. K., . . . & Singh, R. S. (2021). Progress in bioremediation of pesticide residues in the environment. *Environmental Engineering Research, 26*(6).

Gokçe, D. (2016). Algae as an indicator of water quality. *Algae-Organisms for Imminent Biotechnology*, 81–101.

Goutner, V., Frigis, K., Konstantinou, I. K., Sakellarides, T. M., & Albanis, T. A. (2012). Organochlorine pesticide residue concentrations and accumulation patterns in waterbirds and in their prey at Lake Kerkini, a Ramsar wetland, Greece. *Journal of Biological Research, 17*.

Gu, J. D., & Cheung, K. H. (2001). Phenotypic expression of Vogesella indigofera upon exposure to hexavalent chromium, Cr6+. *World Journal of Microbiology and Biotechnology, 17*, 475–480. https://doi.org/10.1023/A:1011917409139

Gupta, V. K., Nayak, A., & Agarwal, S. (2015). Bioadsorbents for remediation of heavy metals: Current status and their future prospects. *Environmental Engineering Research, 20*(1), 1–18.

Gupta, V. K., & Rastogi, A. (2009). Biosorption of hexavalent chromium by raw and acid-treated green alga Oedogonium hatei from aqueous solutions. *Journal of Hazardous Materials, 163*(1), 396–402.

Huo, K., Liu, Y., Huang, R., Zhang, Y., Liu, H., Che, Y., & Yang, C. (2022). Development of a novel promoter engineering-based strategy for creating an efficient para-nitrophenol-mineralizing bacterium. *Journal of Hazardous Materials, 424*, 127672.

Hussain, A., Madan, S., & Madan, R. (2021). Removal of heavy metals from wastewater by adsorption. In: *Heavy metals—Their environmental impacts and mitigation.* IntechOpen.

Jafari, M., Abdollahi, H., Shafaei, S. Z., Gharabaghi, M., Jafari, H., Akcil, A., & Panda, S. (2019). Acidophilic bioleaching: A review on the process and effect of organic— Inorganic reagents and materials on its efficiency. *Mineral Processing and Extractive Metallurgy Review, 40*(2), 87–107.

Jhansi, S. C., & Mishra, S. K. (2013). Wastewater treatment and reuse: Sustainability options. *Consilience,* (10), 1–15.

Kabra, A. N., Ji, M. K., Choi, J., Kim, J. R., Govindwar, S. P., & Jeon, B. H. (2014). Toxicity of atrazine and its bioaccumulation and biodegradation in a green microalga, Chlamydomonas mexicana. *Environmental Science and Pollution Research, 21*(21), 12270–12278.

Karigar, C. S., & Rao, S. S. (2011). Role of microbial enzymes in the bioremediation of pollutants: A review. *Enzyme Research, 2011.*

Khalef, R. N., Hassan, A. I., & Saleh, H. M. (2022). Heavy Metal's Environmental Impact. In IntechOpen eBooks. https://doi.org/10.5772/intechopen.103907

Kharabsheh, H. A., Han, S., Allen, S., & Chao, S. L. (2017). Metabolism of chlorpyrifos by Pseudomonas aeruginosa increases toxicity in adult zebrafish (Danio rerio). *International Biodeterioration & Biodegradation, 121,* 114–121.

Kim, K. H., Kabir, E., & Jahan, S. A. (2017). Exposure to pesticides and the associated human health effects. *Science of the Total Environment, 575,* 525–535.

Kong, L., Zhu, S., Zhu, L., Xie, H., Su, K., Yan, T., . . . & Sun, F. (2013). Biodegradation of organochlorine pesticide endosulfan by bacterial strain Alcaligenes faecalis JBW4. *Journal of Environmental Sciences, 25*(11), 2257–2264.

Kress, N. (2019). Actual impacts of seawater desalination on the marine environment reported since 2001. *Mar. Impacts Seawater Desalin,* 81–133.

Kurade, M. B., Kim, J. R., Govindwar, S. P., & Jeon, B. H. (2016). Insights into microalgae mediated biodegradation of diazinon by Chlorella vulgaris: Microalgal tolerance to xenobiotic pollutants and metabolism. *Algal Research, 20,* 126–134.

Leong, Y. K., & Chang, J. S. (2020). Bioremediation of heavy metals using microalgae: Recent advances and mechanisms. *Bioresource Technology, 303,* 122886.

Lerebours, A., To, V. V., & Bourdineaud, J. P. (2016). Danio rerio ABC transporter genes abcb3 and abcb7 play a protecting role against metal contamination. *Journal of Applied Toxicology, 36*(12), 1551–1557.

Li, J., Margaret Oliver, I., Cam, N., Boudier, T., Blondeau, M., Leroy, E., . . . & Benzerara, K. (2016). Biomineralization patterns of intracellular carbonatogenesis in cyanobacteria: Molecular hypotheses. *Minerals, 6*(1), 10.

Li, F., Zheng, Y., Tian, J., Ge, F., Liu, X., Tang, Y., & Feng, C. (2019). Cupriavidus sp. strain Cd02-mediated pH increase favoring bioprecipitation of Cd2+ in medium and reduction of cadmium bioavailability in paddy soil. *Ecotoxicology and Environmental Safety, 184,* 109655.

Li, W. W., & Yu, H. Q. (2015). Stimulating sediment bioremediation with benthic microbial fuel cells. *Biotechnology Advances, 33*(1), 1–12.

Lin, W., Huang, Z., Li, X., Liu, M., & Cheng, Y. (2016). Bio-remediation of acephate—Pb (II) compound contaminants by Bacillus subtilis FZUL-33. *Journal of Environmental Sciences, 45,* 94–99

Liu, H., Gao, H., Wu, M., Ma, C., Wu, J., & Ye, X. (2020). Distribution characteristics of bacterial communities and hydrocarbon degradation dynamics during the remediation of petroleum-contaminated soil by enhancing moisture content. *Microbial Ecology, 80,* 202–211.

López-Serrano, M. J., Velasco-Muñoz, J. F., Aznar-Sánchez, J. A., & Román-Sánchez, I. M. (2020). Sustainable use of wastewater in agriculture: A bibliometric analysis of worldwide research. *Sustainability, 12*(21), 8948.

Masiá, A., Campo, J., Vázquez-Roig, P., Blasco, C., & Picó, Y. (2013). Screening of currently used pesticides in water, sediments and biota of the Guadalquivir River Basin (Spain). *Journal of Hazardous Materials, 263*, 95–104.

Medfu Tarekegn, M., Zewdu Salilih, F., & Ishetu, A. I. (2020). Microbes used as a tool for bioremediation of heavy metal from the environment. *Cogent Food & Agriculture, 6*(1), 1783174.

Mishra, A., & Malik, A. (2013). Recent advances in microbial metal bioaccumulation. *Critical Reviews in Environmental Science and Technology, 43*(11), 1162–1222.

Mishra, M., Singh, S. K., & Kumar, A. (2021). Environmental factors affecting the bioremediation potential of microbes. In: *Microbe mediated remediation of environmental contaminants* (pp. 47–58). Woodhead Publishing.

Mohan, S. V., Kisa, T., Ohkuma, T., Kanaly, R. A., & Shimizu, Y. (2006). Bioremediation technologies for treatment of PAH-contaminated soil and strategies to enhance process efficiency. *Reviews in Environmental Science and Bio/Technology, 5*(4), 347–374.

Mori, T., Wang, J., Tanaka, Y., Nagai, K., Kawagishi, H., & Hirai, H. (2017). Bioremediation of the neonicotinoid insecticide clothianidin by the white-rot fungus Phanerochaete sordida. *Journal of Hazardous Materials, 321*, 586–590.

Morris, J. C., Georgiou, I., Guenther, E., & Caucci, S. (2021). Barriers in implementation of wastewater reuse: Identifying the way forward in closing the loop. *Circular Economy and Sustainability, 1*(1), 413–433.

Nagvenkar, G. S., & Ramaiah, N. (2010). Arsenite tolerance and biotransformation potential in estuarine bacteria. *Ecotoxicology, 19*(4), 604–613.

Nanda, M., Kumar, V., & Sharma, D. K. (2019). Multimetal tolerance mechanisms in bacteria: The resistance strategies acquired by bacteria that can be exploited to 'clean-up'heavy metal contaminants from water. *Aquatic Toxicology, 212*, 1–10.

Naskar, A., Majumder, R., & Goswami, M. (2020). Bioaccumulation of Ni (II) on growing cells of Bacillus sp.: Response surface modeling and mechanistic insight. *Environmental Technology & Innovation, 20*, 101057.

Nath, S., Deb, B., & Sharma, I. (2018). Isolation of toxic metal-tolerant bacteria from soil and examination of their bioaugmentation potentiality by pot studies in cadmium-and lead-contaminated soil. *International Microbiology, 21*(1), 35–45.

Noroozi, M., Amoozegar, M. A., Pourbabaei, A. A., Naghavi, N. S., & Nourmohammadi, Z. (2017). Isolation and characterization of mercuric reductase by newly isolated halophilic bacterium, Bacillus firmus MN8. *Global Journal of Environmental Science and Management, 3*(4), 427–436.

Odobasic, A., Sestan, I., & Begic, S. (2019). Biosensors for determination of heavy metals in waters'. In: Rinken, T., & Kivirand, K. (Eds.), *Biosensors for environmental monitoring*. IntechOpen. https://doi.org/10.5772/intechopen.84139

Pan, X., Xu, T., Xu, H., Fang, H., & Yu, Y. (2017). Characterization and genome functional analysis of the DDT-degrading bacterium Ochrobactrum sp. DDT-2. *Science of the Total Environment, 592*, 593–599.

Pande, V., Pandey, S. C., Sati, D., Bhatt, P., & Samant, M. (2022). Microbial interventions in bioremediation of heavy metal contaminants in agroecosystem. *Frontiers in Microbiology, 13*, 824084.

Paoli, L., Corsini, A., Bigagli, V., Vannini, J., Bruscoli, C., & Loppi, S. (2012). Long-term biological monitoring of environmental quality around a solid waste landfill assessed with lichens. *Environmental Pollution, 161*, 70–75.

Pervaiz, I., Ahmad, S., Madni, M. A., Ahmad, H., & Khaliq, F. H. (2013). Microbial biotransformation: A tool for drug designing. *Applied Biochemistry and Microbiology, 49*, 437–450.

Povedano-Priego, C., Martín-Sánchez, I., Jroundi, F., Sánchez-Castro, I., & Merroun, M. L. (2017). Fungal biomineralization of lead phosphates on the surface of lead metal. *Minerals Engineering, 106*, 46–54.

Pratush, A., Kumar, A., & Hu, Z. (2018). Adverse effect of heavy metals (As, Pb, Hg, and Cr) on health and their bioremediation strategies: A review. *International Microbiology, 21*(3), 97–106.

Priatni, S., Ratnaningrum, D., Warya, S., & Audina, E. (2018, June). Phycobiliproteins production and heavy metals reduction ability of Porphyridium sp. In *IOP Conference Series: Earth and Environmental Science* (Vol. 160, p. 012006). IOP Publishing.

Qiao, W., Zhang, Y., Xia, H., Luo, Y., Liu, S., Wang, S., & Wang, W. (2019). Bioimmobilization of lead by Bacillus subtilis X3 biomass isolated from lead mine soil under promotion of multiple adsorption mechanisms. *Royal Society Open Science, 6*(2), 181701.

Raklami, A., Meddich, A., Oufdou, K., & Baslam, M. (2022). Plants—Microorganisms-based bioremediation for heavy metal cleanup: Recent developments, phytoremediation techniques, regulation mechanisms, and molecular responses. *International Journal of Molecular Sciences, 23*(9), 5031.

Ramos, J., Naya, L., Gay, M., Abián, J., & Becana, M. (2008). Functional characterization of an unusual phytochelatin synthase, LjPCS3, of Lotus japonicus. *Plant Physiology, 148*(1), 536–545.

Rani, K., & Dhania, G. (2014). Bioremediation and biodegradation of pesticide from contaminated soil and water—A noval approach. *International Journal of Current Microbiology and Applied Sciences, 3*(10), 23–33.

Rani, A., & Goel, R. (2009). Strategies for crop improvement in contaminated soils using metal-tolerant bioinoculants. *Microbial Strategies for Crop Improvement*, 85–104.

Riskuwa-Shehu, M. L., Ismail, H. Y., & Sulaiman, M. (2019). Biosorption of heavy metals by Oscillatoria species. *Microbiology Research Journal International, 27*, 1–8.

Ristić, S. S., Kosanić, M. M., Ranković, B. R., & Stamenković, S. S. (2017). Lichens as biological indicators of air quality in the urban area of Kuršumlija (Southern Serbia). *Kragujevac Journal of Science*, (39), 165–175.

Rodríguez-Tirado, V., Green-Ruiz, C., & Gómez-Gil, B. (2012). Cu and Pb biosorption on Bacillus thioparans strain U3 in aqueous solution: Kinetic and equilibrium studies. *Chemical Engineering Journal, 181*, 352–359.

Romera, E., González, F., Ballester, A., Blázquez, M. L., & Munoz, J. A. (2007). Comparative study of biosorption of heavy metals using different types of algae. *Bioresource Technology, 98*(17), 3344–3353.

Saez, J. M., Álvarez, A., Benimeli, C. S., & Amoroso, M. J. (2014). Enhanced lindane removal from soil slurry by immobilized Streptomyces consortium. *International Biodeterioration & Biodegradation, 93*, 63–69.

Saikia, N., & Gopal, M. (2004). Biodegradation of β-cyfluthrin by fungi. *Journal of Agricultural and Food Chemistry, 52*, 1220–1223.

Sayyed, R. Z., Seifi, S., Patel, P. R., Shaikh, S. S., Jadhav, H. P., & Enshasy, H. E. (2019). Siderophore production in groundnut rhizosphere isolate, Achromobacter sp. RZS2 influenced by physicochemical factors and metal ions. *Environmental Sustainability, 2*(2), 117–124.

Sellami, K., Couvert, A., Nasrallah, N., Maachi, R., Abouseoud, M., & Amrane, A. (2022). Peroxidase enzymes as green catalysts for bioremediation and biotechnological applications: A review. *Science of the Total Environment, 806*, 150500.

Senthil Kumar, P., Femina Carolin, C., & Varjani, S. J. (2018). Pesticides bioremediation. In: *Bioremediation: Applications for Environmental Protection and Management* (pp. 197–222). Springer.

Sher, S., & Rehman, A. (2019). Use of heavy metals resistant bacteria—A strategy for arsenic bioremediation. *Applied Microbiology and Biotechnology, 103*(15), 6007–6021.

Sihag, S., Pathak, H., & Jaroli, D. P. (2014). Factors affecting the rate of biodegradation of polyaromatic hydrocarbons. *Indian Journal of Pure & Applied Biosciences, 2*, 185202.

Silva, V., Mol, H. G., Zomer, P., Tienstra, M., Ritsema, C. J., & Geissen, V. (2019). Pesticide residues in European agricultural soils—A hidden reality unfolded. *Science of the Total Environment, 653*, 1532–1545.

Silva, P. V., Pinheiro, C., Morgado, R. G., Verweij, R. A., van Gestel, C. A., & Loureiro, S. (2022). Bioaccumulation but no biomagnification of silver sulfide nanoparticles in freshwater snails and planarians. *Science of the Total Environment*, 808, 151956.

Singh, R., Misra, V., & Singh, R. P. (2012). Removal of Cr (VI) by nanoscale zero-valent iron (nZVI) from soil contaminated with tannery wastes. *Bulletin of Environmental Contamination and Toxicology, 88*, 210–214.

Taran, M., Fateh, R., Rezaei, S., & Gholi, M. K. (2019). Isolation of arsenic accumulating bacteria from garbage leachates for possible application in bioremediation. *Iranian Journal of Microbiology, 11*(1), 60.

Tarfeen, N., Nisa, K. U., Hamid, B., Bashir, Z., Yatoo, A. M., Dar, M. A., . . . & Sayyed, R. Z. (2022). Microbial remediation: A promising tool for reclamation of contaminated sites with special emphasis on heavy metal and pesticide pollution: A review. *Processes, 10*(7), 1358.

Tayang, A., & Songachan, L. S. (2021). Microbial bioremediation of heavy metals. *Current Science, 120*(6), 00113891.

Thengodkar, R. R. M., & Sivakami, S. (2010). Degradation of chlorpyrifos by an alkaline phosphatase from the cyanobacterium Spirulina platensis. *Biodegradation, 21*(4), 637–644.

Tovar-Sánchez, E., Suarez-Rodríguez, R., Ramírez-Trujillo, A., Valencia-Cuevas, L., Hernández-Plata, I., & Mussali-Galante, P. (2019). The use of biosensors for biomonitoring environmental metal pollution. *Biosensors for Environmental Monitoring*, 1–24.

Tran, K. M., Lee, H. M., Thai, T. D., Shen, J., Eyun, S. I., & Na, D. (2021). Synthetically engineered microbial scavengers for enhanced bioremediation. *Journal of Hazardous Materials, 419*, 126516.

Tsaboula, A., Papadakis, E. N., Vryzas, Z., Kotopoulou, A., Kintzikoglou, K., & Papadopoulou-Mourkidou, E. (2019). Assessment and management of pesticide pollution at a river basin level part I: Aquatic ecotoxicological quality indices. *Science of the Total Environment, 653*, 1597–1611.

Ullah, S., Zuberi, A., Alagawany, M., Farag, M. R., Dadar, M., Karthik, K., . . . & Iqbal, H. M. (2018). Cypermethrin induced toxicities in fish and adverse health outcomes: Its prevention and control measure adaptation. *Journal of Environmental Management, 206*, 863–871.

Van Hamme, J. D., Singh, A., & Ward, O. P. (2003). Recent advances in petroleum microbiology. *Microbiology and Molecular Biology Reviews, 67*, 503–549.

Vasudevan, P., Padmavathy, V., Tewari, N., & Dhingra, S. C. (2001). *Biosorption of heavy metal ions. Journal of Scientific & Industrial Research* 60, 112–120

Velasco-Muñoz, J. F., Aznar-Sánchez, J. A., Batlles-delaFuente, A., & Fidelibus, M. D. (2019). Rainwater harvesting for agricultural irrigation: An analysis of global research. *Water, 11*(7), 1320.

Verma, S., & Kuila, A. (2019). Bioremediation of heavy metals by microbial process. *Environmental Technology & Innovation, 14*, 100369.

Vestola, E. A., Kuusenaho, M. K., Närhi, H. M., Tuovinen, O. H., Puhakka, J. A., Plumb, J. J., & Kaksonen, A. H. (2010). Acid bioleaching of solid waste materials from copper, steel and recycling industries. *Hydrometallurgy, 103*(1–4), 74–79.

Villaverde, J., Rubio-Bellido, M., Merchán, F., & Morillo, E. (2017). Bioremediation of diuron contaminated soils by a novel degrading microbial consortium. *Journal of Environmental Management, 188*, 379–386.

Wierzba, S. (2015). Biosorption of lead (II), zinc (II) and nickel (II) from industrial wastewater by Stenotrophomonas maltophilia and Bacillus subtilis. *Polish Journal of Chemical Technology, 17*(1).

Wołejko, E., Wydro, U., Odziejewicz, J. I., Koronkiewicz, A., & Jabło-ska-Trypuć, A. (2022). Biomonitoring of soil contaminated with herbicides. *Water, 14*(10), 1534.

Wuana, R. A., & Okieimen, F. E. (2011). Heavy metals in contaminated soils: A review of sources, chemistry, risks and best available strategies for remediation. *International Scholarly Research Notices, 2011*.

Xiao, P., Mori, T., Kamei, I., & Kondo, R. (2011). A novel metabolic pathway for biodegradation of DDT by the white rot fungi, Phlebia lindtneri and Phlebia brevispora. *Biodegradation, 22*(5), 859–867.

Yalçin, S., Sezer, S., & Apak, R. (2012). Characterization and lead (II), cadmium (II), nickel (II) biosorption of dried marine brown macro algae Cystoseira barbata. *Environmental Science and Pollution Research, 19*, 3118–3125.

Yang, S., Jin, P., Wang, X., Zhang, Q., & Chen, X. (2016). Phosphate recovery through adsorption assisted precipitation using novel precipitation material developed from building waste: Behavior and mechanism. *Chemical Engineering Journal, 292*, 246–254.

Yatoo, A. M., Ali, M., Baba, Z. A., & Hassan, B. (2021). Sustainable management of diseases and pests in crops by vermicompost and vermicompost tea: A review. *Agronomy for Sustainable Development, 41*(1), 1–26.

Yatoo, A. M., Ali, M., Zaheen, Z., Baba, Z. A., Ali, S., Rasool, S., . . . & Hamid, B. (2022). Assessment of pesticide toxicity on earthworms using multiple biomarkers: A review. *Environmental Chemistry Letters, 1*, 1–24.

Yue, Z. B., Li, Q., Li, C. C., Chen, T. H., & Wang, J. (2015). Component analysis and heavy metal adsorption ability of Extracellular Polymeric Substances (EPS) from sulfate reducing bacteria. *Bioresource Technology, 194*, 399–402.

Zhang, K., Xue, Y., Xu, H., & Yao, Y. (2019). Lead removal by phosphate solubilizing bacteria isolated from soil through biomineralization. *Chemosphere, 224*, 272–279.

Zhang, W. M., & Gu, S. F. (2007). Catalytic effect of activated carbon on bioleaching of low-grade primary copper sulflde ores. *Transactions of Nonferrous Metals Society of China, 17*(5), 1123–1127.

Zouboulis, A. I., Loukidou, M. X., & Matis, K. A. (2004). Biosorption of toxic metals from aqueous solutions by bacteria strains isolated from metal-polluted soils. *Process Biochemistry, 39*(8), 909–916.

8 Urban and Industrial Liquid Waste

Sources, Eco-Toxicological Effects, and Sustainable Management

Malik Ahsaf Aziz, Shayesta Islam,
Zaffar Mahdi Dar, Syed Hujjat ul Baligah,
and Burhan Hamid

8.1 INTRODUCTION

Waste can be delineated as material that is of no use to the user (Basu, 2009). Dijkema et al. (2000) defined waste as the substances for which people pay money to discard. Waste is produced as a byproduct of human actions and as a result of unproductive manufacturing processes leading to the squandering of economic resources (Cheremisinoff, 2003). A material may be labelled as waste by one class and may serve as a resource for another category of people (Dijkema et al., 2000). Besides the character of waste, the constitution of waste is important to define, because categorizing a substance as waste will serve as the basis for the legislation needed to protect the population and the area of processing and discarding of waste (DEFRA, 2011).

Liquid waste comprises wastewater, fats, oils or grease, used oils, liquids, solids, gases or sludges, and hazardous household liquids. This liquid waste is a springing issue globally owing to its tendency to infiltrate watersheds and contaminate drinking and groundwater caused by unscientific disposal and handling operations. The utilization of untreated wastewater has significant harmful impacts on human health, like the emergence of various diseases viz, cholera, plague, tuberculosis hepatitis, and diphtheria. Generally liquid waste is categorized as sewage water and non-sewage water. In general, wastewater has been categorized into two broad types: sewage wastewater and non-sewage wastewater (Schoen and Ashbolt, 2010). Sewage wastewater is generated from domestic activities, which are produced from hospitals, schools, hotels, restaurants, and public toilets, while non-sewage waste includes wastes generated from factories and industries. In addition to this, non-sewage water consists of storm and rainwater.

Liquid waste generated from industries is composed of numerous contaminants, which have hazardous effects on humans and aquatic organisms. The contaminants include metals like chromium, zinc, lead, copper, iron, cadmium, nickel, arsenic, and mercury, produced in the paint and dye (Wang and Yang, 2016; Lokhande et al., 2011), textile, pharmaceutical, paper, and fine chemical industries (Rajkumar and Palanivelu, 2004); phenol and phenolic compounds generated by oil refineries; and phenol–formaldehyde resin from bulk drug manufacturing industries. Petrochemical industries generate numerous pollutants, viz petroleum hydrocarbons, sulphides, aniline, naphthalenic acid, organochlorines, olefins, nitrobenzene, alkanes, and chloroalkanes, which are partly biodegradable owing to their complexity in composition and ineffective and slow biological treatment (Liu et al., 2014). The paper and pulp industry generates organic solids and organic materials as their main liquid pollutants. Suspended solids and highly organic materials are the major water pollutants released by the paper and pulp industry, and the properties of discharge vary according to the quality of the paper formed and pulp processing (Buyukkamaci and Koken, 2010; Lindholm-Lehto et al., 2015) Textile industries produce a number of compounds varying from heavy metals like Cr to surfactants, urea, ammonium nitrogen (NH_4–N), and nitrogenous and phosphorus compounds and bleaching agents comprising of hydrogen peroxide and chlorine, sodium silicate, and alkaline bases (Wang et al., 2011). The surface protectants with extraordinary surface action, stability, and oil–water repellence are perfluoroalkyl acids; among perfluoroalkyl acids, perfluorooctane sulphonate generated from textile treatment, metal plating, and semi-conductor industries and perfluorooctanoic acid generated by the fluoropolymer production and processing industries are hazardous for life (Liu et al., 2017). Wastewater with high salt content is derived from petroleum, leather, food processing, and agro-based industries (Lefebvre and Moletta, 2006).

The vital categories of wastewater are stormwater runoff, domestic, agricultural, and industrial.

8.2 CLASSIFICATION OF LIQUID WASTE

8.2.1 Classification of Liquid Waste Based on Source or Origin

The main categories of wastewater based on source are stormwater runoff, domestic, agricultural, and industrial (Figure 8.1).

1. **Stormwater runoff:** This is the principal source of water pollution as it contains numerous hazardous contaminants like plastics, pesticides, herbicides, heavy metals, and pathogens released from streets, industrial sites, construction sites, and other places.
2. **Domestic wastewater:** Water released as a result of residential activities is known as domestic waste. This can be waste produced from sanitary facilities called toilet waste or generated by other activities like cooking (Mara, 2013) Broadly, it is of three types, viz black, grey, and yellow wastewater (Friedler et al., 2013).
   a. **Black water**: Highly contaminated water having a high risk of disease. It is released from toilets, kitchen dishwashers, and so on. The pollut-

ants present are discarded papers, urine, faecal matter, soaps, and so on.

b. **Grey water**: This is less contaminated than black water, released from laundry and washing. This is black water without faecal matter, urine, and so on (Al-Jayyousi, 2003).

c. **Yellow water**: Wastewater other than black and grey.

3. **Agricultural wastewater**: The prime source of watcr pollution is runoff from agricultural fields. The utilization of fertilizers, herbicides, and pesticides in excess quantity is known to contaminate water bodies (Pedrero et al., 2010).

4. **Industrial wastewater**: Wastewater released from industrial processes, viz manufacturing, cleaning, and other commercial activities. The type of contaminant depends on the type of industry (Munter, 2003). This wastewater is composed of numerous toxic elements, viz Al, Cr, Pb, Zn, and Ni. All these inorganic compounds are noxious and impact health after getting accumulated. Industrial waste is classified into different types based on the type of industry.

a. **Electroplating**. Electroplating is the technique of depositing thin films of metals and metallic alloys on the substrate for decoration, protection from corrosion, and so on. This industry generates solid and liquid toxic chemicals and contains oil and inorganic plating chemicals, viz Ni, Zn, Pb, and Cr (Rastogi, 2019). These substances contaminate the underground water courses if disposed of without treatment.

b. **Leather tanning**: Tanning is the technique of conversion of protein of hide or skin into stable unputrefiable material. Globally in approximately 90% of tanning industries, Cr compounds are utilized as tanning agents. The wastewater generated from this industry contains numerous contaminants like unutilized chemicals, leached proteins, and other deterioration products, which give rise to a number of problems in the environment (Kanagraj et al., 2006).

c. **Batteries**: The industries producing batteries release a number of pollutants like Pb, Cd, Li, and other metals, which are hazardous. The battery industry utilizes various virulent metals like lead and zinc. The improper discarding practices lead to an increase in the amount of e-waste and are hazardous to the health of humans and the environment (Martinez et al., 2021).

d. **Pesticides**: The pesticide utilizes various types of hazardous and non-biodegradable wastes that are toxic to the environment. The hydrocarbons are degraded very slowly and remain intact for a longer duration in the environment. The organophosphate toxicity is very high and dangerous similar to arsenic, strychnine, and cyanide (Theriot and Grunden, 2012).

e. **Plastic industry**: Waste from the plastic industry has caused an environmental crisis worldwide owing to its long life and resistance to decomposition (UNEP, 2018). The treatment of plastic waste is cumbersome and poses many threats to soil health (Jambeck et al., 2014).

f. Textile dyeing: Textile industries produce different types of liquid effluents like chemicals and dyes used in clothing. This waste is characterized by high chemical oxygen demand (COD), biological oxygen demand (BOD), heavy metals, and so on. The quantity of waste generated from different sources is presented in Table 8.1.

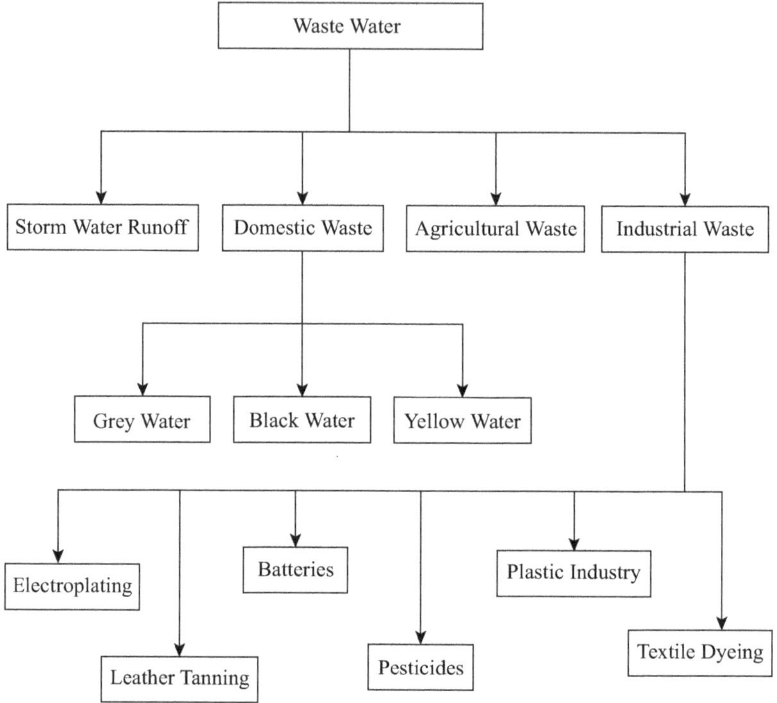

FIGURE 8.1   Flow chart showing different types of water

TABLE 8.1
Different sources of industrial waste and the quantity of waste generated (Chanakya et al., 2006)

Source of waste	Quantity generated per year
Tannery	52,500 m³ wastewater/day
Pulp/paper industry	1,600 m³ wastewater/day
Dairy industry	50–60 million litres/day
Willow dust	30,000 tonnes/year
Municipal liquid	12,145 million litres/day
Distillery	8,057 m³/day
Municipal solid	27.4 million tonnes/year
Pressmud	9 million tonnes/year
Food/fruit processing industry	4.5 million tonnes/year

8.2.2 CLASSIFICATION OF LIQUID WASTE BASED ON NATURE

1. **Physical**: The physical factors of waste include temperature, suspended matter, and colour.
2. **Microbiological**: The microorganisms present in waste are bacteria, viruses, and protozoa.
3. **Chemical**: The chemical components of liquid waste include mineral pollutants like salts, heavy metals, or organic pollutants like pesticides, hydrocarbons, and solvents.

8.2.3 CLASSIFICATION OF LIQUID WASTE BASED ON TEMPORAL DISTRIBUTION

1. **Permanent**: These are wastes which are generated continuously like the percolation of leachates from wastes in landfills.
2. **Accidental**: These are wastes produced accidentally like the fracturing of pipes or tanks.
3. **Seasonal**: These are produced in a particular season due to agricultural practices like spraying of different plant protection chemicals.

8.2.4 CLASSIFICATION OF LIQUID WASTE BASED ON SPATIAL DISTRIBUTION

1. **Diffuse**: These are waste released from non-point sources like agricultural fields and sanitation practices.
2. **Localized**: These wastes are released from point sources like commercial, industrial, and residential areas.
3. **Linear**: These wastes include those generated from roads, railways, water bodies, and so on.

8.3 ECOTOXICOLOGICAL EFFECTS OF WASTE

Ecotoxicology is defined as the study of the baneful influence of pollutants on living organisms in the environment and comprises toxicology, ecology, and chemistry. Truhaut (1977), in his definition, included the influence on humans. Ecotoxicological effects are defined as alterations in the state of organisms or other biological organisms owing to exposure to various chemicals at various stages, viz sub-cellular, cellular, tissues, individuals, populations, communities, ecosystems, landscapes, and the biosphere.

8.3.1 EFFECTS OF HEAVY METALS IN WASTEWATER FROM INDUSTRIES

Heavy metals include metals or metalloids possessing a density greater than 5 g/cm^3, atomic number of more than 20, along with conductivity characteristics and lustre surface. Heavy metal contamination is a significant environmental issue. The unique properties of heavy metals like bioaccumulation, persistent nature, non-biodegradability, contamination of food chains, and being noxious to living organisms make the contamination of the environment a matter of concern (Jaishankar et al., 2014; Ali et al., 2019; Rehman et al., 2021). Heavy metals are one of the critical contaminants that are conservative and endure in the environment for quite a long duration. Ingestion of these metals causes health hazards in lower tropic levels and finally impacts humans via the food chain. Heavy metals are teratogenic and cancer-causing, cause oxidative stress, impair organs, impede the nervous system, and reduce development (Bilal et al., 2013). The effects of various metals are as follows:

Cadmium: Cadmium is released from various industries like plastic, colour pigment, alloys, glass production, and electroplating industries, as well as from mining or smelting processes (Jaishankar et al., 2014; Balali-Mood et al., 2021; Kaur et al., 2019; Al-Lami et al., 2020; Tchounwou et al., 2012; Rashid et al., 2013). It enters into the human body either through ingestion or inhalation (Rehman et al., 2021; Balali-Mood et al., 2021; Aliasgharpour, 2020; Al-Lami et al., 2020; Tchounwou et al., 2012). The toxic effects of Cd rely on the amount of exposure. Divalent Cd is the most dangerous form and interferes with the basic functions of the cell (Rehman et al., 2021; Tchounwou et al., 2012). Cadmium attacks internal organs, particularly kidneys, resulting in their impairment (Rehman et al., 2021; Wu et al., 2016; Al-Lami et al., 2020; Tsutsumi et al., 2014). Long-term exposure results in numerous disorders like emphysema, depletion in olfactory function, anaemia, osteoporosis, renal disorders, anosmia, chronic rhinitis, and nervous damage by altering the levels of serotonin, norepinephrine, or acetylcholine (Rehman et al., 2021; Tchounwou et al., 2012).

Arsenic: Arsenic exists in two forms, viz. trivalent and pentavalent; trivalent form is 2–10-fold more dangerous than the pentavalent form. It is a carcinogenic heavy metal and has resulted in a number of deaths due to the cancer of various internal organs like kidneys and the bladder (Rehman et al., 2021; Kaur et al., 2019; Aliasgharpour, 2020; Tchounwou et al., 2012). The medical condition caused by long-duration exposure is called arsenicosis. It has noxious effects even at lower levels like nausea and vomiting, depletion of RBCs and WBCs, unusual heartbeat, and destruction of blood vessels (Jaishankar et al., 2014). It can affect the foetus owing to its ability to pass the placenta, resulting in abortion, mortality of the foetus, and so on (Aliasgharpour, 2020; Balali-Mood et al., 2021; Quansah et al., 2015). Moreover, it has negative effects on cognitive development, intelligence, and memory (Aliasgharpour, 2020; Tolins et al., 2014).

Mercury: Mercury is utilized in industries like mining, electrical industry, lamp production, measuring equipment, and nuclear reactors (Balali-Mood et al., 2021; Tchounwou et al., 2012; Jaishankar et al., 2014; Pratush et al., 2018; Morais et al., 2012). It enters into the body through inhalation or ingestion (Wu et al., 2016, 22). Elemental and methyl mercury is usually absorbed (Balali-Mood et al., 2021; Tchounwou et al., 2012). Elemental mercury is soluble in fats; 80% is assimilated by the mouth lining and lungs before reaching the cells; it has the tendency to pass the placental and blood–brain barrier and causes neurotoxicity. It is slowly absorbed by the gastrointestinal tract, and thus the toxicity is less there (Balali-Mood et al.,

2021). Methyl mercury is absorbed in ease in contrast to elemental mercury as well as transported in the body, where it attaches to thiol groups, such as cysteine, to form compounds that have the capacity to cross the blood–brain barrier (Balali-Mood et al., 2021; Wu et al., 2016; Bridges and Zalups, 2017). Exposure to methyl mercury causes damage to the central nervous system (Wu et al., 2016).

Zinc: Humans take Zn metal through ingestion. Excess ingestion of Zn impedes the metabolism of other elements like Cu. The toxic effects of Zn are low plasma copper and plasma caeruloplasmin levels, anaemia, changes in serum lipid levels, and impaired immune response (Strachan, 2010; Agnew and Slesinger, 2021).

Chromium: Chromium exists in two ionic forms: Cr3+ and Cr6+. The Cr6+ form, which is inorganic, poses significant health risks, particularly through inhalation and skin contact, which are more toxic than ingestion. When chromium compounds combine with oily scum, they form colloidal matter that blocks sunlight from penetrating water bodies, leading to decreased oxygen levels in the water. Environmental damage from chromium is primarily caused by oil burning, reaction promoters, pigment production, tannery industries, fertilizers, and sewage, due to its extensive use across various industries. (Balali-Mood et al., 2021; Kaur et al., 2019; Tchounwou et al., 2012).

Nickel: Ni intake usually occurs through cigarette smoking and affects respiratory organs, the skin, and the immune system. Exposure through inhalation is riskier than ingestion. The toxicity of Ni varies according to its form. Nickel attacks mostly respiratory organs like the nasal cavities and sinuses; moreover, it is detrimental to the immune system and causes skin diseases (Rehman et al., 2021; Aliasgharpour, 2020). Splitting of DNA strand splitting, oxidation, cross-linking of DNA protection, nucleotide deletion, mutation of genes, chromatid alteration, attachment to enzymes necessary for DNA repair and deterioration of proteins, production of reactive oxygen species (ROS), and so on are some of the effects of Ni at the cellular stage (Rehman et al., 2021).

Lead: Industrial processes like mining, fossil fuel combustion, and smelting are the major sources of lead (Aliasgharpour, 2020; Tchounwou et al., 2012). Organic lead is mainly absorbed through the skin. A major portion of lead absorbed by humans accumulates in bones (nearly 90%) and the rest in the kidneys and liver. Inorganic lead enters into the human body through ingestion, smoking, or inhaling (Tchounwou et al., 2012; Pratush et al., 2018). Malfunctioning of internal organs like the liver and kidney, dehydration, and frailty anaemia are some of the manifestations of Pb toxicity (Aliasgharpour, 2020).

Copper: Copper toxicity primarily occurs through ingestion, leading to long-term effects such as hepatitis, liver impairment, and jaundice. Significant copper accumulation in the liver can be fatal, especially in children. The gastrointestinal tract is the main organ system affected by copper exposure. Humans are exposed to copper through water pipes, plant protectants, cooking and eating utensils, and birth control pills. Copper has a high reduction potential, which results in the generation of reactive oxygen species (ROS). Additionally, copper binds to the thiol (SH) groups of proteins, altering liver enzymes involved in biological processes (Rehman et al., 2021).

8.3.2 Effect of Electroplating Wastewater

This waste is generated from electroplating industries. It carries many toxic metals, viz, Cr, Cd, CN, and other noxious chemicals that contaminate the environment and have pernicious impacts on humans and lower organisms like shrimps and fishes (Cai, 2010; Gao et al., 2013). According to a few reports, liquid waste from electroplating reduces the cellular immunity of mice, reflected by the loss of tendency to fight against disease-causing agents, but the effective component is not clearly known owing to the complexity of waste. The susceptibility of the immune system to electroplating wastewater varies according to the type of component. Cr metal has been reported to escalate the concentration of immunoglobulin M in juvenile rockfish (Kim and Kang, 2016). The amount of heavy metals also alters the immune response of organisms. For instance, a low level of nickel is known to decrease, while a high level is known to escalate immune responses of *Fabricius* larvae (Sun et al., 2011). Furthermore, escalation in the levels of lead in electronic waste recycling points has been reported to diminish the proportion of natural killer cells and blood hepatitis B surface antibody levels in children (Xu et al., 2015; Zhang et al., 2016b). The major pollutants of various industrial sectors are given in Table 8.2.

TABLE 8.2
Major pollutants from different industrial sectors

Industry	Major pollutants	Reference
Dye manufacturing	Copper, colour, salt, sulphides, formaldehydes	Lokhande et al. (2011); Wang et al. (2011)
Paint manufacturing	Chromium, zinc, lead, volatile organic compounds (VOCs)	Lokhande et al. (2011); Datta and Philip (2012)
Textile	Iron, chromium, chlorinated compounds, urea, salts, hydrogen peroxide, high pH NaOH, surfactants	Wang et al. (2011)
Pharmaceutical	Cadmium, nickel, phenolic compounds	Lokhande et al. (2011); Rajkumar and Palanivelu (2004)
Petrochemical	Petroleum hydrocarbons, phenolic compounds, nitrobenzene, alkanes, chloro alkanes, high salt, and so on.	Rajkumar and Palanivelu (2004)
Paper and pulp	Organic and chloro phenolic compounds, suspended solids, alternative oxides lignin, tannins, sterols, colours, biocides, and so on.	Buyukkamaci and Koken (2010); Lindholm-Lehto et al. (2015)
Metal working	Perfluorooctane sulphonate (PFOS), ammonium nitrogen, cyanide, phenol, oil, grease	Liu et al. (2017); Das et al. (2018)
Plastic	Perfluorooctanoic acid (PFOA), lead, mercury, cadmium, diethylhexyl phthalate	Das et al. (2018); Tang et al. (2015); Rowdhwal and Chen (2018).
Agriculture	Fertilizers, pesticides, insecticides	Aktar et al. (2009)

8.3.3 Effect of Liquid Waste on Aquatic Ecosystem

The wastewater from urban areas may contain essential nutrients, particularly N and P, which initiate the development of noxious varieties of phytoplanktons in both freshwater and saline water. These microscopic plankton release toxic substances that achieve undesirable levels during eutrophication and ultimately get concentrated in higher trophic levels when consumed by shellfish and other aquatic organisms. Examples of illnesses induced by noxious algae are paralytic shellfish poisoning, diarrheal shellfish poisoning, and amnesic shellfish poisoning (WHO, 2006). The anthropogenic components in waste liquid have the capacity to disturb the endocrine characteristics. It has been revealed that oestrogenic chemicals in liquid waste can destroy the endocrine functioning of aquatic organisms, thereby resulting in complete variation in the reproductive organs both structurally and functionally (Liney et al., 2006). In aquatic biomes of Europe, North America, and other areas, similar incidents have been reported in amphibians, terrestrial animals, and avis. Levels of metals in squishy parts of periwinkle have been reported from Delta areas of various rivers in Nigeria (Obasi et al., 2015). The recorded malformations in aquatic organisms range from minute to permanent variation, like disruption in differentiation of sex, variation in sexual behaviour, and alteration in immune function (Vos et al., 2000).

8.4 SUSTAINABLE WASTEWATER TREATMENT PROCESS

8.4.1 Biological Treatment

Biological treatment plays a vital role in the treatment of wastewater generated from municipality as well as industries containing dissolvable organic contaminants. Biological treatment of wastewater is of two types, viz aerobic and anaerobic. Aerobic treatment is accomplished in the presence of oxygen and oxygen-consuming microbes referred to as aerobes that utilize molecular oxygen to transform organic contaminants into carbon dioxide, water, and biomass. On the contrary, anaerobic treatment is accomplished devoid of oxygen by anaerobic microbes, leading to the formation of methane and carbon dioxide gas and biomass (Mittal, 2011). The simplified principles of the two processes are shown in Figure 8.2.

8.4.2 Membrane Technology

Membrane processes for the treatment of wastewater are classified into various categories depending on the substance to be segregated and the dimensions of the pores. Microfiltration (MF), ultrafiltration (UF), nanofiltration (NF), forward osmosis (FO), and reverse osmosis (RO) are some of the widely used membrane technologies.

MF: This is considered the first pretreatment of NF and RO to decrease the fouling potential. MF involves the separation of particles of size 0.1–0.2 µm (Qu et al., 2016; Xiaolei et al., 2013). It separates less or no organic matter and fails to remove dissolved substances; moreover, it does not obstruct viruses (Torki et al., 2017).

UF: This process separates substances of the size range 0.005 ≈ 10 µm, between MF and RO (Qu et al., 2014). These membranes utilize less energy for separating

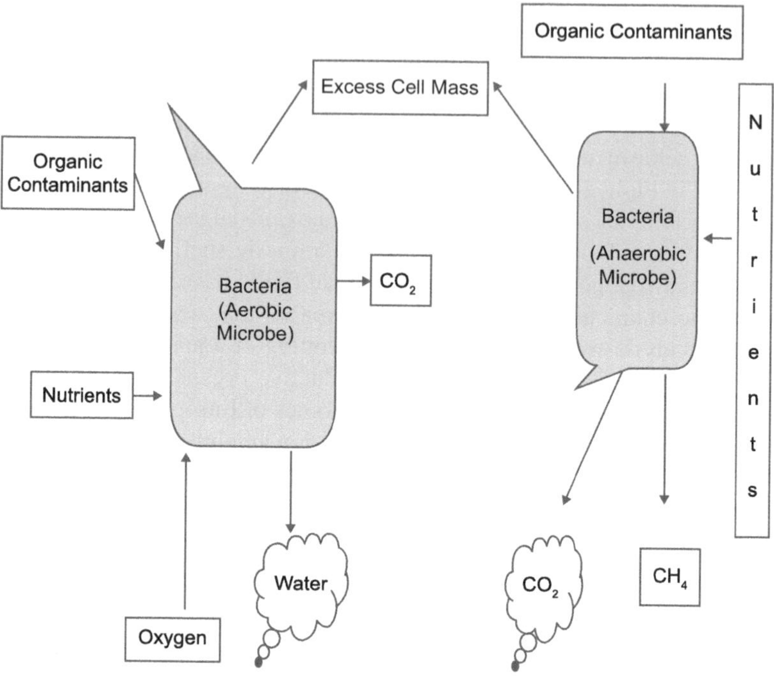

FIGURE 8.2   Principle of aerobic and anaerobic biological treatment of wastewater

pathogenic microorganisms, macromolecules, and suspended particles but cannot separate soluble substances (Krüger et al., 2016; Zhang et al., 2016a). Mocanu et al. synthesized a hybrid UF membrane for liquid waste treatment in which a wet phase inversion procedure was utilized with polysulphone and graphene nanoplatelets amended with poly (styrene). A membrane composed of dissolvable polymers served as a surface for the deposition of ZnO (Mocanu et al., 2017). They used the UF membrane, which was resistant to organic fouling in wastewater treatment technology (Igbinigun et al., 2016).

NF: This process can remove substances that have a remarkable contribution to the osmotic pressure, thus permitting working pressure less than RO; thus, the operational pressure is less than RO. For heavy contamination of liquid waste, efficient treatment is necessary. In the case of highly contaminated water, NF can be successful only if pretreatment is given. These membranes are subtle to free chlorine, and dissolvable substances cannot be isolated from water (Wang et al., 2016). An NF membrane formed by the combination of Poly (m-phenylene isophthalamide) (PMIA) and graphene oxide utilized for water treatment was developed by Yang et al. It has a more hydrophilic surface area in comparison to PMIA alone, thus providing high pure water flux (Yang et al., 2017). Xu et al. synthesized an NF membrane for treating liquid waste from textile plants with high efficiency to separate heavy metal ions, common salts, and dyes (Xu et al., 2016).

FO: This technique involves the passage of the solvent from the region of low concentration to higher concentration through the penetrable membrane (Ong et al., 2017). This is a reliable technique to solve global water problems.

RO: This technique is driven by pressure and can separate soluble solids and small-sized particles through permeable membranes. The pore size of the filter membrane is much smaller than UF and hence can separate particles of size less than 0.1 nm, enables the conversion of hard water to soft water, and has a tendency to separate all particles, bacteria, and organic substances (Wood et al., 2017). The membranes used are economical and susceptible to fouling (Liu et al., 2017). Huang and others revealed the separation of *Escherichia coli* and a decrease in BSA fouling by RO pellicle glazed with azide-functionalized graphene oxide, owing to its smooth, antibacterial, and hydrophilic properties (Huang et al., 2016).

8.4.3 Ion Exchange

Ion exchange treatment of the liquid waste is believed to reduce the levels of nitrate and phosphate; soluble substances are removed by adsorption and suspended by filtration. Elxassen et al. (1965) and Elias and Bennett (1967) used a strong base anion exchanger in the chloride form and pretreated effluent from an activated sludge plant; the results revealed that there was a substantial reduction in nitrate, phosphate, and organic matter (COD), and thus rejuvenation with brine was logically successful. The foul odour of effluents can be removed by periodic cleaning treatment. Ion exchange has been shown to eliminate sulphate from the effluents fully.

8.4.4 Advanced Oxidation Process

This is a promising, environmentally friendly, systematic simple, and cost-effective technique of water treatment to maintain the water quality (Feitz, 2005). It depends on the in-place production of chemical oxidants to sterilize water and deteriorate a variety of noxious organic pollutants (Giannakis et al., 2017a; Marjanovic et al., 2018; Nieto-Juarez and Kohn, 2013; Shabat-Hadas et al., 2017). Advanced oxidation processes commonly used are ozonation, ozonation coupled with H_2O_2 or UV rays, Fenton and similar reactions, photocatalysis activated by semiconductors such as TiO_2, sonolysis, electrochemical oxidation, and several blends of them (Figure 8.3). These can be utilized for pre- or post-treatment to a biological process and work on the principle of generation of ROS (Galeano et al., 2019; Kokkinos et al., 2011; Monteiro et al., 2015). Hydroxyl radical is the main oxidizing agent; other agents are hydroperoxyl and superoxide radicals, and so on (Giannakis et al., 2017a, 2017b, 2017c). This technique has an extraordinary capacity of disinfection against a broad spectrum of microbes, viz bacteria, fungi, viruses, protozoans, and yeasts, particularly through the activity of ROS, viz singlet and triplet oxygen, superoxides, hydroxyl and hydroperoxyl radical, and H_2O_2. ROS oxidizes various types of biomolecules like proteins, lipids, and nucleic acids (Galeano et al., 2019). The redox potential of sulphate radicals is nearly similar to hydroxyl radicals (2.6 V) and thus can be utilized as substitute to hydroxyl radicles, owing to more selective nature against numerous target pollutants. The formation of sulphate radical is

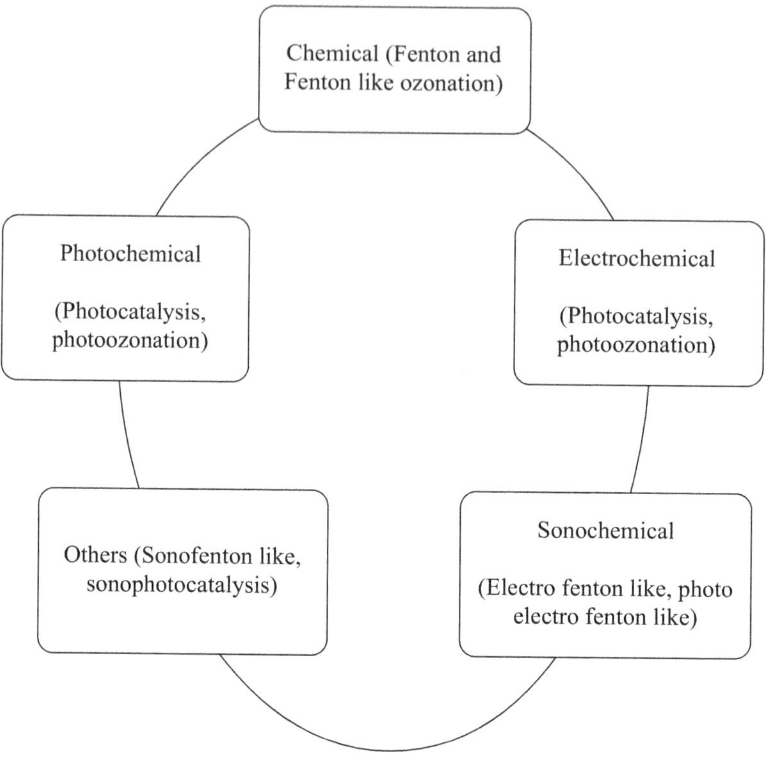

FIGURE 8.3   Different categories of advanced oxidation techniques having the potential for viral disinfection of wastewater

enhanced by factors, viz light, heat, and transition metal chemicals (Marjanovic et al., 2018).

8.4.5 EVAPORATION AND CRYSTALLIZATION

This is an age-old practice, approximately 5,000 years old, when it was utilized for the synthesis of table salt. In addition to this, it is utilized in all fields, viz food, medicine, and so on. Crystallization techniques for the treatment of wastewater are of various types like evaporation crystallization, cooling crystallization, reaction crystallization, drowning-out crystallization, and membrane distillation crystallization. These are commonly used for desalination, salt recovery, and water recovery. In this process the solution is heated, and the solute gets separated from the solvent in the form of solid crystals. In the case of unsaturated solution at a certain temperature, the heat is given from outside to evaporate the solvent; due to the evaporation of the solvent, the solution finally attains the saturation stage. This is a traditional technique for treating wastewater, particularly with high salt concentrations containing the following ions like Cl^-, SO_4^{2-}, Na^+, and Ca^{2+}. During the present time, owing to the maturity and high efficiency of multieffect evaporation crystallization, it is a widely used technique (Haijiao et al., 2017).

8.4.6 ZERO LIQUID DISCHARGE

This is the waste treatment technology that removes waste liquid leaving the plant periphery, and most of the water is recovered for reuse. This process maintains an equilibrium between overutilization of freshwater and the protection of aquatic sources by evading the pollution hazard associated with the release of wastewater and increasing the water use efficiency (Oren et al., 2010). This is a very economic process and requires a great amount of energy and thus is not regarded as a feasible approach and has restricted applications. The two global challenges, viz water shortage and pollution, in the present scenario, have driven interest towards zero liquid discharge (ZLD). More stern regulations, escalation in water dispersal expenditure, and growing demand for freshwater have led ZLD to be the essential and needful alternative management of wastewater (Oren et al., 2010).

8.4.7 WASTE STABILIZATION PONDS

These are large but shallow lagoons of distinct depths, in which the wastewater is treated by utilizing microorganisms including bacteria and algae. This is a commercial technique for treating wastewater in tropical and temperate areas and is regarded as an easy, authentic, and cost-effective method (Abdullahi et al., 2014). Since it requires only sunlight for operation, the intensity of solar radiation and duration of light hours in tropical areas provide outstanding opportunities for its extraordinary efficiency and performance. Moreover, it needs less management, which involves cleaning of inlet and outlet. These lagoons are more efficient in removing faecal coliform and are most suitable for low-income tropical countries where conventional wastewater treatment is not possible owing to a lack of a reliable energy source. It has been reported that anaerobic, facultative, and maturation ponds together were successful in reducing BOD, COD, and total soluble salts (TSS) to 50.65%, 48.95%, and 44.3%, respectively, of wastewater (Ghazy et al., 2008).

8.5 CONCLUSION

Wastewater from industries and urban areas is an impending issue owing to its tendency to enter water courses and contaminate ground- and drinking water when handled in improper manner. The wastewater, besides polluting the environment, causes numerous diseases in humans and lower organisms because of the presence of noxious substances in it. The oxygen concentration of water bodies gets depleted due to the influx of waste, leading to anoxic conditions that ultimately result in the death of aquatic flora and fauna. Thus, the scientific and sustainable management and disposal of liquid waste is the need of the hour. Sustainable waste management techniques have the potential to make a useful change in sanitation and hygiene behaviour and will assist in income generation through fish culture, and water can be reused for irrigation.

REFERENCES

Abdullahi, I., Naguri, I., Saminu, A., Sagir, L., & Charity, E. (2014). Design of waste stabilization pond for sewage at Nigerian defense academy staff quarter, permanent site Mando Kaduna. *International Journal of Engineering and Applied Sciences, 1*(2), 9–15.

Agnew, U. M., & Slesinger, T. L. (2021). *Zinc toxicity.* StatPearls Publishing.

Aktar, W., Sengupta, D., & Chowdhury, A. (2009). Impact of pesticides use in agriculture: Their benefits and hazards. *Interdisciplinary Toxicology, 2,* 1–12.

Ali, H., Khan, E., & Ilahi, I. (2019). Environmental chemistry and ecotoxicology of hazardous heavy metals: Environmental persistence, toxicity and bioaccumulation. *Journal of Chemistry,* 6730305.

Aliasgharpour, M. (2020). Trace elements in human nutrition (II)—An update. *International Journal of Preventive Medicine, 11*(2), 1–17.

Al-Jayyousi, O. R. (2003). Grey water reuse: Towards sustainable water management. *Desalination, 156,* 181–192.

Al-Lami, A. M. A., Khudhaier, S. R., & Aswad, O. A. (2020). Effects of heavy metals pollution on human health. *Annals of Tropical Medicine and Public Health, 23,* 1–4.

Balali-Mood, M., Naseri, K., Tahergorabi, Z., Khazdair, M. R., & Sadeghi, M. (2021). Toxic mechanisms of five heavy metals: Mercury, lead, chromium, cadmium and arsenic. *Frontiers in Pharmacology, 12,* 643–972.

Basu, R. (2009). Solid waste management-A model study. *Sies Journal of Management, 6,* 20–24.

Bilal, M., Shah, J. A., Ashfaq, T., Gardazi, S. M. H., Tahir, A. A., Pervez, A., Haroon, H., & Mahmood, Q. (2013). Waste biomass adsorbents for copper removal from industrial wastewater—A review. *Journal of Hazardous Materials, 263,* 322–333.

Bridges, C. C., & Zalups, R. K. (2017). Mechanisms involved in the transport of mercuric ions in target tissues. *Archives of Toxicology, 91*(1), 63–81.

Buyukkamaci, N., & Koken, E. (2010). Economic evaluation of alternative wastewater treatment plant options for pulp and paper industry. *Science of the Total Environment, 408,* 6070–6078.

Cai, Y. (2010). Effect of electroplating wastewater on human health and its centralized processing. *Agricultural Journal of Environmental Science, 29,* 205–208.

Chanakya, H. N., Ramachandra, T. V., & Vijayachamundeeswari, M. (2006). Anaerobic digestion and reuse of digested products of selected components of urban solid waste, CES and CST, IIS Bangalore. *Technical Reports, 114,* 1–109.

Cheremisinoff, N. P. (2003). *Handbook of solid waste management and waste minimization technologies [electronic resource].* Butterworth-Heinemann.

Das, P., Mondal, G. C., Singh, S., Singh, A. K., Prasad, B., & Singh, K. K. (2018). Effluent treatment technologies in the iron and steel industry-A state of the art review. *Water Environment Research, 90,* 395–408.

Datta, A., & Philip, L. (2012). Biodegradation of volatile organic compounds from paint industries. *Applied Biochemistry and Biotechnology, 167,* 564–580.

DEFRA. (2011). *Business waste prevention evidence review: WR1403.* http://randd.defra.gov.uk/Default.aspx?Menu=Menu&Module=More&Location=None&Completed=0&ProjectID=17499

Dijkema, G. P. J., Reuter, M. A., & Verhoef, E. V. (2000). A new paradigm for waste management. *Waste Management, 20*(8), 633–638.

Elias, S. R., & Bennett, G. E. (1967). Anion exchange and filtration techniques for waste water renovation. *Journal of the Water Pollution Control Federation, 39,* R82–R91.

Elxassen, R., Wyckoff, B. M., & Tonkm, C. D. (1965). Ion exchange for reclamation of reusable supplies. *Journal American Water Works Association, 57*, 1113–1122.

Feitz, A. (2005). Advanced oxidation processes and industrial wastewater treatment. *Water, 32*, 59–65.

Friedler, E., Butler, D., & Alfiya, Y. (2013). Wastewater composition. In: *Source separation and decentralization for wastewater management* (pp. 241–257). IWA Publishing.

Galeano, L. A., Guerrero-Flórez, M., Sánchez, C. A., Gil, A., & Vicente, M. Á. (2019). Disinfection by chemical oxidation methods. In: *Applications of Advanced Oxidation Processes (AOPs) in drinking water treatment* (Vol. 67, pp. 257–295). The Handbook of Environmental Chemistry. Springer.

Gao, C., Liu, Z., & Zhou, P. (2013). Electroplating industry environment analysis and pollution control counter Z. Q. Ma et al. 320 measures. *Environmental Protection, 5*, 206–207.

Ghazy, M. M. E. D., El-Senousy, W. M., Abdel-Aatty, A. M., & Kamel, M. (2008). Performance evaluation of a waste stabilization pond in a rural area in Egypt. *American Journal of Environmental Sciences, 4*(4), 316–325.

Giannakis, S., Liu, S., Carratala, A., Rtimi, S., Bensimon, M., & Pulgarin, C. (2017a). Effect of Fe (II)/Fe(III) species, pH, irradiance and bacterial presence on viral inactivation in wastewater by the photo-Fenton process: Kinetic modeling and mechanistic interpretation. *Applied Catalysis B: Environmental, 204*, 156–166.

Giannakis, S., Liu, S., Carratalà, A., Rtimi, S., Talebi Amiri, M., Bensimon, M., & Pulgarin, C. (2017b). Iron oxide-mediated semiconductor photocatalysis vs. heterogeneous photo-Fenton treatment of viruses in wastewater: Impact of the oxide particle size. *Journal of Hazardous Materials, 339*, 223–231.

Giannakis, S., Rtimi, S., & Pulgarin, C. (2017c). Light-assisted advanced oxidation processes for the elimination of chemical and microbiological pollution of wastewaters in developed and developing countries. *Molecules, 22*(7), 1070.

Haijiao, L., Jingkang, W., Ting, W., Wang, N., Bao, Y., & Hao, H. (2017). Crystallization techniques in wastewater treatment: An overview of applications. *Chemosphere, 173*, 474–484.

Huang, X., Marsh, K. L., McVerry, B. T., Hoek, E. M., & Kaner, R. B. (2016). Low-fouling antibacterial reverse osmosis membranes via surface grafting of graphene oxide. *ACS Applied Materials & Interfaces, 8*(23), 14334–14338.

Igbinigun, E., Fennell, Y., Malaisamy, R., Jones, K. L., & Morris, V. (2016). Graphene oxide functionalized polyethersulfone membrane to reduce organic fouling. *Journal of Membrane Science, 514*, 518–526.

Jaishankar, M., Tseten, T., Anbalagan, N., Mathew, B. B., & Beeregowda, K. (2014). Toxicity, mechanism and health effects of some heavy metals. *Interdisciplinary Toxicology, 7*(2), 60–72.

Jambeck, J. R., Geyer, R., Wilcox, C., Siegler, T. R., Perryman, M., Andrary, A, & Law, K. L. (2014). Plastic waste inputs from land into the ocean. *Science*, 1655–1734.

Kanagraj, J., Velappen, K. C., Chandra babu, N. K., & Sadulla, S. (2006). Solid waste generation in leather industry and its utilization for cleaner environment: A review. *Journal of Scientific and Industrial Research, 65*, 541–548.

Kaur, R., Sharma, S., & Kaur, N. (2019). Heavy metals toxicity and the environment. *Journal of Pharmacognosy and Phytochemistry, SP1*, 247–249.

Kim, J. H., & Kang, J. C. (2016). The toxic effects on the stress and immune responses in Juvenile Rockfish, sebastes schlegelii exposed to hexavalent chromium. *Environmental Toxicology and Pharmacology, 43*, 128–133.

Kokkinos, P. A., Ziros, P. G., Mpalasopoulou, G., Galanis, A., & Vantarakis, A. (2011). Molecular detection of multiple viral targets in untreated urban sewage from Greece. *Virology Journal, 8*, 195.

Krüger, R., Vial, D., Arifin, D., Weber, M., & Heijnen, M. (2016). Novel ultrafiltration membranes from low-fouling copolymers for RO pretreatment applications. *Desalination and Water Treatment: Wastewater Treatment Using Membrane Technology, 57*(48–49), 23185–23195.

Lefebvre, O., & Moletta, R. (2006). Treatment of organic pollution in industrial saline wastewater: A literature review. *Water Research, 40*, 3671–3682.

Lindholm-Lehto, P. C., Knuutinen, J. S., Ahkola, H. S., & Herve, S. H. (2015). Refractory organic pollutants and toxicity in pulp and paper mill wastewaters. *Environmental Science and Pollution Researc, 22*, 6473–6499.

Liney, K. E., Hagger, J. A., Tyler, C. R., Depledge, M. H., Galloway, T. S., & Jobling, S. (2006). Health effects in fish of long-term exposure to effluents from wastewater treatment works. *Environmental Health Perspectives, 114*(S-1), 81–89.

Liu, G., Han, K., Ye, H., Zhu, C., Gao, Y., Liu, Y., & Zhou, Y. (2017). Graphene oxide/triethanolamine modified titanate nanowires as photocatalytic membrane for water treatment. *Chemical Engineering Journal, 320*, 74–80.

Liu, S., Ma, Q., Wang, B., Wang, J., & Zhang, Y. (2014). Advanced treatment of refractory organic pollutants in petrochemical industrial wastewater by bioactive enhanced ponds and wetland system. *Ecotoxicology, 23*, 689–698.

Liu, Z., Lu, Y., Wang, P., Wang, T., Liu, S., Johnson, A. C., Sweetman, A. J., & Baninla, Y. (2017). Pollution pathways and release estimation of Perfluorooctane Sulfonate (PFOS) and Perfluorooctanoic Acid (PFOA) in central and eastern China. *Science of Total Environment, 580*, 1247–1256.

Lokhande, R. S., Singare, P. U., & Pimple, D. S. (2011). Toxicity study of heavy metals pollutants in waste water effluent samples collected from taloja industrial estate of Mumbai, India. *Resources, Environment and Sustainability, 1*, 13–19.

Mara, D. (2013). *Domestic wastewater treatment in developing countries.* Routledge.

Marjanovic, M., Giannakis, S., Grandjean, D., de Alencastro, L. F., & Pulgarin, C. (2018). Efect of µM Fe addition, mild heat and solar UV on sulfate radical-mediated inactivation of bacteria, viruses, and micropollutant degradation in water. *Water Research, 140*, 220–231.

Martinez, E. M., Garbett, R. M., Becerra, A. M., Iqbal, H. M. N., Hernandez, J. E. S., & Saldivar, R. P. (2021). Environmental impact of emerging contaminants from battery waste; A mini review. *Case Studies in Chemical and Environmental Engineering, 3*, 100–104.

Mittal, A. (2011). Biological wastewater treatment. *Water today, 1*, 32–44.

Mocanu, A., Rusen, E., Diacon, A., Damian, C., Dinescu, A., & Suchea, M. (2017). Electrochemical deposition of zinc oxide on the surface of composite membrane polysulfone-graphene-polystyrene in the presence of water soluble polymers. *Journal of Nanomaterials,* 11. https://doi.org/10.1155/2017/1401503

Monteiro, G. S., Staggemeier, R., Klauck, C. R., Bernardes, A. M., Rodrigues, M. A. S., & Spilki, F. R. (2015). Degradation and inactivation of adenovirus in water by photo-electro-oxidation. *Brazilian Journal of Biology, 75*(4), S37–S42.

Morais, S., Costa, F. G., & Pereira, M. L. (2012). Heavy metals and human health. In: Oosthuisen, J. (Ed.), *Environmental health—Emergencies issues and practice* (pp. 227–246). IntechOpen.

Munter, R. (2003). *Industrial wastewater characteristics* (pp. 185–194). The Baltic University Programme (BUP).

Nieto-Juarez, J. I., & Kohn, T. (2013). Virus removal and inactivation by iron (hydr)oxide-mediated Fenton-like processes under sunlight and in the dark. *Photochemical & Photobiological Sciences, 12*(9), 1596–1605.

Obasi, K. O., Chinedu, K., Udebuani, A. C., Okereke, J. N., Ezeji, E. U., & Anyadoh, S. O. (2015). Study on concentrations of selected heavy metals: Cadmium, lead, arsenic and mercury in the soft tissue of periwinkle (Tympanotonos fuscatus Var Radula) from Brass Island River, Bayelsa State, Nigeria. *International Journal of Science and Technology, 4*(10), 1–7.

Ong, C. S., Al-anzi, B., Lau, W. J., Goh, P. S., Lai, G. S., Ismail, A. F., & Ong, Y. S. (2017). Anti-fouling doubleskinned forward osmosis membrane with zwitterionic brush for oily wastewater treatment. *Scientific Reports, 7*(1), 1–11.

Oren, Y., Korngold, E., Daltrophe, N., Messalem, R., Volkman, Y., Aronov, L., & Gilron, J. (2010). Pilot studies on high recovery 541 BWRO-EDR for near zero liquid discharge approach. *Desalination, 261*(3), 321–330.

Pedrero, F., Kalavrouziotis, I., Alarcón, J. J., Koukoulakis, P., & Asano, T. (2010). Use of treated municipal wastewater in irrigated agriculture: Review of some practices in Spain and Greece. *Agricultural Water Management, 97*, 1233–1241.

Pratush, A., Kumar, A., & Hu, Z. (2018). Adverse effect of heavy metals (As, Pb, Hg, and Cr) on health and their bioremediation strategies: A review. *International Microbiology, 21*, 97–106.

Qu, F., Liang, H., Zhou, J., Nan, J., Shao, S., Zhang, J., & Li, G. (2014). Ultrafiltration membrane fouling caused by Extracellular Organic Matter (EOM) from Microcystis aeruginosa: Effects of membrane pore size and surface hydrophobicity. *Journal of Membrane Science, 449*, 58–66.

Qu, X., Alvarez, P., Werber, J. R., Deshmukh, A., & Elimelech, M. (2016). The critical need for increased selectivity, not increased water permeability, for desalination membranes. *Environmental Science & Technology Letters, 3*(4), 112–120.

Quansah, R., Armah, F. A., Essumang, D. K., Luginaah, I., Clarke, E., & Marfoh, K. (2015). Association of arsenic with adverse pregnancy outcomes/infant mortality: A systematic review and meta-analysis. *Environmental Health Perspectives, 123*, 412–421.

Rajkumar, D., & Palanivelu, K. (2004). Electrochemical treatment of industrial wastewater. *Journal of Hazardous Materials, 113*, 123–129.

Rashid, K., Sinha, K., & Sil, P. C. (2013). An uppdate on oxidative stress-mediated organ pathophysiology. *Food and Chemical Toxicology, 62*, 5.

Rastogi, R. (2019). Waste management process in electroplating industries. *International Journal of Advanced Research, 7*, 1611–1614.

Rehman, A. U., Nazir, S., Irshad, R., Tahir, K., Ur Rehman, K., & Ul Islam, R. (2021). Toxicity of heavy metals in plants and animals and their uptake by magnetic iron oxide nanoparticles. *Journal of Molecular Liquids, 321*, 114–455.

Rowdhwal, S., & Chen, J. (2018). Toxic effects of Di-2-ethylhexyl Phthalate: An overview. *BioMed Research International*, 1750368–1750378.

Schoen, M. E., & Ashbolt, N. J. (2010). *Assessing pathogen risk to swimmers at non-sewage impacted recreational beaches.* ACS Publications.

Shabat-Hadas, E., Mamane, H., & Gitis, V. (2017). Rhodamine B in dissolved and nano-bound forms: Indicators for light-based advanced oxidation processes. *Chemosphere, 184*, 1020–1027.

Strachan, S. (2010). Trace elements. *Current Anaesthesia & Critical Care, 21*, 44–48.

Sun, H. X., Dang, Z., Xia, Q., Tang, W. C., & Zhang, G. R. (2011). The effect of dietary nickel on the immune responses of spodoptera litura fabricius larvae. *Journal of Insect Physiology, 57*, 954–961.

Tang, Z., Zhang, L., Huang, Q., Yang, Y., Nie, Z., Cheng, J., & Chai, M. (2015). Contamination and risk of heavy metals in soils and sediments from a typical plastic waste recycling area in North China. *Ecotoxicology and Environmental Safety, 122*, 343–351.

Theriot, C. M., & Grunden, A. M. (2010). Hydrolysis of organophosphorus compounds by microbial enzymes. *Applied Microbiology and Biotechnology*, 2010, 89, 35–43.

Tchounwou, P. B., Yedjou, C. G., Patlolla, A. K., & Sutton, D. J. (2012). Heavy metals toxicity and the environment. *EXS, 101*, 133–164.

Tolins, M., Ruchirawat, M., & Landrigan, P. (2014). The developmental neurotoxicity of arsenic: Cognitive and behavioral consequences of early life exposure. *Annals of Global Health, 80*, 303–314.

Torki, M., Nazari, N., & Mohammadi, T. (2017). Evaluation of biological fouling of RO/MF membrane and methods to prevent it. *European Journal of Advances in Engineering and Technology, 4*(9), 707–710.

Truhaut, R. (1977). Ecotoxicology: Objectives, principles and perspectives. *Ecotoxicology and Environmental Safety, 1*, 151–173.

Tsutsumi, T., Ishihara, A., Yamamoto, A., Asaji, H., Yamakawa, S., & Tokumura, A. (2014). The potential protective role of lysophospholipid mediators in nephrotoxicity induced by chronically exposed cadmium. *Food and Chemical Toxicology, 65*, 52–62.

UNEP, State Plastics World Environment Day Outlook. (2018). https://www.Unenvironment. org/resources/report/state-plastics-world-environmentday-outlook-2018.

Vos, J. G., Dybing, E., Greim, H. A., Ladefoged, O., Lambre, C., Tarazona, J. V., . . . & Vethaak, A. D. (2000). Health effects of endocrine-disrupting chemicals on wildlife, with special reference to the European situation. *Critical Reviews in Toxicology, 30*(1), 71–133.

Wang, N., Liu, T., Shen, H., Ji, S., Li, J. R., & Zhang, R. (2016). Ceramic tubular MOF hybrid membrane fabricated through in situ layer-by-layer self-assembly for nanofiltration. *AIChE Journal, 62*(2), 538–546.

Wang, Q., & Yang, Z. (2016). Industrial water pollution, water environment treatment, and health risks in China. *Environmental Pollutution, 218*, 358–365.

Wang, Z., Xue, M., Huang, K., & Liu, Z. (2011). Textile dyeing waste water treatment. *Advances in Treating Textile Effluent, 5*, 91–116.

WHO. (2006). *Guidelines for the safe use of wastewater, excreta and greater* (Vol. 3). World Health Organisation Press.

Wood, A. R., Justus, K., Parigoris, E., Russell, A., & LeDuc, P. (2017). Biological inspiration of salt exclusion membranes in mangroves toward fouling-resistant reverse osmosis membranes. *The FASEB Journal, 31*(1 Supplement), 942–949

Wu, X., Cobbina, S. J., Mao, G., Xu, H., Zhang, Z., & Yang, L. (2016). A review of toxicity and mechanism of individual and mixtures of heavy metals in the environment. *Environmental Science and Pollution Research, 23*, 8244–8259.

Xiaolei, Q., Alvarez, P. J. J., & Li, Q. (2013). Applications of nanotechnology in water and wastewater treatment. *Water Research, 47*(12), 3931–3946.

Xu, X., Chen, X., Zhang, J., Guo, P., Fu, T., Dai, Y., . . . & Huo, X. (2015). Decreased blood hepatitis B surface antibody levels linked to E-waste lead exposure in preschool children. *Journal of Hazardous Materials, 298*, 122–128.

Xu, Y. C., Wang, Z. X., Cheng, X. Q., Xiao, Y. C., & Shao, L. (2016). Positively charged nanofiltration membranes via economically mussel-substance-simulated co-deposition for textile wastewater treatment. *Chemical Engineering Journal, 303*, 555–564.

Yang, M., Zhao, C., Zhang, S., Li, P., & Hou, D. (2017). Preparation of graphene oxide modified poly (m-phenylene isophthalamide) nanofiltration membrane with improved water flux and antifouling property. *Applied Surface Science, 394*, 149–159.

Zhang, L., Zhang, P., Wang, M., Yang, K., & Liu, J. (2016a). Research on the experiment of reservoir water treatment applying ultrafiltration membrane technology of different processes. *Journal of Environmental Biology, 37*(5), 1007.

Zhang, Y., Huo, X., Cao, J., Yang, T., Xu, L., & Xu, X. (2016b). Elevated lead levels and adverse effects on natural killer cells in children from an electronic waste recycling area. *Environmental Pollution, 213*, 143–150.

9 Role of Ornamental Plants in Clean-Up of Waste Disposal Sites

Shabeer Ahmad Dar, Sumira Tyub, Fahima Gul, Burhan Hamid, Irshad Ahmad Nawchoo, and Azra N. Kamili

9.1 INTRODUCTION

When a product or material is no longer useful for its intended purpose, it is referred to as waste. The waste created by human activity has the same physical components as the valuable output (White et al., 1995). Garbage also includes any components or goods that are useless to the producer (Basu, 2009). Dijkema et al. (2000) claim that trash is an item that people would still want to get rid of even if they had to pay. Although trash is a necessary by-product of human activity, it also comes from inefficient manufacturing methods, whose constant output results in a loss of essential resources (Cheremisinoff, 2003). Human-produced waste is typically used as food or a reactant, unlike waste from natural ecosystems, which is frequently very strong and takes an extremely long time to decompose (e.g., oxygen, carbon dioxide, and deceased natural matter). A substance can only be labelled as waste if its proprietor thinks it is, according to Dijkema et al. (2000). Consequently, a substance that is regarded as trash by one person might be a resource for another. Therefore, a product or substance that cannot be used in the future is considered trash. The trash from human activities is regularly very vigorous and takes many years to disintegrate in accepted environments, while waste from daily activities is usually used as foodstuff or a reactant (e.g., oxygen, carbon dioxide, and dead organic matter). To offer proper and effective waste management, it is crucial for legislators and governments to define and categorize garbage based on dangers to the environment and public health. Determining whether a substance is a waste or not is crucial for the manufacturer or holder when deciding whether waste regulations need to be followed. Definitions are important for domestic and international reporting requirements and for the gathering and analysis of waste data.

The choice to label a substance as trash will decide the regulations that must be followed to safeguard the public and the environment while the waste is processed or disposed of (DEFRA, 2009). On a more specific level, there is a substantial variation in classification methods and terminologies utilized globally.

DOI: 10.1201/9781003359326-9

Since the producer or holder discards materials and substances that are intended for recycling or reuse, they are frequently (but not always) recognized as garbage and won't stop being waste until specific procedures are followed and documented. Sometimes, defining waste also requires making case-by-case decisions. For instance, in certain circumstances, industrial by-products may be classed as non-waste.

The Waste Management and Survey Guide define waste as any substance that is unwanted by the producer. By-products of a manufacturing process, such as fly ash from a furnace, may also be considered waste. Such a substance might be removed from the refuse stream by a waste management company because it might be useful to another party. For instance, the apple might be used by a composting facility, or the newspaper might be used as an input at a pulp and paper plant. The material is treated in a way that allows the market to once again accept it as a desirable product, giving it value once more. Metropolitan areas and large towns appear to have the worst solid waste management issues due to the enormous quantity of solid waste generated by residential and commercial activities. Solid garbage is not only piled high on refuse dumps in the majority of the world's cities and large towns, but it is also frequently dumped and left to accumulate in street piles and small illegal dumps in any vacant area. The majority of third-world countries experience worse situations than industrialized nations, which have the resources, know-how, and public sentiment to partially control and manage their waste (Igbinomwanhia, 2011). A novel method that effectively lowers the metal burden in the food chain is the accumulation of heavy metals (HMs) by ornamental plants (OPs) from contaminated agricultural soils (Awad et al., 2021).

Urban waste is defined by the European Union as "household waste as well as any other waste that, by type or content, is comparable to domestic garbage" for legal purposes under the Directive on the Landfill of Waste 1999/31/EC. This broader meaning means that waste generated by businesses is classified as municipal waste when its makeup is comparable to that of household waste. Due to population shifts from rural to urban areas brought on by the Industrial Revolution, there was a major uptick in the production of trash (Wilson, 2007). Population growth brought on by the flow of people into cities increased both the quantity and type of waste created there (Amasuomo and Baird, 2016). Around that time, the quantity of metals and glass in the urban waste stream began to rise (Williams, 2005). Government officials started managing garbage disposal in the nineteenth century as a result of protecting public health (Theisen and Vigil, 1993).

9.2 CLASSIFICATION AND TYPES OF WASTE

According to White et al. (1995), trash can be broadly categorized into three kinds based on their physical states, including solid, liquid, and gaseous. There are obviously many categories, but the ones that are most frequently applied are those based on physical state, source, and environmental effect (Figure 9.1)

Based on the various legal and policy frameworks that are typically in place, two broad categories of trash can be identified: hazardous and non-hazardous or solid waste. The Basel Convention also employs this classification system.

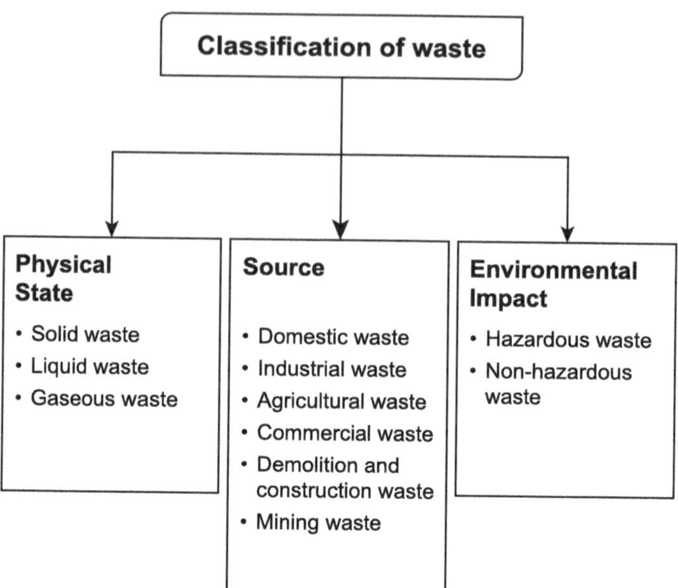

FIGURE 9.1   Classification of wastes on the basis of physical state, source, and environmental impact

9.2.1   NON-HAZARDOUS OR SOLID WASTE

Non-hazardous or solid waste includes all debris that has not been marked as hazardous, including paper, plastic, glass, metal, and beverage cans. Although solid waste is not dangerous, it can harm the ecosystem if it is not gathered and processed. Ordinarily, non-hazardous waste is coordinated at the regional or local (municipal) level, whereas hazardous waste is typically controlled at the federal level. Although it is theoretically possible to reuse or recycle a sizeable amount of solid waste, selective waste collection, which is required for reuse and recycling, presents one of the major waste management issues.

9.2.2   HAZARDOUS WASTE

The hazardous wastes, which have the potential to harm the environment, need stricter environmental standards. The liquid wastes are disposed of through sewer networks or lost to groundwater. At landfill sites, only non-hazardous solid waste is discharged. According to Theisen and Vigil (1993), disparate other wastes, solid wastes are difficult to eliminate. They asserted that they will always be discovered where they are found today. Chemical and physical characteristics decide how to collect and recycle things exactly. In flammability, corrosiveness, toxicity, ecotoxicity, and explosiveness are the most common characteristics of hazardous waste. Waste that is liquid, gaseous, or powdery must always need exacting treatment to avoid dissemination. To prevent contact with non-hazardous trash, separate collection and handling procedures are typically developed. Chemical, high-temperature, incineration, secure storage, recovery, and recycling processes can all be used to treat hazardous waste. The majority of hazardous garbage is produced industrially. Hazardous wastes that are unique include:

E-waste: Waste from electric and electronic devices, such as outdated computers, phones, and house appliances, is known as "e-waste." Generally speaking, e-waste is considered hazardous since it contains damaging components (e.g., PCB and various metals).

Medical waste: Bandages, reusable medical devices, drugs, chemicals, pharmaceuticals, bodily fluids, and body parts make up the majority of the waste from the human and animal healthcare systems. Radioactive, infectious, poisonous, or radioactively tainted waste from the medical industry is possible (including those that are drug-resistant).

Radioactive waste has radioactive components in it. The way radioactive waste is managed is very different from how normal waste is managed.

9.3 WASTE MANAGEMENT

Waste has traditionally been produced as a result of interactions between people and the environment (human activities). According to Giusti (2009), before people started living in communities, trash production and control were not a significant problem. Vergara and Tchobanoglous (2012) claim that as the world's population and purchasing power grows, more goods are made to meet the demand, which leads to the creation of more trash. As a result, careful planning and management are required to reduce the impact of trash on the ecosystem in a study by Marchettini et al. (2007). It was observed that the environment is overburdened by the constant fluxes of trash brought on by human activity. Ghiani et al. (2014) came to the conclusion that maintaining an effective method for managing solid waste is essential for maintaining the ecosystem. According to Beranek (1992), having an effective solid waste management system is now just as important as having electricity, airports, and roads, which are other necessities. According to a statement made by Basu (2009), processing waste is a necessary action to protect the public's health because it is impractical to continuously dump trash in landfills. There is little information available about the use of decorative plants (OPs) in phytoremediation, and little is known about how HMs affect OPs. The OP-mediated HMs remediation can enhance the site's aesthetics and remove contaminants at the same time (Khan et al., 2021).

Waste managing is the procedure of collecting, moving, and getting rid of wastes in the best way possible to reduce or completely eradicate their negative effects. A variety of strategies for supervision and getting rid of garbage is known as "waste management." Wastes may be eliminated, processed, recycled, reused, or managed. Waste management's primary objectives are lowering garbage production and averting risks to human health and the ecosystem. The management of the environment is just as crucial as other public services or infrastructure, without which modern man's existence would be very challenging. It also discusses the legal and administrative framework for garbage management, including recycling standards. In contrast to other types of remediation plants, OPs cultivated for decorative reasons in gardens and landscape design projects have become a significant source of remediation plants in recent years.

9.3.1 Types of Waste Management

The most widespread waste managing methods are recycling, incinerating, land filling, biological reprocessing, and animal feeding, among others (Figure 9.2).

The protection of the ecosystem is greatly enhanced by recycling. Recycling is one of the main waste-running techniques that avoids the disposal of garbage in landfills or waterways by converting it into recyclable litter components. Many organizations and municipalities have made it easier to recycle things by requiring labelling to say whether or not a substance is recyclable. This method of trash management is excellent because it has positive effects on both the economics and the environment. In addition to producing a sizeable quantity of income and thousands of new jobs, it saves the government money that would otherwise be spent on garbage programmers.

9.4 WASTE DISPOSAL SITES AND LANDFILLS

Municipal solid waste (MSW) landfills or dumping sites are designated areas of land or excavations that accept residential garbage. To put it another way, a landfill

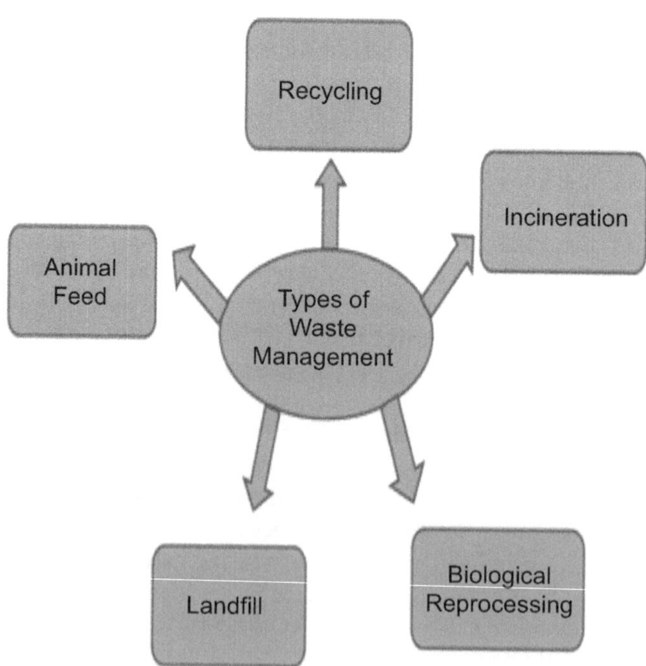

FIGURE 9.2 Various types of waste management

is a mechanism for disposing of rubbish and garbage in which the waste is buried between layers of soil to raise low-lying land. Along with profitable solid waste, non-hazardous sludge, tiny amounts of producer refuse, and industrial non-hazardous solid waste, non-hazardous waste is disposed of in landfills. As can be seen from the variety of garbage composition with respect to location and time, a landfill is an extremely varied and diverse ecosystem. Landfills store refuse that includes a variety of organic molecules from both natural and xenobiotic sources. It is well known that disposing of municipal solid refuse in landfills has negative effects on the environment and public health. MSWs are disposed of in sanitary landfills that have been professionally planned in many modern nations. They are frequently illegally dumped in impoverished nations without any safeguards in place to deal with the development of leachate and gas emissions, which are hazardous to the environment.

Today's landfills are engineered pits where solid refuse is layered, compacted, and covered before being disposed of. The floor is lined to prevent groundwater contamination. Engineered landfills have a lined floor in addition to a system for collecting and treating leachate, tracking groundwater, extracting gas (which is flared or used to produce energy), and a cap system. Using an analysis of environmental risk assessment, the capability is designed and the location is selected (UNEP, 2002). In addition, there are landfills specifically created to promote anaerobic biodegradation of the organic refuse component for the creation of biogas by monitoring the oxygen levels and moisture content.

Faeces sludge can be disposed of in landfills along with city solid waste. Even the cleanest landfill will ultimately fill up and, after a number of years, probably begin to leak. Landfills that have been properly built and kept are safer than open dumps, though. Therefore, only garbage that cannot be recycled should be disposed of in landfills.

9.5 PHYTOREMEDIATION OF THE WASTE DISPOSAL SITES OR LANDFILLS

A developing method called phytoremediation uses plants to collect and biomagnified pollutants from their surroundings and transform them into other types of biomolecules in their tissues (Pant et al., 2011). Green plants are employed to clean up contaminated environments since phytoremediation refers to plant-based approaches (Sadowsky, 1999). It is one of the most cost-effective and environmentally friendly technologies used for the renewal of waste disposal sites (Wu et al., 2016). For the phytoremediation of metals and organics in polluted soils, two of the more developed subcategories of phytoremediation are phytoextraction and phytodegradation (Lee, 2014). Phytoextraction, as defined by Mani et al. (2015) is the process of extracting or concentrating metals or organics using plants, particularly hyperaccumulator plants, into a biomass that can be harvested, whereas phytodegradation is the process of utilizing plant root exudates to promote microbial activity. Increased mineralization at the root–soil contact is believed to be caused by microorganisms that are present on the root surface.

The natural or cultivated vegetation on a landfill plays a significant role in erosion control and pollutant removal and also adds aesthetic value. Additionally, it can be utilized for leachate remediation (Maurice, 1998). The vegetation surrounding landfills demonstrates damage frequently brought on by the presence of landfill gas in the root zone. Rebuilding a suitable media for landfill re-vegetation should have as favourable of a capping for root growth as is required to achieve optimum plant performance (Vogel, 1987). It is known that plants increase the availability of nutrients by secreting certain enzymes, such as phosphates, cationic chelators, organic acids, or cationic chelators into the soil systems. One way that plants could encourage biodegradation is through increased nutrient availability brought on by plant growth (Nagendran et al., 2006).

According to the types, forms, and media of the contaminants, various plants use various methods or combinations of them to remediate soil and water. Contaminations caused by dirt, sediments, or sludge are cleaned up using phytoextraction, phytodegradation, Phytostabilization, rhizodegradation, or phytovolatilization (Figure 9.3). The effectiveness of remediation is significantly influenced by plant biomass and metabolism in addition to soil pH, electric conductivity, the quantity of organic matter, microbial activity, and other soil amendments (Nissim et al., 2018). The phytoremediation capacity of the plants is assessed using the bioconcentration factor (BCF). The ratio of elemental accumulation in a plant's shoot to that in its base is known as the BCF (Kafle et al., 2022).

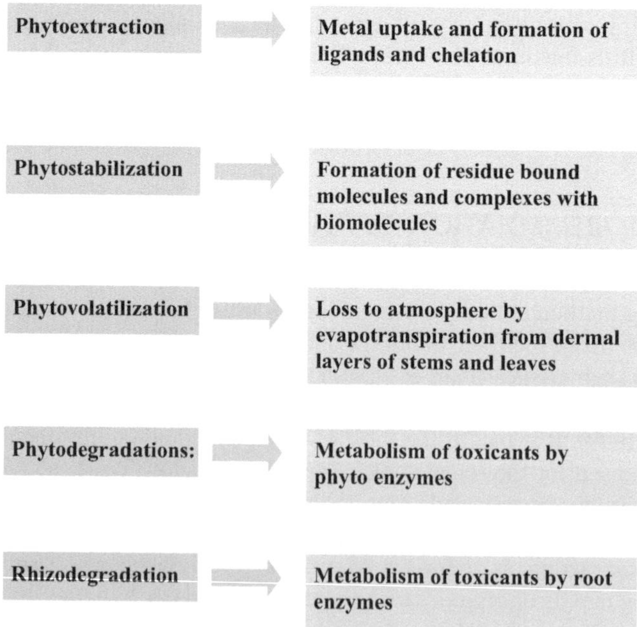

FIGURE 9.3 Schematic representation of phytoremediation process

9.6 ORNAMENTAL PLANTS

Plants can be found in a wide range of sizes, shapes, and aesthetics all over the globe. While others provide us with food, shelter, or building supplies, others simply offer us aesthetic pleasure. OPs are those that are utilized by humans for a number of purposes. In general, OPs include a variety of plant kinds, including herbaceous, woody, lower-level, aquatic, and terrestrial plants (Liu et al., 2018). The main reason that OPs are produced is to show off their flowers or for aesthetic, screening, accent, specimen, or colour purposes. Leaves, fragrances, berries, stems, and bark are examples of common ornamental elements. Because of their rapid growth, viability, and abundance, OPs may be particularly interesting and are therefore best selected from the local environment (Awad et al., 2021). Numerous species of OPs have demonstrated a high capacity to withstand and accumulate HMs and organic contaminants, making them exceptionally excellent prospects for phytoremediation (Liu et al., 2018; Khan et al., 2021). They could therefore improve the ecosystem in addition to remediating damaged soils. OPs have significant economic and ecological value due to the fact that their non-food biomass frequently includes contaminants. According to Liu et al. (2018), the pollutants that have accumulated in OPs can be kept out of food chains, which reduce the risk to human health to some degree. Compared to crop plants, they present a lower danger of biomagnifications (Khan et al., 2021).

9.6.1 PHYTOREMEDIATION OF MSW DISPOSAL SITES BY OPs

There are many applications for OPs in environmental management, but the most obvious ones are those that have a direct impact on how humans fit into the natural world, including reducing erosion caused by wind and water. The phytoremediation of landfills or waste disposal sites is a recent concept developed in a few countries. Wildflowers in particular offer a wide variety of plant heights, root depths, and aesthetic options. Grasses and flower plants are typically herbaceous and thrive in fallow areas. Some factors to take into account when choosing these plants for phytoremediation include their life duration, resistance to invasive species, root depth, and seeding cycle, and whether they need to be reseeded. OPs have most of these characteristic features so have a good potential to restore waste disposal sites. The goal of employing OPs at garbage disposal sites is to integrate people, buildings, and places in a practical and aesthetically pleasing manner, using plants and space as its primary tools. It is essential to landscape architecture for long-term, effective management of the environment's rapid change. Additionally, OPs have aesthetic functions by attracting people's attention and acting as complements, emphasizes, diverters, and markers.

When updating such facilities in developing countries, landfills and dumpsites utilized for the disposal of MSWs occasionally need rehabilitation. Environmental considerations, such as groundwater contamination brought on by the migration of leachate, transmigration of pollutants, and aesthetics must always be taken into consideration while planning rehabilitation plans and solutions. Phytoremediation has several uses in restoring garbage disposal sites and offers practical answers to many environmental issues connected to landfill rehabilitation. The kind of plant, the age of the contaminant, and numerous other physical and chemical properties of the soil can all affect how much organic matter plants can absorb. Eighty-eight types of plants and trees were shown to absorb and collect more than seventy organic chemicals from one study, representing a wide range of chemical groups (Paterson et al., 1990). When root tips sprout and begin to push through the soil, the soil aggregates physically break up. Especially as they deteriorate, roots have the potential to create huge gaps in the soil called macrospores, which can affect the soil's aeration and water status and facilitate the movement of contaminants, gases, and water. In addition to adding visual value, OPs on a landfill play a significant role in preventing erosion and cleaning up contaminants. Using OPs for phytoremediation is an appealing option to many current methods for a number of reasons.

9.6.2 PHYTOREMEDIATION OF HAZARDOUS WASTES CONTAMINATED SITES BY OPs

Agricultural and industrial operations include wastewater irrigation, the use of pesticides and chemical fertilizers, metal smelting, and mining for metalliferous minerals have all resulted in significant sections of land becoming contaminated (Damodaran et al., 2014; Shao et al., 2014; Hagmann, 2021). The most significant metallic pollutants for the environment are the metalloids arsenic (As) and selenium (Se), as well as the metals mercury (Hg), cadmium (Cd), lead (Pb), nickel (Ni), copper (Cu), and zinc (Zn) (Zhang et al., 2015). Even though some metals, like Zn, Cu, and Ni, are required for many physiological functions, they can become toxic when present in excess, particularly given that the window between metal deficit and toxicity can be quite small (Touceda-Gonzalez et al., 2015). The People's Republic of China's Ministries of Environmental Protection and Land and Resources jointly published a study on the condition of soil contamination in China (MLR). The state of the soil environment was not good: the percentages of inorganic contaminants that were above the limits were 7.0%, 1.6%, 2.7%, 2.1%, 1.5%, 1.1%, 0.9%, and 4.8% respectively, for Cd, Hg, As, Cu, Pb, Cr, Zn, and Ni (Luo and Chen, 2021). Excessive amounts of HMs can deteriorate the quality of the soil, lower crop yields, produce agricultural products of poor quality, contaminate surface water and groundwater, and risk human and animal health through the food chain (Vamerali et al., 2010; Motuzova et al., 2014). A promising technique for dealing with widespread but low-level radionuclide pollution is phytoremediation, which employs plants to remove, transfer, or immobilize radionuclides found in contaminated soil, water, or sludge. Numerous uses for radioactive elements exist in the fields of science, medicine, agriculture, industry, and energy production. They are essential to daily existence in human society. Therefore, it is expected that such a variety of operations will result in the production of radioactive waste. In this method, harmful metals are removed from the environment, decomposed, or detoxified using green plants. It is known that plants may absorb, decompose, and accelerate the decomposition of organic pollutants in soil (Anderson et al., 1993). An efficient way to lessen HM contamination in soil is by the sequestration of HMs by plants (Cunningham et al., 1996). Research on phytoremediation frequently focuses on the sequestration of toxicants by plants. OPs can also improve the aesthetics of dumping sites, and the majority of them are herbs with short life cycles and short roots that can thrive there with ease. A number of mechanisms, including phytostabilization, rhizodegradation, phytoextraction, phytodegradation, phytoaccumulation, and phytovolatilization, are involved in the uptake of HMs by phytoremediation technology (Tangahu et al., 2011). The plants chosen for phytoremediation have the capacity to absorb large amounts of HMs into their roots, can withstand various metals, and have the physiological parameters to adapt to various settings (Guerra et al., 2021). The Lamiaceae family, which includes plants like Anethum graveolens, Lavandula vera, and Lavandula angustifolia, has reportedly been used to remove HMs from soil. The biomass so generated is reportedly used to make essential oils (Angelova et al., 2021; Barouchas et al., 2019). According to the findings, Phragmites australis may accumulate the highest levels of Cd and Ni, but Echinochloa stagnina can accumulate a lot of Pb in its tissues. Consequently, phytoremediation is a practical and affordable method for removing HMs from soil and water (Eid et al., 2020). As model plants for the removal of HMs, two genera, Petunia and Nicotiana, of the Solanaceae family, have been used widely in research (Khan et al., 2021; Khan et al., 2021). Plants belonging to Asteraceae family including Tagetes erecta, Calendula officinalis, and Chrysanthemum indicum have also been reported to perform well in HM-contaminated soils (Coelho et al., 2023; Goswami and Das, 2016; Mani et al., 2015). It is well known that each of these plants has commercial uses in horticulture and floriculture. Therefore, it is possible to predict that phytoremediation through OPs brings about two advantages. These include the removal of HMs and concurrent site beautification.

Only shallow depths where their roots can engage with metals are suitable for using small herbaceous OPs for phytoremediation. Dendroremediation is the process of using plants to remove contaminants (Khan et al., 2021). Using ornamental trees and bushes to remove HMs from lower soil levels has a selective benefit over using ornamental herbs (Saxena et al., 2020). According to a research by Pilipovic et al. (2023), Zn and Cd were phytoextracted from the aerial parts of Populus deltoides and Salix viminalis while Cr, Ni, Pb, and Cu were phytostabilized in the roots. Morus alba was found to phytoextract Cd and Pb from its shoot, according to Jiang et al. (2019), who came to the conclusion that these trees might serve as an alternative plant in paddy fields that have been contaminated with HMs.

On the basis of the metal uptake mechanism OPs are classified into three groups namely excluder plants, accumulators/indicator plants, and hyperaccumulator plants. The ability of excluder plants to endure HMs in the soil up to a threshold concentration is due to their ability to stop the buildup of metal in their cells. Either active (energy-dependent) efflux pumps or blocking the uptake in the roots are used to accomplish this exclusion. HM sensitivity is typically present in indicator plants. According to Leitenmaier and Küpper (2013), hyper accumulators are plants that can actively take up high concentrations of particular elements and accumulate them to a number of percent of the dry mass of their above-ground parts in addition to tolerating high concentrations of those elements. Some of the representative groups of OPs involved in HM accumulation is given in Figure 9.4.

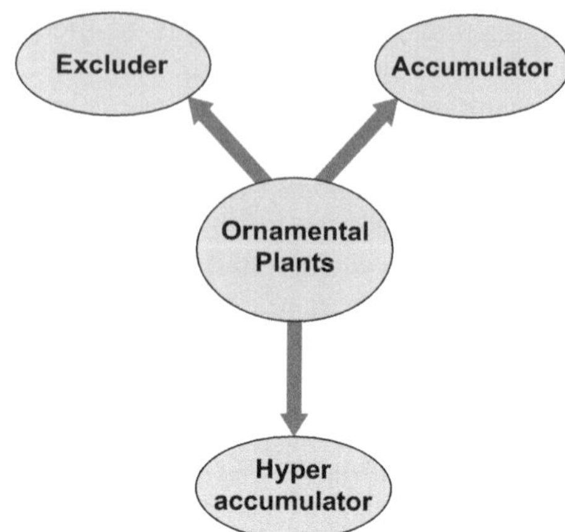

FIGURE 9.4   Groups of the ornamental plants involved in heavy metal uptake

9.7   VEGETATION OF OPS AT DUMPSITES

Plants are known to increase the abundance of nutrients in the soil by secreting cationic chelators, organic acids, or specific enzymes like phosphates. The degree of competition between degrading and non-degrading species for these nutrients will determine the quantity of pollution that is degraded (Steffensen and Alexander, 1995). Additionally, atrazine degradation by an inoculation consortium in maize grown with fertilizer and corn grown without fertilizer was comparable (Alvay and Crowley, 1996). Plants may improve the bioavailability of the pollutant in addition to enhancing the availability of nutrients. In terms of landfill vegetation, this characteristic is important. Biodegradation is frequently impeded by contaminant bioavailability, and stimulation of degradation might increase it (Siciliano and Germida, 1998). By competing with the pollutant for binding sites on the soil matrix, root exudates can improve the bioavailability of contaminants.

Employing indigenous species is a good place to start when choosing the right plant species for the repair and restoration of dumpsites. Even though they have a little surface area, landfills frequently provide a wide variety of environmental niches for wildlife. Many fluxes of trash and cover materials from many sources wind up in landfills where they form microhabitats that give one type of vegetation a competitive edge and allow it to flourish while making other species rare. Reviewing plant species found at various landfills makes it easier to choose plants that can handle a variety of toxins at once. It's noteworthy to note that the origin of the wastes, the vegetation in the area, and the circumstances at the dump all have an impact on species diversity. Therefore, it is impossible to use a single species as a general indicator, and plant selection should be dependent on the local climate and native plants that grow in a particular dump.

9.8 VEGETATED FILTER STRIPS FOR AGRICULTURAL WASTE TREATMENT

Vegetated filter strips are those that have vegetation growing downward from a farm or an animal production centre where animal waste has been spread. From the contributing regions, the strips can filter viruses, organics, agrichemicals, nutrients, silt, and runoff. There are four steps involved in purging the run-on water of its components. The first procedure is the deposition of sediment (solid substance) in the strip. Grass or other dense vegetation makes up a vegetated filter strip, which provides resistance to shallow overland flow. The capacity to carry sediment is drastically reduced by the drop in flow velocity at the upslope border of the vegetated filter strip, and suspended particles are deposited as a result. In the second process, vegetation makes it possible for water that has run off the surface to get into the soil profile. The elements are absorbed into the soil and changed into plant nutrition or organic soil components by chemical, physical, and biological processes. In the third and fourth processes, solid particles carried by runoff are physically filtered out by the dense, upright vegetation. In the third phase, some soluble nutrients that are travelling with the runoff water can be directly absorbed through the plant leaves and stems. All of the processes convert a potential contaminant into vegetative biomass, which can be used as feed, fibre, or mulch material, by the plants utilizing nutrients from the run-on water. According to recent research findings, vegetated filter strips are highly effective in a variety of situations (Adam et al., 1986; Dillaha et al., 1988; Doyle et al., 1977; Schwer and Clausen, 1989; Young et al., 1980). Individual site conditions, including the contributing area and the vegetated filter strip location, can vary in effectiveness. Site factors that affect effectiveness include land slope, soils, land use and management, climate, vegetation type and density, application rates for sites that are regularly loaded, and concentration and features of constituents. The capacity of the system as a whole to lower the concentration and volume of contaminants present in the runoff from the site depends on how the contributing area is operated and managed as well as how well the vegetated filter strip is maintained. Before making decisions regarding how effectively vegetated filter strips perform during planning, it is crucial to be aware of site characteristics.

Planning a vegetated filter strip should take into account the following qualities that research and operation sites exhibit:

- The sheet flow must be preserved. Except when low-velocity Grass Rivers are utilized, concentrated flow should be avoided.
- To keep the proper depth of flow, hydraulic loading must be carefully managed.
- To give the vegetated filter strip a chance to recover, wastewater must be applied on a regular basis. For rest periods and climate factors, wastewater storage is crucial.
- The amount of soluble components removed from the run-on water will be small unless infiltration occurs.
- For correctly constructed and maintained strips, removal of suspended solids and associated constituents from the run-on can be significant, ranging from 60% to 80%.

- Vegetated filter strips shouldn't be employed in place of other suitable management and structural procedures. They often do not exist in isolation.
- It is necessary to do maintenance, which includes taking good care of the plants and removing any accumulated solids.
- To ensure that regular slopes can be installed and maintained along and perpendicular to the flow channel, proper siting is crucial.

9.9 ADVANTAGES AND DISADVANTAGES OF PHYTOREMEDIATION

9.9.1 ADVANTAGES

1. Early estimates of the costs of phytoremediation indicate that the use of OPs in the restoration of waste disposal sites will be a lower-cost technology.
2. Phytoremediation of wastes using OPs seems to be environment-friendly with high aesthetic value. So the public reception could be greater.
3. OPs can be applied to remediate degraded shallow soils.
4. Regardless of the level of contaminant reduction, the presence of plants is likely to enhance the soil's overall condition.
5. Vegetation can also reduce or prevent erosion and fugitive dust emissions.

9.9.2 DISADVANTAGES

1. The depth limitation resulting from the usually shallow distribution of plant roots is a major drawback of phytoremediation by ornamentals.
2. Phytoremediation in general and remediation by ornamentals is a slow process and will take a much extended time period.
3. The harvested plant material containing radioactive load again becomes a problem for disposal.
4. In some areas having prolonged winter season phytoremediation may not be so effective especially with ornamentals due to lack of suitable growth conditions.
5. Cultivation of plants often needs great care which is not possible in dumping sites.

9.10 CONCLUSION

OPs are great alternatives to other types of plants for the reinstatement of polluted urban soil, and their usage in this regard has gained extra interest in modern days. Numerous decorative plant species have shown a great capability to withstand and collect HMs and natural pollutants, which enables them to both clean up contaminated soils and enhance the environment. OPs' ecological and commercial value is greatly increased by the presence of contaminants in their non-food biomass, which is common. More research is required to comprehend the molecular processes underlying contaminant tolerance, accumulation/degradation by OPs, and its practical application because the majority of the literature presently available focuses on laboratory findings. Current research should focus on analysing the molecular mechanisms through which hazardous substances are accumulated and transferred in OPs to optimize the phytoremediation of OPs and discover more environmentally friendly strengthening methods. In the end, getting away from decorative clean-up plants should also make sense.

REFERENCES

Adam, R., Lagace, R., & Vallieres, M. (1986). Evaluation of beef feedlot runoff treatment by a vegetative filter. ASAE paper, 86–208.

Alvey, S., & Crowley, D. E. (1996). Survival and activity of an atrazine-mineralizing bacterial consortium in rhizosphere soil. *Environmental Science and Technology, 30,* 1596–1603.

Amasuomol, E., & Baird, J. (2016). The concept of waste and waste management. *Journal of Management and Sustainability, 6*(4), 88–96.

Anderson, J. L. (1993). *The hazardous waste land.* Va. Envtl. LJ, 13, 1.

Angelova, G., Brazkova, M., Stefanova, P., Blazheva, D., Vladev, V., Petkova, N., & Krastanov, A. (2021). Waste rose flower and lavender straw biomass—An innovative lignocellulose feedstock for mycelium bio-materials development using newly isolated Ganoderma resinaceum GA1M. *Journal of Fungi, 7*(10), 866.

Awad, M., El-Desoky, M. A., Ghallab, A., Kubes, J., Abdel-Mawly, S. E., Danish, S., & Sabagh, A. (2021). Ornamental plant efficiency for heavy metals phytoextraction from contaminated soils amended with organic materials. *Molecules, 26*(11), 3360.

Barouchas, P. E., Akoumianaki-Ioannidou, A., Liopa-Tsakalidi, A., & Moustakas, N. K. (2019). Effects of vanadium and nickel on morphological characteristics and on vanadium and nickel uptake by shoots of mojito (Mentha× villosa) and lavender (Lavandula anqustifolia). *Notulae Botanicae Horti Agrobotanici Cluj-Napoca, 47*(2), 487–492.

Basu, R. (2009). Solid waste management – A model study. *Sies Journal of Management, 6*(2).

Beranek Jr, W. (1992). Solid waste management and economic development. *Economic Development Review, 10*(3), 49.

Cheremisinoff, N. P. (2003). *Handbook of solid waste management and waste minimisation technology* (p. 477). Elsevier Science.

Coelho, D. G., da Silva, V. M., Gomes Filho, A. A. P., Oliveira, L. A., de Araújo, H. H., dos Santos Farnese, F., & de Oliveira, J. A. (2023). Bioaccumulation and physiological traits qualify Pistia stratiotes as a suitable species for phytoremediation and bio indication of iron-contaminated water. *Journal of Hazardous Materials, 446,* 130701.

Cunningham, S. D., Anderson, T. A., Schwab, A. P., & Hsu, F. C. (1996). Phytoremediation of soils contaminated with organic pollutants. *Advances in Agronomy, 56*(1), 55–114.

Damodaran, D., Shetty, K. V., & Mohan, B. R. (2014). Uptake of certain heavy metals from contaminated soil by mushroom—Galerina vittiformis. *Ecotoxicology and Environmental Safety, 104,* 414–422.

DEFRA, U. (2009). *Household waste prevention evidence review.* http://randd. defra. gov. uk/Default. Project ID= 16161.

Dijkema, G. P. J., Reuter, M. A., & Verhoef, E. V. (2000). A new paradigm for waste management. *Waste Management, 20*(8), 633–638.

Dillaha, T. A., Sherrard, J. H., Lee, D., Mostaghimi, S., & Shanholtz, V. O. (1988). Evaluation of vegetative filter strips as a best management practice for feed lots. *Journal of Water Pollution Control Federation,* 1231–1238.

Doyle, R. C., Stanton, G. C., & Wolf, D. C. (1977). Effectiveness of forest and grass buffer strips in improving the water quality of manure polluted runoff. *American Society of Agricultural Engineers.*

Eid, E. M., Galal, T. M., Sewelam, N. A., Talha, N. I., & Abdallah, S. M. (2020). Phytoremediation of heavy metals by four aquatic macrophytes and their potential use as contamination indicators: A comparative assessment. *Environmental Science and Pollution Research, 27*, 12138–12151.

Ghiani, G., Laganà, D., Manni, E., Musmanno, R., & Vigo, D. (2014). Operations research in solid waste management: A survey of strategic and tactical issues. *Computers & Operations Research, 44*, 22–32.

Giusti, L. (2009). A review of waste management practices and their impact on human health. *Waste Management, 29*(8), 2227–2239.

Goswami, S., & Das, S. (2016). Copper phytoremediation potential of Calandula officinalis L. and the role of antioxidant enzymes in metal tolerance. *Ecotoxicology and Environmental Safety, 126*, 211–218.

Guerra Sierra, B. E., Muñoz Guerrero, J., & Sokolski, S. (2021). Phytoremediation of heavy metals in tropical soils an overview. *Sustainability, 13*(5), 2574.

Hagmann, D. (2021). *Effect of abiotic factors on enzyme activity in brown field soils.* Montclair State University.

Igbinomwanhia, D. I. (2011). Status of waste management. *Integrated Waste Management, 2*, 11–34.

Jiang, M., Liu, S., Li, Y., Li, X., Luo, Z., Song, H., & Chen, Q. (2019). EDTA-facilitated toxic tolerance, absorption and translocation and phytoremediation of lead by dwarf bamboos. *Ecotoxicology and Environmental Safety, 170*, 502–512.

Kafle, A., Timilsina, A., Gautam, A., Adhikari, K., Bhattarai, A., & Aryal, N. (2022). Phytoremediation: Mechanisms, plant selection and enhancement by natural and synthetic agents. *Environmental Advances, 8*, 100203.

Khan, A. H. A., Kiyani, A., Mirza, C. R., Butt, T. A., Barros, R., Ali, B., & Yousaf, S. (2021). Ornamental plants for the phytoremediation of heavy metals: Present knowledge and future perspectives. *Environmental Research, 195*, 110780.

Khan, D., Yadav, S., Kapley, A. (2021). Impact of COVID-19 pandemic on the current status of solid waste management in India. *Journal of Clinical Images and Medical Case Reports, 2*(4), 1261.

Lee, W. S., Chua, A. S. M., Yeoh, H. K., & Ngoh, G. C. (2014). A review of the production and applications of waste-derived volatile fatty acids. *Chemical Engineering Journal, 235*, 83–99.

Leitenmaier, B., & Küpper, H. (2013). Compartmentation and complexation of metals in hyper accumulator plants. *Frontiers in Plant Science, 4*, 55114.

Liu, J., Xin, X., & Zhou, Q. (2018). Phytoremediation of contaminated soils using ornamental plants. *Environmental Reviews, 26*(1), 43–54.

Luo, L., Wang, B., Jiang, J., Fitzgerald, M., Huang, Q., Yu, Z., & Chen, S. (2021). Heavy metal contaminations in herbal medicines: Determination, comprehensiv risk assessments, and solutions. *Frontiers in Pharmacology, 11*, 595335.

Mani, D., Kumar, C., & Patel, N. K. (2015). Integrated micro-biochemical approach for phytoremediation of cadmium and zinc contaminated soils. *Ecotoxicology and Environmental Safety, 111*, 86–95.

Mani, D., Kumar, C., & Patel, N. K. (2015). Integrated micro-biochemical approach for phytoremediation of cadmium and zinc contaminated soils. *Ecotoxicology and Environmental Safety, 111*, 86–95.

Marchettini, N., Ridolfi, R., & Rustici, M. (2007). An environmental analysis for comparing waste management options and strategies. *Waste Management, 27*(4), 562–571.

Maurice, C. (1998). Landfill gas emission and landfill vegetation [Doctoral dissertation]. Luleå tekniska universitet.

Motuzova, G. V., Minkina, T. M., Karpova, E. A., Barsova, N. U., & Mandzhieva, S. S. (2014). Soil contamination with heavy metals as a potential and real risk to the environment. *Journal of Geochemical Exploration, 144*, 241–246.

Nagendran, R., Selvam, A., Joseph, K., & Chiemchaisri, C. (2006). Phytoremediation and rehabilitation of municipal solid waste landfills and dumpsites: A brief review. *Waste Management, 26*(12), 1357–1369.

Nissim, W. G., Cincinelli, A., Martellini, T., Alvisi, L., Palm, E., Mancuso, S., & Azzarello, E. (2018). Phytoremediation of sewage sludge contaminated by trace elements and organic compounds. *Environmental Research, 164*, 356–366.

Pant, M., Dolma, S., Gahlot, M., Sharma, A., & Mundepi, S. (2011). Phytoremediation of Heavy Metals 15. *Heavy Metal Toxicity: Environmental Concerns, Remediation and Opportunities, 313*.

Paterson, S., Mackay, D., Tam, D., & Shiu, W. Y. (1990). Uptake of organic chemicals by plants: A review of processes, correlations and models. *Chemosphere, 21*(3), 297–331.

Pilipović, A., Kováčik, J., Beronská, N., Opálková Šišková, A., Dvorák, T., & Rujnić Havstad, M. (2023). Comparison of the influence of carbon obtained from kitchen waste and synthetic carbon on the tensile properties of polyolefins. *Materials Science Forum, 1108*, 11–16.

Sadowsky, M. J. (1999). Phytoremediation: Past promises and future practices. In: *Proceedings of the 8th international symposium on microbial ecology* (pp. 1–7). Halifax.

Saxena, N., Sarkar, B., & Singh, S. R. (2020). Selection of remanufacturing/production cycles with an alternative market: A perspective on waste management. *Journal of Cleaner Production, 245*, 118935.

Schwer, C. B., & Clausen, J. C. (1989). Vegetative filter treatment of dairy milk house wastewater. American *Society of Agronomy, Crop Science Society of America, and Soil Science Society of America, 18* (4), 446–451.

Shao, X., Huang, B., Zhao, Y., Sun, W., Gu, Z., & Qian, W. (2014). Impacts of human activities and sampling strategies on soil heavy metal distribution in a rapidly developing region of China. *Ecotoxicology and Environmental Safety, 104*, 1–8.

Siciliano, S. D., & Germida, J. J. (1998). Degradation of chlorinated benzoic acid mixtures by plant—bacteria associations. *Environmental Toxicology and Chemistry: An International Journal, 17*(4), 728–733.

Steffensen, W. S., & Alexander, M. (1995). Role of competition for inorganic nutrients in the biodegradation of mixtures of substrates. *Applied and Environmental Microbiology, 61*(8), 2859–2862.

Tangahu, B. V., Sheikh Abdullah, S. R., Basri, H., Idris, M., Anuar, N., & Mukhlisin, M. (2011). A review on heavy metals (As, Pb, and Hg) uptake by plants through phytoremediation. *International journal of chemical engineering, 2011*(1), 939161.

Theisen, H., & Vigil, S. A. (1993). *Integrated solid waste management: Engineering principles and management issues.* McGraw-Hill.

Touceda-Gonzalez, M., Brader, G., Antonielli, L., Ravindran, V. B., Waldner, G., Friesl-Hanl, W., & Sessitsch, A. (2015). Combined amendment of immobilizers and the plant growth-promoting strain Burkholderia phytofirmans PsJN favours plant growth and reduces heavy metal uptake. *Soil Biology and Biochemistry, 91*, 140–150.

UNEP. 2002. *Environmental impact assessment training resource manual* (2nd Ed).

Vamerali, T., Bandiera, M., & Mosca, G. (2010). Field crops for phytoremediation of metal-contaminated land. A review. *Environmental Chemistry Letters, 8*, 1–17.

Vergara, S. E., & Tchobanoglous, G. (2012). Municipal solid waste and the environment: A global perspective. *Annual Review of Environment and Resources, 37*, 277–309.

Vogel, W. G. (1987). A manual for training reclamation inspectors in the fundamentals of soils and revegetation. *Soil and Water Conservation Society.*

White, P. R., Franke, M., & Hindle, P. (1995). *Integrated solid waste management: A lifecycle inventory: A lifecycle inventory.* Springer Science & Business Media.

Williams, P. T. (2005). *Waste treatment and disposal.* John Wiley & Sons.

Wilson, D. C. (2007). Development drivers for waste management. *Waste Management & Research, 25*(3), 198–207.

Wu, H., Wang, J., Duan, H., Ouyang, L., Huang, W., & Zuo, J. (2016). An innovative approach to managing demolition waste via GIS (geographic information system): A case study in Shenzhen city, China. *Journal of Cleaner Production, 112*, 494–503.

Young, R. A., Huntrods, T., & Anderson, W. (1980). Effectiveness of vegetated buffer strips in controlling pollution from feedlot runoff. *American Society of Agronomy, Crop Science Society of America, and Soil Science Society of America, 9*(3), 483–487.

Zhang, J., Wang, L. H., Yang, J. C., Liu, H., & Dai, J. L. (2015). Health risk to residents and stimulation to inherent bacteria of various heavy metals in soil. *Science of the Total Environment, 508*, 29–36.

10 Low Cost and Economical Alternatives for Remediation of Industrial Waste

S. Karan and B.S. Anusha

10.1 INTRODUCTION

Industries significantly contribute to the worldwide economies of most of the countries, although they are also, regrettably, the primary sources of environmental pollution. There are several benefits and drawbacks associated with industrialization. The production of waste is one of industrialization's drawbacks. Waste is generated by human activities in general, including the mining and processing of basic resources. Industrial waste could be solid, gaseous, or liquid, and each type is managed and disposed of differently. Industrial waste management addresses all industrial wastes, including domestic, biological, and industrial byproducts, before, during, and after production as well as trash left over after consumer use. Industrial waste sometimes poses a concern to human health. Among the different sources (soil and water), industrial wastewater released from various enterprises is recognized as the main cause of environmental contamination (Goutam et al., 2018; Maszenan et al., 2011).

Waste collection services are frequently depleted in regions with growing economies, and dumpsites are poorly supervised and managed. The issues are becoming worse (Chinaza et al., 2020). The situation becomes increasingly challenging as a result of governance issues. Due to these nations' poor institutions, rising urbanization, and continued underfunding, waste management in their cities and countries continues to be a concern. These difficulties, along with a lack of knowledge of the various elements that subsidize to the steps of management of waste, have an impact on waste treatment (Aparcana, 2017). Despite advances, controlling industrial waste is still problematic in developed countries. Recent trend in methods and techniques for managing and treating solid waste, electronic waste, and industrial wastewater, as well as the consequences of inappropriate management of industrial waste for human health and the ecosystem.

The huge effects of the entire waste management operations on the economy and ecology have been found by various recent researches. Significant amounts of studies have concentrated on taking into account the cost of the environmental damage caused by the air contaminants that are produced by each treatment facility,

DOI: 10.1201/9781003359326-10

"

even though some of these have created various scenarios involving waste control measures and have analysed these scenarios using pre-treatment techniques. There hasn't been any research, however, that incorporates a quantitative cost analysis of all treatment paths that accounts for waste transportation in each treatment phase as well as the costs related to environmental degradation from air pollution emissions during treatment.

As the importance of wastewater treatment plants has come to light, research experts are under pressure to develop innovative, affordable, and eco-friendly technologies that can remove harmful chemicals from wastewater while also protecting the general public from dangerous health conditions. Low-cost technologies to clean up industrial wastes are covered in this chapter.

10.2 INDUSTRIAL WASTE

Any substance that is made unfeasible during a manufacturing process, such as that of mineral extraction, factories, or mills is called industrial waste. Everything from soil and gravel to masonry and concrete to chemicals, scrap metal, solvents, oil, and even leftover food from restaurants can be considered industrial waste. Groundwater, rivers, lakes, coastal waterways, streams, surrounding soil, and other water bodies can all be contaminated by industrial pollution (Maczulak, 2010). Due to the frequent mixing of industrial waste and municipal trash, accurate assessments are difficult to make. The majority of nations have passed legislation to address the problems of industrial waste, although the level of rigour and compliance requirements varies. A constant issue is enforcement.

10.2.1 Types

Industrial waste comes in two forms, solid and liquid, and is frequently created as a result of industrial activity. It is made up of any elements that are rendered unusable during the process of making a product in an industry.

10.2.1.1 Solid Waste

Solid trash produced by production processes like manufacturing, transportation, and resource development is referred to as industrial solid waste. They consist of liquid and gaseous wastes in tanks that are not allowed to discharge into the environment, as well as solid, semi-solid, and liquid wastes. Industrial solid wastes are divided into hazardous wastes and common wastes depending on pollution characteristics, as well as into inorganic wastes and organic wastes based on their constituents and solid, semi-solid, and liquid wastes depending on their species. They typically receive extra care since many industrial solid wastes have hazardous qualities.

Pollution caused by industrial solid waste has grown much worse all across the world. The developing nations must face it specifically. Huge amounts of industrial solid waste are generated annually by expanding industries. However, many poor nations lack suitable treatment and disposal facilities as well as trained people. All of these have negatively impacted both human health and the environment along with the development of the industries in these countries.

Controlling the contamination of industrial solid wastes has been paid much attention nowadays. Significant advancement has been made in creating treatment and disposal technologies, establishing related management and legislative frameworks, and putting the findings of research into use in industry. Because of this, although the situation has not entirely altered, the major problem of hazardous waste contamination has been somewhat lessened in affluent nations. But there are a lot of issues that need to be fixed in emerging nations regarding the management of industrial solid wastes (Jinhui, 2004).

At numerous landfill sites, tonnes of solid trash are disposed inappropriately leading to global warming, and other ecological implications (Yatoo et al. 2024). The majority of these wastes are frequently generated by industrial sources and are either disposed of directly or after usage. The wastes are generated by numerous agricultural-related activities as well as garbage from residences, businesses, and enterprises. If the waste is not stored appropriately and handled, the landfill sites emit unpleasant odours and can pollute the air around them, having a harmful impact on humans, the environment, and wildlife (Rinkesh, 2018).

10.2.1.2 Sources of Solid Waste

Agricultural: Farm wastes, dairy farms, orchards, and vineyards all produce solid waste. They generate trash such as rotten food, pesticide bottles, agricultural waste, and other potentially harmful substances.

Commercial: Buildings and commercial facilities also produce solid waste. In this context, the terms "commercial facilities" and "buildings" refer to establishments like restaurants, motels, marketplaces, shops, and office buildings. Plastic garbage is one type of solid waste produced by these facilities (Awuchi and Awuchi, 2019a, 2019b; Chinaza et al., 2020). Wood, glass, food waste, cardboard, special wastes, metals, and other hazardous wastes are only a few examples.

Institutional: Organized facilities, including universities, military barracks schools, prisons, and other government facilities, generate solid waste. wastes, metals, wood, cardboard plastics, electronics, glass, food waste, paper, rubber, and various hazardous wastes are only a few of the common solid wastes produced in these locations.

Demolition and construction sites: The creation of solid garbage is increased by construction and demolition projects. Road repair sites, new building, and road construction sites, and building demolition and renovation sites are construction sites (Rinkesh, 2018). Plastic, steel, glass, rubber, copper wires, and steel are just a few of the solid waste generated in these sites.

Biomedical: Solid waste comes from hospitals, chemical manufacturing companies, and biomedical equipment. Syringes, medications, paper, plastics, bandages, soiled gloves, food scraps, chemicals, and plastics are just a few of the various solid wastes that are generated in hospitals. All of these need to be properly disposed of to avoid creating major issues for both humans and the environment.

Residential: Paper, leather, glass, food wastes, ashes, metals, plastics, wastes, cardboard yards, and special wastes such as large households. Objects like old mattresses, electronics, used tyres, oil, and batteries are among the garbage from these locations. Most homes have trash cans where residents can dispose of their solid wastes. A company or individual who collects trash will then empty the can and take it for treatment.

Industrial: The largest source of solid waste is the industrial sector. They consist of both heavy and light manufacturing businesses, canning and fabrication facilities, building sites, power plants, and others. These businesses generate solid wastes in the form of ashes, medical wastes, special wastes, food wastes, packaging wastes, plastic wastes, household wastes, and other hazardous wastes.

Municipal services: In most nations, the generation of solid waste is significantly influenced by industrialization and urbanization in metropolitan areas (Yatoo et al. 2024). Wastewater treatment facilities, street cleaning, waste from parks, landscaping wastes, garbage from recreational areas and beaches, and sludge are a few examples of solid wastes produced by municipal activities.

10.2.1.3   Liquid Waste (Industrial Wastewater)

Given that liquid waste (different wastewaters produced by various businesses) contains a variety of potentially dangerous chemicals, it is regarded as being extremely hazardous to both living things. However, the type of industry, production methods used, and product quality all affect the nature and features of industrial effluent. Industries frequently release wastewater with a high COD, BOD, TDSs, and TSSs, as well as several inorganic and organic contaminants (Saxena and Bharagava, 2017).

10.2.1.4   Sources of Industrial Wastewater

Food industry: The wastewater produced by agricultural and food processing techniques differs from the typical municipal wastewater treated by private or public sewage treatment plants worldwide in that it is biodegradable and non-toxic but contains high levels of BOD and TSS (European Environment Agency, 2001). Due to the seasonality of post-harvesting and food processing as well as differences in BOD and pH and between the wastewater from horticultural, and meat products, the components of food and agriculture wastewater are typically difficult to anticipate. Large amounts of water of good quality are needed for food processing starting with basic materials. Simply washing vegetables produces water that is highly contaminated with both dissolved organic matter and particle debris. Surfactants could also be present. The factories that process dairy release conventional pollutants (SS, BOD). Aspartame and other beverage flavouring sweeteners, like sugars and sugar alcohols, may wash off during production and end up in wastewater (Awuchi and Echeta, 2019; Awuchi, 2017).

Iron and steel industries: There is no doubt that the cooling waters are tainted with chemicals, particularly cyanide and ammonia. Naphthalene, benzene, anthracene, ammonia, phenols, cyanide, cresols, and polycyclic aromatic hydrocarbons commonly known as Hazardous organic compounds are among the gasification products that contaminate waste streams (Environmental Protection Agency, 2002). Before being sold into manufacturing, iron and steel goods undergo their last treatment. Sulphuric acid (H_2SO_4) and hydrochloric acid (HCl) are the two regularly utilized acids, also referred to as soluble oil, that have contaminated numerous wastewater from the steel sector.

Mines and quarries: Slurries of rock particles in water are the main wastewaters produced by mines and quarries. This frequently includes extra metallic elements like zinc and other substances like arsenic for metal mining. High-value metal extraction processes, like those used to extract silver and gold, may result in slimes with incredibly small particles that make it challenging to physically remove contaminants. One of the numerous negative effects of mining on the environment is acid mine drainage, which is frequently enriched in heavy metals.

Battery manufacturing: Cadmium, cobalt, oil iron, silver, zinc, lead, chromium, manganese, cyanide, nickel, copper, mercury, and grease are among the contaminants produced by battery manufacturing companies, that require attention to get recycled (U.S. Environmental Protection Agency, 2017).

Organic chemicals industries: The following organic substances may be released: naphthalene, benzene, phenols, chloroform, vinyl chloride, and toluene. Metal pollutant discharges can include zinc, lead, chromium, nickel, and copper.

Electric power plants: Many electric power plants in the form of wastewater add significant amounts of metals like chromium, cadmium, lead, mercury, selenium, and arsenic are released into the environment. Flue-gas desulfurization, bottom ash, flue-gas mercury controls, and fly ash are examples of wastewater streams. Wet scrubbers and other air pollution control facilities typically add the seized contaminants to the wastewater stream.

Textile industries: Pollutants produced by textile mills include suspended solids, grease and oil, BOD, sulphide, chromium, and phenols (U.S. Environmental Protection Agency, 2017). Due to pesticide residues in fleece, the treatment of wastewater created during the manufacture of wool is particularly challenging. Natural and manufactured colours, gum thickener (guar), pH buffers, various wetting agents, and dye are all present in wastewater from textile dyeing operations. COD, BOD, grease and oil, colour (ADMI), phenol, sulphide, heavy metals, and TSS are typical monitoring indicators after subsequent treatment with settling agents and polymer-based flocculants (copper, chromium, lead).

Petroleum refining and petrochemicals industries: The distinctive pollutants (SS, BOD, oil, COD, and grease), phenols, ammonia, sulphides, and chromium are among the pollutants released by petroleum refineries and petrochemical plants.

Paper and pulp industries: BOD and suspended solids are often high in the effluents from the paper and pulp industries. Chloroform, COD, furans, and phenols may be produced by plants that bleach wood pulp used to make paper, including 2,3,7,8-TCDD.

Industrial oil contaminations: Workshops, car wash bays, fuel depots, transportation, and power generation facilities hubs are examples of industrial applications where oil can enter the effluent stream. Solvents, lubricants, grits, detergents, and hydrocarbons are examples of common contaminants.

Wood preserving: Traditional and harmful pollutants, such as COD, chromium, copper, arsenic, excessively high or low pH, phenols, SS, and oil and grease, are produced by wood preservation operations (Environmental Protection Agency, 2018).

10.3   LOW-COST ALTERNATIVES FOR REMEDIATION OF INDUSTRIAL WASTE

10.3.1   Low-Cost Adsorbents Used for Heavy Metal Removal

Due to the recognized effects of heavy metals on the environment and living things, removing heavy metals from municipal wastewater, industrial wastewater, water supplies, and other wastewater sources has been the topic of substantial research. The most popular methods for removing dangerous metals from wastewater include coagulation, ion exchange, electro-dialysis, flocculation, electrochemical treatment, chemical precipitation, membrane separation, and adsorption (Carlos et al., 2013). The bulk of the systems have several disadvantages, such as higher running costs and problems with sludge disposal brought on by these methods. The most successful method for treating wastewater is adsorption since it is inexpensive, easy to use, and requires little upkeep (Bhattacharyya and Gupta, 2008).

10.3.1.1   Zeolite

Zeolites are hydrated 3D alumina silicate minerals that exist naturally. They are members of the "tecto silicates" mineral class. Molecular sieves are the names given to the structures created when alumina and silica interact. By swapping the cations, zeolites can offer heavy metal ion adsorption sites. In addition to ion exchange and adsorption, surface precipitation, zeolite structure dissolution, and degradation are additional processes that contribute to zeolite's removal of heavy metals. Zeolites, because of their huge surface areas, which range from 600 to 800 m^2g^{-1}. They also have the ability to exchange ions. Depending on the zeolites' grade, zeolite is thought of as being very inexpensive.

The choice of a heavy metal is greatly influenced by adsorption parameters like pH. While at low pH, Cr^{+6} was better adsorbed (Barakat, 2008). According to Ouki and Kavanagh (1997), zeolites, both in their clinoptilolite and in their chabazite forms, are highly effective at removing pollutants like Cu^{+2}, Pb^{+2}, Co^{+2}, Cd^{+2}, Ni^{+2}, and Zn^{+2} from effluent. As a result, there is a better chance for both the ions to interact with the Na ions in the zeolite. At a wide range of pH 5 to 11, magnetically modified synthetic zeolite with Fe_2O_3 shown significant Pb^{+2} adsorption capabilities for and good chemical resistance (Nah et al., 2006). In general, zeolite has a considerable the capacity to extract metal ions from many industrial wastes.

10.3.1.2   Clay

Clays have a number of advantages over other adsorbents, including their widespread availability, low cost, exceptional adsorption ability, strong ion-exchange potential, high specific surface area, and non-toxic nature. Clay minerals are generally very efficient and regularly used to adsorb metal cations because metal cations are typically negatively charged by a number of adsorption mechanisms, including metal cations, ion exchange, and clay surface bonding. There are several different kinds of clays, including mectites like saponite and montmorillonite and mica like illite, vermiculite, kaolinite, and serpentine (Shichi and Takagi, 2000).

It was proposed that the adsorption process for nickel and copper removal by utilizing raw montmorillonite depended more on internal solid pore diffusion than on external mass transfer (Ijagbemi et al., 2009). As studied in other research, sodium–montmorillonite may be more successful than calcium-montmorillonite at adsorbing Cu^{+2}, Zn^{+2}, Pb^{+2}, Cd^{+2}, and Co^{+2}ions (Chen et al., 2015). The ion-exchange capacity of kaolinite clay is low. Electrolyte concentrations ranged from 0.01 to 0.1 molar, and the amount of metal ions Pb^{+2}, Ni^{+2}, Cu^{+2}, and Cd^{+2} that could be absorbed was dramatically reduced. According to Jiang et al. (2010), the quantity of nickel that could be maximally absorbed by kaolinite was 140.84 mg g^{-1}. The naturally available form of vermiculite demonstrated outstanding potential for the adsorption-based removal of metal ions. pH had an impact on Pb adsorption by vermiculite significantly (Liu et al., 2007). Environmental risks must be well assessed before utilizing goethite to remove heavy metals. The first accurate prediction is produced by the extended triple layer model, was tested using As^{+3} and As^{+5} based data (Kanematsu et al., 2013).

Numerous researches have been conducted using clay minerals as a low-cost method, easily accessible, and effective adsorbent material. It may be said that montmorillonite, demonstrates higher adsorption of mercury and chromium contrasted with the alternative heavy metals, based on comparative studies on adsorption capabilities among various clay minerals. Lead, zinc, and nickel can be more effectively adsorbed by kaolin, while cadmium can be effectively adsorbed by bentonite and impressive amounts of arsenic by goethite clay.

10.3.1.3   Chitosan

Chitin is a widely available natural adsorbent that can be found in the cell walls of some fungi and the exoskeletons of insects (Ren et al., 2008). After cellulose, chitosan is the second most prevalent biopolymer and a deacetylated derivative of chitin. Chitosan is not only effective and plentiful but also affordable. Chitosan offers easy derivatization and a number of unique properties, including biocompatibility, hydrophilicity, biodegradability, and non-toxicity (Huo et al., 2009). Because heavy metals include amino and hydroxyl groups, with these, they can create chelate complexes (Hu et al., 2011). Despite these benefits, chitosan has certain drawbacks, including the fact that it is mechanically feeble, pH sensitive, and soluble in an acidic medium (Hu et al., 2011). Therefore, several cross-linkers such as glutaraldehyde have been utilized to increase its stability and adsorption efficiency (Wang et al., 2011).

The inverse emulsion method was used to create chitosan and gelatin spherical hydrogel particles, and the results showed that the Hg^{+2} ion had the highest adsorption effectiveness for single ions (98%) and multiple ions (73–94%) for multiple ions. For the same aim, a team of scientists also created "ion imprint technology," for Cd^{+2} removal which uses urea-modified magnetic ion embossed chitosan@TiO2 composite. According to reports, the order of Cd > Cu > Pb for metal ion 23 adsorption on chitosan (Li et al., 2015).

10.3.1.4   Peat Moss

One of the naturally occurring, less expensive adsorbent materials with a surface area of more than 200 m^2 g^1 and porosity of more than 95% is peat moss (McLellan and Rock, 1988). Lignin and cellulose are two important peat moss constituents that can chemically bond with metal ions by using functional groups including phenolic hydroxides, –OH, ketones, acids, –CHO, and ether. As a result, peat moss functions as a potent adsorbent for the straightforward cation-exchange approach of eliminating dangerous heavy metals. Eutrophic and oligotrophic peats are just two types of peats. According to Babel and Kurniawan (2003), eutrophic peat has a lower cellulose concentration but a higher humic acid content, whereas oligotrophic peat, like sphagnum, is high acidic.

10.3.1.5 Coal

Coals are readily available and reasonably priced everywhere. They serve as a very calorie-dense fuel and retain their qualities even after use. So, raw, hard coal is one of the promising adsorbents. There aren't many papers that talk about using raw hard coals as cheap adsorbents to purify water to eliminate dyes, heavy metal ions, and other impurities (Lakatos et al., 2001; Venkata Mohan et al., 2002). Two of the most important variables affecting the metal ion adsorption by coal are ionic strength and pH.

10.3.1.6 Bio-Adsorbent

Bio adsorption, a recent advancement in the arena of wastewater treatment, is highly effective at removing heavy metals. Because they are made from natural materials, bioadsorbents are cheap and can be used in this procedure quickly and reversibly. Microbial biomass, algal biomass, and non-living materials like shrimp shells, bark, crab shells, and lignin are the three main sources of bio-adsorbent. Algae are the biosorbent that are most readily available in the crust of the earth. The three different types of marine algae are brown, green, and red algae. It could be used as a promising bio-adsorbent because of its exceptional adsorption capacity and renewable availability. They are made up of certain functional active groups, such as hydroxyl and amine on the cell wall's surface, which has a chemical connection with metals. Ion exchange, uptake, chelation, and precipitation are the mechanisms behind heavy metal adsorption.

10.3.1.7 Agricultural Waste

To address the global problem of reducing environmental contamination by "treating waste by waste," researchers are now very interested in the use of agricultural waste, which can serve as a respectable replacement for activated carbon (AC) adsorbent that is already available commercially. Agricultural waste has a large supply, is cost-effective, has good adsorption ability and selectivity, and is simple to process, separate, and regenerate adsorbent, making it a viable biosorbent for heavy metals from wastewater. Heavy metals are bound by functional groups like hydroxyl, carbonyl, and ether by chelating, coordinating, and hydrogen bonding.

An easily available, inexpensive adsorbent made of cellulose, lignin, hemicellulose, and mineral ash is rice husk. The existence of functional groups in rice husk, such as hydroxyl, silanol, and siloxane, indicates a high affinity for the cations of heavy metals. Researchers found that rice husk works well as a cheap adsorbent to remove metals like Zn, Ni, Pb, Cr, and Cu. A potential adsorbent for the adsorption of lead is rice husk. because of the following mechanisms: ion exchange with calcium and magnesium, surface complexation with—OH groups and—COOH, and forces of attraction. With a 95% adsorption capability, biochar, a byproduct of the pyrolysis of rice husk, Chromium may be removed from water using this.

Coconut shell is a well-known, widely used, inexpensive adsorbent that is found in large quantities in tropical countries as agricultural waste. Heavy metals may be effectively removed using coconut shells from wastewater, including Zn, Cu, Pb, Ni, and Cr. According to an experimental investigation, Pb is more readily absorbed by activated coconut shells as compared to unactivated coconut shells (Sharaf El-Deen and Sharaf El-Deen, 2016). Researchers have found that peanut shells are an effective adsorbent for heavy metals removal. Researchers also accomplished an efficient Cr^{+6} removal at low pH values (Ahmad et al., 2017). Fruit peel waste material exhibits impressive heavy metal adsorption the possibility of treating of various wastewaters. Using orange peel, various heavy metals like Cu, Ni, Zn, Cd, and Pb were eliminated. According to studies, chemically altered orange peel improves the adsorption of metals like Cu, Pb, Ni, and Cd. Sweet lime or mosambi peel dust had the ability to adsorb Cr^{+6} ions out of an aqueous solution (Saha et al., 2013).

10.3.1.8 Industrial Waste

Another accessible and affordable source of adsorbent of heavy metals from industrial wastewater. With relatively little processing, adsorbent can be produced from industrial waste, primarily byproducts. For the purpose of removing heavy metals from wastewater, industrial wastes such as lignin, fly ash, grape-stalk trash, red mud, battery industry waste, and blast furnace sludge can be employed.

Fly ash, a byproduct of coal combustion, has a high capacity for adsorption and is employed in a variety of applications. The primary usage of fly ash in research on the adsorption of Hg, Cr, Cd, Pb, Ni, As, and Cu. The maximal metal removal relies on the content of Zn^{+2} and Ni^{+2} and the pH of the solution, according to research on the ability of AC and fly ash to remove Zn^{+2} and Ni^{+2} (Cetin and Pehlivan, 2007). Red mud, a fairly common industrial waste product, is also useful as a low-cost adsorbent. It is a byproduct of the Bayer's method, also known as caustic leaching, used to produce alumina from bauxite ore. Red mud can be used to eliminate the harmful element arsenic. Red mud can be made more adsorbent by thermal and chemical treatment. Red mud activity depends on pH just like the activity of other adsorbents.

10.3.2 LOW-COST ADSORPTION TECHNOLOGIES FOR REMOVAL OF DYES FROM INDUSTRIAL WASTEWATER

Dyes are important types of organic macromolecules. These dyes play a crucial role in our existence and are used in a variety of industrial fields, such as textile, greases, plastics, sensors, paints, optics, leather, pharmaceuticals, waxes, cosmetics, dye-sensitized solar cells, fur, and hair. Some of the characteristics used to categorize dyes include source, colour, molecular structure, and colour index. They can be identified mostly based on the chromophores present in these compounds. Industrial dye wastewater is also treated using practices that are more or less the same, but no one treatment approach was used for all varieties of dye effluent. Although there are numerous methods for dye removal, some of them are less effective due to their flaws (Katheresan et al., 2018).

10.3.2.1 Adsorbent from Agricultural Wastes

Cellulosic compounds, often known as agricultural wastes, are mostly composed of cellulose, lignin, and hemicelluloses (Salleh et al., 2011). Recently, several research teams have started using agricultural wastes to make AC since they are low in inorganics, need less mechanical strength, have low levels of ash content, and are inexpensive (Chowdhury et al., 2011; Zhou et al., 2015). This process has been tested on a variety of agricultural wastes, including bagasse, grape seeds, straw, nut shells, almond shells, coconut husks, maize stover, apricot hulls, oat hulls, peanut cherries, and stones (Aldegs et al., 2008; El Qada et al., 2008).

Natural polymer lignin is primarily made of aromatic molecules. It is a macromolecule with an erratic three-dimensional structure made up of alkylphenols. Due to hardwoods, lignin is covalently linked to xylans, and in softwoods, galacto glucomannans (Parvathi et al., 2010). It has a variety of functional groups, including methoxyl, hydroxyl, and carbonyl, which provide the lignin macromolecule with high extremities (Young, 1986). A byproduct of the bamboo pulp mill was converted into a spherical sulphonic lignin adsorbent. This adsorbent was tested for the removal and recovery of cationic dyes from fluid arrangements, including cationic red GTL, cationic turquoise GB, and cationic yellow X-5GL (Liu et al., 2005).

Among agricultural waste products, sawdust is one of the most interesting materials for removing pollutants like colours, salts, and heavy metals from water and wastewater (Shukla et al., 2002). Lignin, cellulose, and hemicellulose make up the substance, with polyphenolic groups playing a crucial role in binding colours through diverse mechanisms. For the removal of dyes from wastewater, several researchers have used both the natural and modified forms of this saw dust. The researchers have published numerous adsorption kinetics and thermodynamics isotherm models for the removal of dye by sawdust. Moreover, the thermodynamics of how sawdust removes dyes has been described.

Rice husk is a byproduct of the rice processing industry and is a type of agricultural trash. Generally speaking, these industries produce more than 100 million tonnes of trash, 96% of which is produced by developing nations (Savova et al., 2001). The main components of rice husk are silica (15–22% SiO_2) and carbon (Vadivelan and Kumar, 2005). To effectively remove colours from wastewater, many studies have been employing rice husk as an adsorbent (Chakraborty et al., 2011). Trash peels are inexpensive, environmentally beneficial, and natural sources of adsorbents. These wastes can be employed to remove coloured wastewater. Grass waste, guava leaves, plant leaf powder, rice straw, garlic peel waste, barley straw, papaya seeds, jackfruit peel waste, pumpkin seed palm kernel fibre, pumpkin seed hull, banana waste, castor seed shell, Citrus peel waste, pineapple stem, cassava peel waste, coffee husks, coconut husk, dehydrated peanut hull, and other agricultural wastes have all been used successfully to remove colour from wastewater (Kumar et al., 2019).

10.3.2.2   Adsorbent Derived from Industrial Wastes

Typically, strong waste materials are produced in vast quantities as a byproduct of industrial activities. While some byproducts are recycled, the rest are dumped in landfills. Thus, the likelihood of reuse in adsorption forms points to an interesting arrangement, especially given that these industrial waste materials are freely available and pose serious disposal problems. Among the industrial wastes are wastes from the fertilizer business, fly ash, leather industry, paper industry, steel industry, and aluminium industry (Bhatnagar and Sillanpaa, 2010). An unwanted byproduct of ignition operations is fly ash. Ash is mostly used in the construction of buildings, concrete, and other structures. The fly ash debris is a good candidate for use as a modest mass-market adsorbent because to the high concentration of silica and alumina in it (Bhatnagar and Sillanpaa, 2010; Gisi et al., 2016). The materials that have undergone the greatest testing throughout the adsorption process are the blast furnace lag, dust, and sludge. When the bauxite ore is treated to caustic leaching, red mud (aluminium industrial waste) is a waste product that is created (Bhatnagar et al., 2011). Red mud particles present a real contamination risk due to their toxic nature and colloidal structure. There are several ideas for using red mud, including making red mud blocks, filling up black-top streets, making iron, and using it as a source of other minerals (Bhatnagar and Sillanpaa, 2010).

Utilizing industrial byproducts to successfully remove dye contaminants from wastewater. Complex regeneration activities are not necessarily due to the low effort and high accessibility of these wastes. However, as these adsorbents often have low adsorption capabilities, more research and development into additional novel adsorbents is still being done. The goal of the current research is to develop inexpensive and practical alternatives to the current adsorbents using agricultural waste.

10.3.3   TECHNOLOGIES USING INEXPENSIVE ADSORBENTS TO REMOVE PHENOLIC COMPOUNDS

In fact, cost is a crucial factor to consider when contrasting different adsorbent materials. If an adsorbent is plentiful and needs little processing, it might be regarded as low-cost (Bailey et al., 1999). It could also be described as an industry waste product or a byproduct. Natural materials, bioadsorbents, and some industrial and agricultural waste products all have the potential to be more cost-effective alternatives to traditional adsorbents. Many of them have been investigated and put forth for the elimination of phenolic chemicals.

10.3.3.1   Natural Materials

10.3.3.1.1   Clay

A negative charge on the structure of clay minerals is what gives them their adsorption properties. Clay can absorb positively charged substances thanks to its negative charge. Their vast surface area and high porosity also contribute to their adsorption properties (Babel and Kurniawan, 2003). Due to their ability to adsorb both inorganic and organic molecules, clay minerals like kaolinite, diatomite, bentonite, and fullerenes earth have attracted more attention in recent years.

The adsorption of trichlorophenols (TCP) on calcium-montmorillonite (pure clay), manufactured montmorillonite-humic complexes, and natural soils was investigated (Isam and Menahem, 1997). Solutions of TCP, dichlorophenol (DCP), and pentachlorophenol were eliminated utilizing mesoporous alumina aluminium phosphates and alumina pillared montmorillonite. Shen (2004) also used additional organosmectites that were created by removing phenol from cation-exchange smectites by adding benzyltrimethylammonium and hexadecyltrimethyl ammonium ions. The exchanged organic cations' impact on the smectite's adsorptive behaviour in the clay appeared to be related to their size and molecular organization. According to Arellano-Cárdenas et al. (2005), DCP and phenol were adsorbable from water using a porous clay heterostructure.

Recently, clay adsorption and ultrafiltration (clay-UF) have both been employed as part of a hybrid method to remove phenolic chemicals from water (Lin et al., 2006). By using cation exchange to place quaternary ammonium cations on smectite, several clay-organic complexes were created. Then, their ability to absorb phenol and some of its chlorinated congeners has been investigated (Mortland et al., 1986). Additionally, bentonite and perlite were used to study the adsorption of p-chlorophenol from aqueous solutions (Koumanova and Peeva-Antova, 2002). Both kaolinite and montmorillonite may adsorb phenolalone, but when surfactant molecules are present, a significant portion of the adsorption is caused by the solubilization of the phenol molecule within surfactant surface aggregates. A comparison is made between the partition coefficients estimated for bulk micelles and those between an aqueous solution and surface aggregates. It was discovered that phenol had a stronger affinity for surface micelles than for heavy micelles.

10.3.3.1.2 Siliceous Materials

Because of their availability, affordability, and abundance, natural siliceous adsorbents such as silica beads, perlite, alunite, dolomite, and glasses are being used to treat wastewater. Given that their hydrophilic surface is chemically reactive, caused by the presence of silanol groups on silica beads, which is one of the inorganic materials, silica beads demand special attention (Krysztafkiewicz et al., 2002). They are desirable as adsorbents for decontamination applications because of their porous texture, mechanical stability, and high surface area (Ahmed and Ram, 1992). Utilizing silane coupling agents with the amino functional group, the silica surface can be changed to enhance their interaction with phenolic compounds (Krysztafkiewicz et al., 2002). To adsorb phenol isoalumina, another adsorbent made of siliceous materials is used. Alunite, one of the jarosite group's minerals, has a silica content of about 50%. Other siliceous materials have been suggested for phenol elimination, including dolomite, perlite, and glass. Dolomite is both a rock and a mineral. According to a theory, phenolic chemicals are adsorbable onto the perlite. For decontamination purposes, perlite makes a suitable adsorbent.

10.3.3.1.3 Zeolite

Zeolites are sought-after adsorbents because of their great ion-exchange capacity, comparatively high specific surface areas, and, most importantly, their reasonably inexpensive cost (Babel and Kurniawan, 2003). Zeolites have another benefit over resins: their stiff porous architectures, which produce high ion selectivity. Zeolites are increasingly exploited as substitute materials in fields that call for adsorptive functions. They have recently received a great deal of attention due to their potential for eliminating traces of contaminants like phenols and heavy metal ions. Using natural zeolite from Indonesia, phenol was eliminated when microbial growth was present (Syamsiah and Hadi, 2004). *Pseudomonas putida* was used to inoculate the bed, and it was discovered that the microbial mass had retarded adsorption because it had increased mass transfer resistance. The remaining phenol might either be irreversibly adsorbed or transformed into microbial mass and dissolved products.

10.3.3.2 Bioadsorbents

The phrase "bioadsorption" refers to the process of employing biological materials to collect and concentrate pollutants from aqueous solutions. In this case, biological chelating and complexing adsorbents are utilized to remove the chitin, peat, yeast, fungus, chitosan, or bacterial biomass. Phenolic compounds are more difficult to work with since these bioadsorbents and their derivatives include a wide range of functional groups.

10.3.3.2.1 Chitin and Chitosan

One of the well-known bioadsorption methods for the elimination of phenols, even at low concentrations, is the adsorption of phenolic compounds utilizing biopolymers like chitin and chitosan. Estimates place the annual production of crustacean shells at 6–8 million tonnes, and the chitin and protein recovery from these shells provides an extra revenue stream (Teng et al., 2001). In contrast to AC, chitosan is less expensive and has a greater abundance of amino and hydroxyl functional groups. This biopolymer offers an appealing alternative to conventional biomaterials due to its physico-chemical properties, chemical stability, high reactivity, and improved chelatability (Guibal, 2004). Chitosans have a high affinity for phenolic compounds. To boost phenol's ability to bind to matrix *P. putida*, chitosan was immobilized. Researchers have investigated *P. putida* (NICM 2174), a suspected phenol biodegradant, for its capacity to break down phenol in a variety of environmental circumstances (Annadurai et al., 2000).

10.3.3.2.2 Peat

The main components of raw peat are cellulose, lignin, humic acid, and fulvic acid. Because they contain polar functional groups including ketones, alcohols, carboxylic acids, aldehydes, phenolic hydroxides, and ethers, which are particularly present in lignin and humic acid, these substances can form chemical bonds. Raw peat has a number of limitations when used as an adsorbent directly, including limited mechanical strength, poor chemical stability, a strong affinity for water, a tendency to shrink and/or swell, and a tendency to leach fulvic acid (Couillard, 1994). Higher PCP reductions are possible at higher dissolved oxygen (DO) levels, according to batch adsorption tests with peat at various water DO levels (Tanjore and Viraraghavan, 1997).

10.3.3.2.3 Biomass

To successfully remove phenolic compounds from solutions, bioadsorbents made from suitable microbial biomass can be used. This is because some phenolic compounds have a particular tendency for interacting with microbial species. Because it is readily available in big numbers and at a low cost, biomass is being used more frequently to treat wastewater. In fermentation operations, microbial biomass is created to create essential compounds like antibiotics and enzymes. Such methods result in the production of several byproducts that can be employed for the bioadsorption of contaminants. A considerable potential for adsorption exists in biomass because of its physico-chemical properties.

Certain species, such as *Pseudomonas* sp., can break down phenol under tightly controlled pH, temperature, and nutritional conditions. *Pseudomonas aeruginosa*, a strain of *Pseudomonas* capable of digesting pentachlorophenol, has been isolated from tannery soil (Bogan and Sullivan, 2003). *Phanerochaetechrysosporium* can be reused up to ten times without significantly losing its adsorption capacity. The removal of phenol from an aqueous solution using non-viable pre-treated *Aspergillus niger* cells (Rao and Viraraghavan, 2002). The results lead to the notion that the considerable pH-dependent adsorption of phenol is purely a spatially organized pattern of hydroxyl groups.

10.3.3.3 Waste from Agriculture and Industry

It is reasonable to suppose that waste items and byproducts from industry and agriculture serve as non-expensive adsorbents. due to their natural abundance, low cost, and need for little processing.

10.3.3.3.1 ACs from Solid Wastes

In the production of carbon adsorbents, ACs are frequently generated from organic materials including coconut shells, wood, lignite, or coal. However, almost any carbonaceous substance can be used as a precursor. Because it is easily available and affordable, coal is the most widely employed antecedent for the creation of AC (Illán-Gómez et al., 1996). Coal is created when carbonaceous materials and mineral matter mix when plants decay. Each coal's adsorption properties vary depending on the initial vegetation's features and the extent of the physical-chemical changes that took place after deposition (Karaca et al., 2004). It has been studied how aqueous monochlorophenols, DCPs, and TCPs cling to carbons derived from wood and lignite. The extent of 2,4-DCP was found to depend on pH (Colella et al., 1998).

10.3.3.3.2 Agricultural Solid Wastes

Sawdust, rice husk, and bark are examples of unprocessed waste products from the agricultural and forestry sectors that have been used as adsorbents. Sawdust is used to eliminate phenolic components from wastewater. Bark, a common forest byproduct, is helpful in removing phenols from aqueous solutions. Because it is readily available and inexpensive, bark is highly sought after as an adsorbent. Researchers studied the ability of thermally treated rice husk to adsorb DCP from aqueous solutions (Akhtar et al., 2006).

10.3.3.3.3 Fly Ash

Large amounts of fly ash are a waste product that is created during combustion processes. Nonetheless, the sugar industry's bagasse fly ash has been widely used for the adsorption of phenolic compounds and does not contain appreciable amounts of harmful metals (Mohan et al., 2002). According to recognized isotherms, phenol, 2,4-DCP, and 3-chlorophenol were adsorbable from water on fly ash (Akgerman and Zardkoohi, 1996). Using a fly ash column, it was also possible to successfully remove chlorophenols from wastewater; breakthrough times were inversely correlated with flow rates. By employing coal fly ash as an adsorbent, 2,4-dimethylphenol was successfully extracted from aqueous solutions.

10.3.3.3.4 Red Mud

Red mud is a byproduct of the bauxite processing technique used to create alumina. The completed product has been applied to the wastewater treatment process to remove 2,4-DCP, 4-chlorophenol, 2-chlorophenol, and phenol (Gupta et al., 2004). Red mud, a byproduct of the aluminium industry, has the potential to be a cheap adsorbent. The discovered adsorbent can adsorb 2,4-DCP and 4-chlorophenol up to 94–97%, whereas 2-chlorophenol and phenol can be removed up to 50%–81%. At a flow rate of 0.5 mL/min, column tests were able to remove phenols and their derivatives up to 98% of the time. Red mud as an adsorbent to remove 2,4-dinitrophenol (Gupta and Ali, 2006).

10.4 CONCLUSION

To give the reader an understanding of the many low-cost materials utilized for the remediation of industrial wastes, this chapter has made an effort to cover a widespread of low-cost adsorbents and several economically viable treatment methodologies. Commercial AC could be replaced with inexpensive, easily accessible, and efficient alternatives to remove hazardous chemicals from aqueous solutions. The adsorbents utilized in this technique are made from straightforward materials including sawdust, lignin, agricultural wastes, peels from waste fruit and vegetables, industrial waste, and AC, all of which are readily available and inexpensive. Affordable adsorbents unquestionably have a variety of intriguing commercial applications in the future. To determine whether the scenario for using and discarding adsorbents can be repeated, the life cycle analysis of the created material must be examined. The regeneration process for the used adsorbent must be meticulously planned to precisely predict the necessary regenerating strategies.

REFERENCES

Ahmad, A., Ghazi, Z. A., Saeed, M., Ilyas, M., Ahmad, R., MuqsitKhattak, A. M., & Iqbal, A. (2017). A comparative study of the removal of Cr(VI) from synthetic solution using natural biosorbents. *New Journal of Chemistry*, *41*(19), 10799–10807. https://doi.org/10.1039/C7NJ02026K

Ahmed, M. N., & Ram, R. N. (1992). Removal of basic dye from waste-water using silica as adsorbent. *Environmental Pollution*, *77*(1), 79–86. https://doi.org/10.1016/0269-7491(92)90161-3

Akgerman, A., & Zardkoohi, M. (1996). Adsorption of phenolic compounds on fly ash. *Journal of Chemical and Engineering Data, 41*(2), 185–187. https://doi.org/10.1021/je9502253

Akhtar, M., Bhanger, M. I., Iqbal, S., & Hasany, S. M. (2006). Sorption potential of rice husk for the removal of 2, 4-dichlorophenol from aqueous solutions: Kinetic and thermodynamic investigations. *Journal of Hazardous Materials, 128*(1), 44–52. https://doi.org/10.1016/j.jhazmat.2005.07.025

Aldegs, Y. S., Elbarghouthi, M. I., Elsheikh, A. H., & Walker, G. M. (2008). Effect of solution pH, ionic strength, and temperature on adsorption behavior of reactive dyes on activated carbon. *Dyes and Pigments, 77*(1), 16–23. https://doi.org/10.1016/j.dyepig.2007.03.001

Annadurai, G., Rajesh Babu, S., Mahesh, K. P. O., & Murugesan, T. (2000). Adsorption and bio-degradation of phenol by chitosan-immobilized Pseudomonas putida (NICM 2174). *Bioprocess Engineering, 22*(6), 493–501. https://doi.org/10.1007/s004499900092

Aparcana, S. (2017). Approaches to formalization of the informal waste sector into municipal solid waste management systems in low-and middle-income countries: Review of barriers and success factors. *Waste Management, 61*, 593–607. https://doi.org/10.1016/j.wasman.2016.12.028

Arellano-Cárdenas, S., Gallardo-Velázquez, T., Osorio-Revilla, G., López-Cortéz, M., & Gómez-Perea, B. (2005). Adsorption of phenol and dichlorophenols from aqueous solutions by Porous Clay Heterostructure (PCH). *Journal of the Mexican Chemical Society, 49*(3), 287–291.

Awuchi, C. G. (2017). Sugar alcohols: Chemistry, production, health concerns and nutritional importance of mannitol, sorbitol, xylitol, and erythritol. *International Journal of Advanced Academic Research, 3*(2), 31–66. ISSN: 2488–9849.

Awuchi, C. G., & Awuchi, C. G. (2019a). Physiological effects of plastic wastes on the endocrine system (Bisphenol A, Phthalates, Bisphenol S, PBDEs, TBBPA). *International Journal of Bioinformatics and Computational Biology, 4*(2), 11–29. www.aascit.org/journal/archive?journalId=809

Awuchi, C. G., & Awuchi, C. G. (2019b). Impacts of plastic pollution on the sustainability of seafood value chain and human health. *International Journal of Advanced Academic Research, 5*(11), 46–138. ISSN: 2488–9849.

Awuchi, C. G., & Echeta, K. C. (2019). Current developments in sugar alcohols: Chemistry, nutrition, and health concerns of Sorbitol, Xylitol, Glycerol, Arabitol, Inositol, Maltitol, and Lactitol. *International Journal of Advanced Academic Research. Babel, 5*(11), 1–33. ISSN: 2488–9849, S, and Kurniawan, T. A. (2003). Low-cost adsorbents for heavy metals uptake from contaminated water: A review. *Journal of Hazardous Materials B, 97* 3894(02)00263–7, 219–243. https://doi.org/10.1016/S0304

Babel, S., & Kurniawan, T. A. (2003). Low cost adsorbents for heavy metals uptake from contaminated water: A review. *Journal of Hazardous Materials, B97*, 219–243. http://dx.doi.org/10.1016/S0304-3894(02)00263-7

Bailey, S. E., Olin, T. J., Bricka, R. M., & Adrian, D. D. (1999). A review of potentially low-cost sorbents for heavy metals. *Water Research, 33*(11), 2469–2479. https://doi.org/10.1016/S0043-1354(98)00475-8

Bhatnagar, A., & Sillanpää, M. (2010). Utilization of agro-industrial and municipal waste materials as potential adsorbents for water treatment—A review. *Chemical Engineering Journal, 157*(2–3), 277–296. https://doi.org/10.1016/j.cej.2010.01.007

Bhatnagar, A., Vilar, V. J. P., Botelho, C. M. S., & Boaventura, R. A. R. (2011). A review of the use of redmud as adsorbent for the removal of toxic pollutants from water and wastewater. *Environmental Technology, 32*(3–4), 231–249. https://doi.org/10.1080/09593330.2011.560615

Bhattacharyya, K. G., & Gupta, S. S. (2008). Adsorption of a few heavy metals on natural and modified kaolinite and montmorillonite: A review. *Advances in Colloid and Interface Science, 140*(2), 114–131. https://doi.org/10.1016/j.cis.2007.12.008

Bogan, B. W., & Sullivan, W. R. (2003). Physicochemical soil parameters affecting sequestration and mycobacterial biodegradation of polycyclic aromatic hydrocarbons in soil. *Chemosphere, 52*(10), 1717–1726. https://doi.org/10.1016/S0045-6535(03)00455-7

Carlos, L., Einschlag, F. S. G., González, M. C., & Martire, D. O. (2013). *Waste water—treatment technologies and recent analytical developments.* Intech-Open Access Publisher. https://doi.org/10.5772/3443

Cetin, S., & Pehlivan, E. (2007). The use of fly ash as a low cost, environmentally friendly alternative to activated carbon for the removal of heavy metals from aqueous solutions. *Colloids and Surfaces A: Physicochemical and Engineering Aspects, 298*(1–2), 83–87. https://doi.org/10.1016/j.colsurfa.2006.12.017

Chakraborty, S., Chowdhury, S., & Saha, P. D. (2011). Adsorption of crystal violet from aqueous solution onto NaOH-modified rice husk. *Carbohydrate Polymers, 84*, 1533–1541.

Chen, C., Liu, H., Chen, T., Chen, D., & Frost, R. L. (2015). An insight into the removal of Pb(II), Cu(II), Co(II), Cd(II), Zn(II), Ag(I), Hg(I), Cr(VI) by Na(I)-montmorillonite and Ca(II) montmorillonite. *Applied Clay Science, 118*, 239–247. https://doi.org/10.1016/j.colsurfa.2006.12.017

Chinaza, G., Awuchi, A., Chibueze, G., Amagwula, I., Otuosorochi, Igwe, V., & Somtochukwu. (2020). Industrial and community waste management: Global perspective. *American Journal of Physical Sciences, 1*, 1–16. www.iprjb.org/journals/index.php/AJPS/article/view/1043

Chowdhury, S., Mishra, R., Saha, P., & Kushwaha, P. (2011). Adsorption thermodynamics, kinetics and isosteric heat of adsorption of malachite green onto chemically modified rice husk. *Desalination, 265*(1–3), 159–168. https://doi.org/10.1016/j.desal.2010.07.047

Colella, L. S., Armenante, P. M., Kafkewitz, D., Allen, S. J., & Balasundaram, V. (1998). Adsorption isotherms for chlorinated phenols on activated carbons. *Journal of Chemical and Engineering Data, 43*(4), 573–579. https://doi.org/10.1021/je970217h

Couillard, D. (1994). The use of peat in wastewater treatment. *Water Research, 28*(6), 1261–1274. https://doi.org/10.1016/0043-1354(94)90291-7

El Qada, E. N., Allen, S. J., & Walker, G. M. (2008). Adsorption of basic dyes from aqueous solution onto activated carbons. *Chemical Engineering Journal, 135*(3), 174–184. https://doi.org/10.1016/j.cej.2007.02.023

Environmental Protection Agency (2002). *"Textile Mills Effluent Guidelines".* chapters 7. Wastewater characterization. In *The development document for final effluent limitations guidelines and standards for the iron and steel manufacturing point source category. EPA, Environmental protection Agency, 2002* (pp. 7–1ff). EPA 821-R-02–004. EPA. 2017–06–30.

Environmental Protection Agency (2018). *Timber products processing effluent guidelines.* EPA. 2018–03–13.

European Environment Agency (2001). Copenhagen, Denmark. *Indicator: Biochemical oxygen demand in rivers (2001).*

Gisi, S. D., Lofrano, G., Grassi, M., & Notarnicola, M. (2016). Characteristics and adsorption capacities of low-cost sorbents for wastewater treatment: A review. *Sustainable Materials and Technologies, 9*, 10–40.

Goutam, S. P., Saxena, G., Singh, V., Yadav, A. K., Bharagava, R. N., & Thapa, K. B. (2018). Green synthesis of TiO2 nanoparticles using leaf extract of Jatropha curcas L. for photocatalytic degradation of tannery wastewater. *Chemical Engineering Journal, 336*, 386–396. https://doi.org/10.1016/j.cej.2017.12.029

Guibal, E. (2004). Interactions of metal ions with chitosan-based sorbents: A review. *Separation and Purification Technology, 38*(1), 43–74. https://doi.org/10.1016/j.seppur.2003.10.004

Gupta, V. K., & Ali, I. (2006). Removal of 2, 4-dinitrophenol from wastewater by adsorption technology: A batch and column study. *International Journal of Environment and Pollution, 27*(1/2/3), 104–120. https://doi.org/10.1504/IJEP.2006.010457

Gupta, V. K., Ali, I., & Saini, V. K. (2004). Removal of chlorophenols from wastewater using red mud: An aluminum industry waste. *Environmental Science and Technology, 38*(14), 4012–4018. https://doi.org/10.1021/es049539d

Hu, X. J., Wang, J. S., Liu, Y. G., Li, X., Zeng, G. M., Bao, Z. L., . . . & Long, F. (2011). Adsorption of chromium(VI) by ethylenediamine-modified cross-linked magnetic chitosan resin: Isotherms, kinetics and thermodynamics. *Journal of Hazardous Materials, 185*(1), 306–314. https://doi.org/10.1016/j. jhazmat.2010.09.034

Huo, H., Su, H., & Tan, T. (2009). Adsorption of Ag+ by a surface molecular-imprinted biosorbent. *Chemical Engineering Journal, 150*(1), 139–144. https://doi.org/10.1016/j.cej.2008.12.014

Ijagbemi, C. O., Baek, M. H., & Kim, D. S. (2009). Montmorillonite surface properties and sorption characteristics for heavy metal removal from aqueous solutions. *Journal of Hazardous Materials, 166*(1), 538–546. https://doi.org/10.1016/j.jhazmat.2008.11.085

Illán-Gómez, M. J., Garcia-Garcia, A., Salinas-Martinez de Lecea, C., & Linares-Solano, A. (1996). Activated carbons from Spanish coals. 2. Chemical activation. *Energy and Fuels, 10*(5), 1108–1114.

Isam, S., & Menahem, R. (1997). Adsorption-desorption of trichlorophenol in water-soil systems. *Water Environment Research, 69*(5), 1032–1038. https://doi.org/10.2175/106143097X125731

Jiang, M. Q., Jin, X. Y., Lu, X. Q., & Chen, Z. L. (2010). Adsorption of Pb(II), Cd(II), Ni(II) and Cu(II) onto natural kaolinite clay. *Desalination, 252*(1–3), 33–39. https://doi.org/10.1016/j.desal.2009.11.005

Kanematsu, M., Young, T. M., Fukushi, K., Green, P. G., & Darby, J. L. (2013). Arsenic(III, V) adsorption on a goethite-based adsorbent in the presence of major coexisting ions: Modeling competitive adsorption consistent with spectroscopic and molecular evidence. *Geochimica et Cosmochimica Acta, 106*, 404–428. https://doi.org/10.1016/j.gca.2012.09.055

Karaca, S., Gürses, A., & Bayrak, R. (2004). Effect of some pre-treatments on the adsorption of methylene blue by Balkaya lignite. *Energy Conversion and Management, 45*(11–12), 1693–1704. https://doi.org/10.1016/j.enconman.2003.09.026

Katheresan, V., Kansedo, J., & Lau, S. Y. (2018). Efficiency of various recent wastewater dye removal methods: A review. *Journal of Environmental Chemical Engineering, 6*(4), 4676–4697. https://doi.org/10.1016/j.jece.2018.06.060

Koumanova, B., & Peeva-Antova, P. (2002). Adsorption of p-chlorophenol from aqueous solutions on bentonite and perlite. *Journal of Hazardous Materials, 90*(3), 229–234. https://doi.org/10.1016/s0304-3894(01)00365-x

Krysztafkiewicz, A., Binkowski, S., & Jesionowski, T. (2002). Adsorption of dyes on a silica surface. *Applied Surface Science, 199*(1–4), 31–39. https://doi.org/10.1016/S0169-4332(02)00248-9

Kumar, P. S., Joshiba, G. J., Femina, C. C., Varshini, P., Priyadharshini, S., Karthick, M. A., & Jothirani, R. (2019). A critical review on recent developments in the low-cost adsorption of dyes from wastewater. *Desalination and Water Treatment, 172*, 395–416.

Lakatos, J., Brown, S. D., & Snape, C. E. (2001). Application of coals as sorbents for the removal of Cr from aqueous waste streams. *Environmental Geochemistry and Health, 23*(3), 287–290. https://doi.org/10.1023/A:1012252423675

Li, X., Zhou, H., Wu, W., Wei, S., Xu, Y., & Kuang, Y. (2015). Studies of heavy metal ion adsorption on chitosan/sulfydryl-functionalized graphene oxide composites. *Journal of Colloid and Interface Science, 448*, 389–397. https://doi.org/10.1016/j.jcis.2015.02.039

Lin, S. H., Hsiao, R. C., & Juang, R. S. (2006). Removal of soluble organics from water by a hybrid process of clay adsorption and membrane filtration. *Journal of Hazardous Materials, 135*(1–3), 134–140. https://doi.org/10.1016/j.jhazmat.2005.11.030

Liu, M. H., Hong, S. N., Huang, J. H., & Zhan, H. Y. (2005). Adsorption/desorption behavior between a novel amphoteric granular lignin adsorbent and reactive red K-3B in aqueous solutions. *Journal of Environmental Sciences, 17*(2), 212–214.

Liu, Y., Xiao, D., & Li, H. (2007). Kinetics and thermodynamics of lead(II) adsorption on vermiculite. *Separation Science and Technology, 42*(1), 185–202. https://doi.org/10.1080/01496390600998045

Maczulak, A. E. (2010). *Pollution: Treating environmental toxins* (p. 120). ISBN 9781438126333.

Maszenan, A. M., Liu, Y., & Ng, W. J. (2011). Bioremediation of wastewaters with recalcitrant organic compounds and metals by aerobic granules. *Biotechnology Advances, 29*(1), 111–123. https://doi.org/10.1016/j.biotechadv.2010.09.004

McLellan, J. K., & Rock, C. A. (1988). Pretreating land-fill leachate with peat to remove metals. *Water, Air, and Soil Pollution, 37*(1–2), 203–215. https://doi.org/10.1007/BF00226492

Mohan, D., Singh, K. P., Singh, G., & Kumar, K. (2002). Removal of dyes from wastewater using fly ash, a low-cost adsorbent. *Industrial and Engineering Chemistry Research, 41*(15), 3688–3695.

Mortland, M. M., Shaobai, S., & Boyd, S. A. (1986). Clay-organic complexes as adsorbents for phenol and chlorophenols. *Clays and Clay Minerals, 34*(5), 581–585. https://doi.org/10.1346/CCMN.1986.0340512

Nah, I. W., Hwang, K. Y., Jeon, C., & Choi, H. B. (2006). Removal of Pb ion from water by magnetically modified zeolite. *Minerals Engineering, 19*(14), 1452–1455. https://doi.org/10.1016/j.mineng.2005.12.006

Ouki, S. K., & Kavannagh, M. (1997). Performance of natural zeolites for the treatment of mixed metal-contaminated effluents. *Waste Management and Research: The Journal for a Sustainable Circular Economy, 15*(4), 383–394. https://doi.org/10.1177/0734242X9701500406

Parvathi, C., Maruthvanan, T., Sivamani, S., Prakash, C., & Koushik, C. V. (2010). Role of Tapioca Peel Activated Carbon (TPAC) in decolourisation of Red Brown C4R reactive dye. *Indian Journal of Science and Technology, 3*(3), 290–292. https://doi.org/10.17485/ijst/2010/v3i3.14

Rao, J. R., & Viraraghavan, T. (2002). Biosorption of phenol from an aqueous solution by Aspergillus niger biomass. *Bioresource Technology, 85*(2), 165–171. https://doi.org/10.1016/s0960-8524(02)00079-2

Ren, Y., Zhang, M., & Zhao, D. (2008). Synthesis and properties of magnetic Cu(II) ion imprinted composite adsorbent for selective removal of copper. *Desalination, 228*(1–3), 135–149. https://doi.org/10.1016/j.desal.2007.08.013

Rinkesh, W. (2018). *What is solid waste management? Sources and methods of solid waste management—Conserve energy future.*

Saha, R., Mukherjee, K., Saha, I., Ghosh, A., Ghosh, S. K., & Saha, B. (2013). Removal of hexavalent chromium from water by adsorption on mosambi (Citrus limetta) peel. *Research on Chemical Intermediates, 39*(5), 2245–2257. https://doi.org/10.1007/s11164-012-0754-z

Salleh, M. A. M., Mahmoud, D. K., Karim, W. A. W. A., & Idris, A. (2011). Cationic and anionic dye adsorption by agricultural solid wastes: A comprehensive review. *Desalination, 280*(1–3), 1–13. https://doi.org/10.1016/j.desal.2011.07.019

Savova, D., Apak, E., Ekinci, E., Yardim, F., Petrov, N., Budinova, T., . . . & Minkova, V. (2001). Biomass conversion to carbon adsorbents and gas. *Biomass and Bioenergy, 21*(2), 133–142. https://doi.org/10.1016/S0961-9534(01)00027-7

Saxena, G., & Bharagava, R. N. (2017). Organic and inorganic pollutants in industrial wastes: Ecotoxicological effects, health hazards, and bioremediation approaches. In: *Environmental pollutants and their bioremediation approaches* (pp. 23–56). CRC Press.

Sharaf El-Deen, G. E., & Sharaf El-Deen, S. E. A. (2016). Kinetic and isotherm studies for adsorption of Pb (II) from aqueous solution onto coconut shell activated carbon. *Desalination and Water Treatment, 57*(59), 28910–28931. https://doi.org/10.1080/194 43994.2016.1193825

Shen, Y. H. (2004). Phenol sorption by organoclays having different charge characteristics. *Colloids and Surfaces A: Physicochemical and Engineering Aspects, 232*(2–3), 143–149. https://doi.org/10.1016/j.colsurfa.2003.10.014.

Shichi, T., & Takagi, K. (2000). Clay minerals as photochemical reaction fields. *Journal of Photochemistry and Photobiology C, 1*(2), 113–130. https://doi.org/10.1016/S1389-5567(00)00008-3

Shukla, A., Zhang, Y. H., Dubey, P., Margrave, J. L., & Shukla, S. S. (2002). The role of saw-dust in the removal of unwanted materials from water. *Journal of Hazardous Materials, 95*(1–2), 137–152. https://doi.org/10.1016/s0304-3894(02)00089-4

Syamsiah, S., & Hadi, I. S. (2004). Adsorption cycles and effect of microbial population on phenol removal using natural zeolit. *Separation and Purification Technology, 34*(1–3), 125–133. https://doi.org/10.1016/S1383-5866(03)00186-2

Tanjore, S., & Viraraghavan, T. (1997). Effect of oxygen on the adsorption of pentachlorophe-nol by peat from water. *Water, Air, and Soil Pollution, 100*, 151–162.

Teng, W. L., Khor, E., Tan, T. K., Lim, L. Y., & Tan, S. C. (2001). Concurrent production of chitin from shrimp shells and fungi. *Carbohydrate Research, 332*(3), 305–316. https://doi.org/10.1016/s0008-6215(01)00084-2

United States Environmental Protection Agency (2017). *Battery manufacturing effluent guidelines.* US EPA. 2017–06–12.

Vadivelan, V., & Kumar, K. V. (2005). Equilibrium, kinetics, mechanism, and process design for the sorption of methylene blue onto rice husk. *Journal of Colloid and Interface Science, 286*(1), 90–100. https://doi.org/10.1016/j.jcis.2005.01.007

Venkata Mohan, S., Chandrasekhar Rao, N., & Karthikeyan, J. (2002). Adsorptive removal of direct azo dye from aqueous phase onto coal based sorbents: A kinetic and mecha-nistic study. *Journal of Hazardous Materials, 90*(2), 189–204. https://doi.org/10.1016/s0304-3894(01)00348-x

Wang, J. S., Peng, R. T., Yang, J. H., Liu, Y. C., & Hu, X. J. (2011). Preparation of ethylenediamine-modified magnetic chitosan complex for adsorption of uranyl ions. *Carbohydrate Polymers, 84*(3), 1169–1175. ttps://doi.org/10.1016/j.carbpol.2011.01.007

Yatoo, A. M., Hamid. B., Sheikh, T. A., Ali, S., Bhat, S. A., Ramola, S., Baba, Z. A., & Kumar, S. (2024). Global perspective of municipal solid waste and landfill leachate: Generation, composition, ecotoxicity, and sustainable management strategies. *Environmental Science and Pollution Research, 31*(16), 23363–23392

Young, R. A. (1986). Structure, swelling and bonding of cellulose fibers. In: *Cellulose, Modification, and Hydrolysis* (pp. 91–128). Wiley & Sons.

Zhou, Y., Zhang, L., & Cheng, Z. (2015). Removal of organic pollutants from aqueous solu-tion using agricultural wastes: A review. *Journal of Molecular Liquids, 212*, 739–762. https://doi.org/10.1016/j.molliq.2015.10.023

11 Waste Management in Past, Present, and Future

Fayaz A. Malla, Mukhtar Ahmed,
Afaan A. Malla, Suhaib A. Bandh,
and Nazir A. Sofi

11.1 INTRODUCTION

The evolution of waste management and the evolution of the language used to describe it are inextricably linked. A cursory examination reveals that three distinct vocabularies have developed to define what we now term "garbage." Déchet (French) comes from the verb choir (to fall); refuse (also known as garbage) in English (usually referring to animal offal); rifiuti (Italian); residuo (Spanish); and Abfall (German) all have connotations of waste and inefficiency. In the second group, words like "immondice" (French) and "immondizia" (Italian), derived from the Latin word for "clean," mundus, and ordure (French), derived from the Latin word for "awful," horridus, highlight the filthy or repulsive quality of these materials. Finally, the names in the third group explain the components of the garbage: "boues" in French, "spazzatura" in Italian, "Müll" and "Schmutz" in German, and "rubbish" in English, from rubble (Harpet, 1998). The word "waste" certainly fits into the first group. Derived from the mediaeval French word "vastum," which meant "empty" or "desolate," the term was originally used to describe a barren, destroyed, or abandoned landscape. In later times, it came to indicate frivolous or unnecessary spending (much as déchet does in French). In the 15th century, it started being used to imply what it means now. That "trash," originally described as a location, suggests that its original meaning was not neutral, much like "spazzatura," from the verb "spazzare" (to make room, eliminate clutter) also has a spatial component. It's also true of the extensive jargon used to define different kinds of trash, which we've just scratched the surface of here. Managing urban pee and excrement has long been intertwined with the waste problem, and the two issues are sometimes thought of as synonymous.

After a brief introduction to the time between Antiquity and the beginning of the Industrial Revolution, we will examine the time between 1770 and 1860, a time when the value of excreta, especially urban excreta, rose due to its agricultural and industrial significance. Then, from the 1870s through the 1960s, we'll show the earliest examples of trash in the form of discarded objects and materials. Finally, we'll show how the environmental catastrophe of the 1960s and the 1970s morphed into a garbage crisis for which only partial remedies have been developed. Though garbage is a global problem, this book focuses on the history

DOI: 10.1201/9781003359326-11

of garbage in Europe and North America (Melosi, 2005; Strasser, 1999; Barles, 2005; Giuntini, 2006) and avoids discussing garbage in the countries of the former Eastern Bloc (Gille, 2007).

At the current time, the primary goal of waste management in advanced economies is to prevent pollution. The situation is not so straightforward in terms of waste management goals for a developing country like India. Waste management is crucial for a variety of reasons, including but not limited to environmental protection, productivity enhancement, job creation, resource recovery, and meeting the welfare requirements of a large population. Determining waste management's goals in such situations is a complex issue with numerous angles to consider. Waste management can have a wide variety of aggregate and sector-level goals, all of which are interconnected and, as a result, structured in some way. As a result, structure alone is often insufficient to understand the nuances of many interactions. They call for additional investigation. Keeping all the temporal developments in waste management in view, this chapter tries to present how waste management strategies have changed over the period.

11.2 HISTORY OF WASTE MANAGEMENT

11.2.1 History in Early Times

Because of the long history of using streets and other urban open spaces as dumps for everything from human and animal waste to organic materials from the home and workshop to demolition rubble and mineral debris, the make-up of the soils in these areas can be used to reconstruct the city's past. Since many cities were built on low, even marshy ground, and because streets and squares were often unpaved, the impregnation of waste into the soil was especially important.

The famous Roman cloaca maximum was constructed during the reign of Tarquin the Proud (seventh to 6th century BC) to remove excess water from the Velabrum and the lowlands between Capitoline Hill and Palatine Hill to clean up the dirty metropolitan area.

It discharged sewage and other waste from cities into the Tiber through an underground sewage system and an aboveground canal network.

Cleaning up Rome was aided by subsidiary lines, such as ditches from homes, that connected to the cloaca maxima. Thus, Pliny the Elder called Rome a "hanging city" (urbs pensilis) because of its perch on these tunnels below ground. The cloacarium tax was based on the work of ex-cons cleaning and maintaining sewers. This sort of public centre might be seen in a lot of different cities back then.

In Europe, the Middle Ages saw a gradual drop in the usage of these underground pipelines to drain and clean urban areas. This was followed by the rise of surface runoff for rainwater and drainage fluids. The brooks and moats of cities served as sewage systems (in France, these waterways are still known as Merdereau and Merderet). To facilitate the growth of their artisan operations, many cities diverted, canalized, and developed networks of drainage systems (at first, this water served a mechanical role; subsequently, depending on its composition, it served a chemical and biological role). Although these canals were mostly thought of as sewers due to their role in draining, they played a much more important part in the establishment of urban prosperity (Guillerme, 1983, 1988).

The classification of human waste also changed over time and between different locations. Some cities kept using the same mixed sewerage system that had been in place since Roman times, while others switched to using pit privies—at first just pits, and then subterranean reservoirs put under dry latrines—during the Middle Ages or the Renaissance.

An entirely new occupation, that of the cesspool emptier, arose in response to the rising popularity of these facilities (although in some cities local growers did this job). In addition, the need for sanitation caused many towns to ban the dumping of human waste into sewers and waterways; in Paris, the Great Sewer ("Grand Égout"), a former backwater of the Seine River, has been draining the Right Bank of the city since the 13th century. Residents of metropolitan areas, however, frequently disregarded these prohibitions as well as any potential sweeping and cleaning duties (Boudriot, 1990).

A significant elevation of the ground caused the ground floors, and in some cases, the second floors, of houses to become buried in these cities because only a small fraction of the streets were paved, the slope of the streets was not regulated, the human and animal populations were extremely dense, and cart and other tip-cart traffic contributed to the formation of a putrid mud. To some extent, the piling of urban waste and rubbish near the surface led to a rise in ground level. It happened at different periods and usually picked up speed after demolitions during the war. Artificial embankment buildings, which turned marshy areas into developable land and whose construction material was often a type of trash (such as excavated debris from moats, demolition rubble, and urban mud), also contributed to an overall upward trend.

Some cities in the Middle Ages had dumpsites where large amounts of trash were dumped after being collected after pit latrines had been drained or street cleaners had been dispatched (Chevallier, 1849). As the city expanded, the dumps were moved to areas outside the new metropolitan limits, where they frequently reached the height of real hills. In Paris, for example, these mounds have become an integral part of the cityscape; since they are elevated above the surrounding area, they can house windmills with more efficiency.

The maze-like structure of the Jardin des Plantes is yet another remnant of a former dump. The Monceau Saint-Gervais, which used to be positioned just behind the Hôtel de Ville, and the motta papellardorum, which was on the westernmost point of the Île de la Cité, are two other mounds that have since vanished (Belgrand, 1887). From the 15th to the 18th centuries, European urban centres experienced a general decline in sanitation standards (Leguay, 1999). Two movements began in the 18th century and led to a new perspective on how human waste should be handled in urban areas.

11.2.2 History in Modern Era

The arrival of industrialization and the continual growth of large population centres in England led to rapid deterioration in urban areas in terms of cleanliness and quality of life. As there were no regulations in place for trash pickup, the streets quickly became overrun with litter. According to Londoner Corbyn Morris in 1751,

> the cleaning of this city, should be put under one uniform public supervision, and all the dirt be . . . conveyed by the Thames to proper distance in the country, as the maintenance of the health of the people is of vital importance.

This was one of the first proposals to establish a government agency with the power to collect and dispose of garbage (Herbert, 2013). The rubbish management system established in London in the latter part of the 18th century was revolutionary (Velis et al., 2009). The "dust yards" were surrounded by a waste management and recycling system. Coal ash, or "dust," was the most common substance found in landfills because of its widespread use as a construction material and soil amendment. After the "rag-and-bone men" of the informal economy had stolen useful products, dust contractors swept all of the rubbish off the streets due to the large profit margins. Thus, this was one of the earliest examples of citywide trash collection and disposal. The dust-yard system was effective until the middle of the 1850s when the market value of "dust" dropped dramatically. This was a crucial factor in England's transition to a centralized, government-run waste management system (Velis et al., 2009).

The horrific impact of cholera outbreaks prompted the first comprehensive public health legislation to be drafted in the middle of the 19th century (Chadwick, 1842). The Sanitary Condition of the Labouring Population was a major impetus for this change. Chadwick argued that better sanitation would have a profound effect on the health and well-being of city dwellers. Chadwick's views were debunked in the early 20th century since they were based on the miasmatic notion of disease transmission.

Since the passage of the Nuisance Removal and Disease Prevention Act in 1846, London's waste regulations have undergone constant revision. Households were required to place their weekly rubbish in "moveable receptacles" (the first dust bins) for disposal with the passage of the Public Health Act in 1875, and the Metropolitan Board of Works was the first municipal authority to centralize sanitary control for the rapidly growing metropolis. The Ashanti Empire's Public Works Department provided essential services to Kumasi and the surrounding area. Daily, they swept and weeded the streets and mandated that residents clean up their yards. Because of the ever-increasing volume of trash that needed to be disposed of, incinerators (or "destructors" as they were once called) were invented (Maier, 1979). Alfred Fryer developed the first incinerator, which was built by Manlove, Alliott & Co. Ltd. in Nottingham in 1874. Nonetheless, they encountered widespread opposition from their neighbours (Herbert, 2007) due to the massive volumes of ash they produced (Gandy, 1994). Throughout the early 1900s, several large cities in Europe and North America installed municipal rubbish disposal systems comparable to those used today. In 1895, New York became the first U.S. city to offer a public waste collection service (Oatman-Stanford, 2013).

In the beginning, garbage trucks were little more than open-sided dump trucks pulled by sleighs. The first close-body trucks with a dumping lever mechanism to decrease odours were introduced in Britain in the 1920s, and this mechanization occurred in the early 20th century. Soon after, "hopper mechanisms" were incorporated, enabling the scooper to be loaded at ground level before being lifted and emptied into the truck. The Garwood Load Packer, introduced in 1938, was the first vehicle to feature a hydraulic compactor.

In the 19th century, horse excrement was a common source of odour in American cities. Even though 19th-century noses were able to tolerate the odour, bare feet on the sidewalks were a terrible idea. Since trash wasn't being picked up, pigs and dogs feasted on the garbage and left behind offensive odours. Dead horses in particular were a major source of the disease outbreak (Tarr, 1971).

11.2.3 WASTE MANAGEMENT IN THE CONTEMPORARY WORLD

Waste is produced as a byproduct of human activity and must be managed, stored, collected, and disposed of in ways that might be harmful to the environment and human health (Saxena et al., 2010; Zhu et al., 2008). Growth in the economy increased urbanization, and higher urban resident incomes all contribute to an increase in solid waste production and its increasingly complicated nature (Yatoo et al., 2024; Gidde et al., 2008; Rathi, 2007). In the context of discussing solid waste, particular types of waste are universally acknowledged as being of high importance. Some types of solid waste are: residential, commercial, industrial (from building and tearing down), agricultural, institutional, and other. Wastes from homes and businesses are often lumped together and referred to as "urban wastes" (Syed, 2006).

Household and business trash generated by a human community make up the bulk of what is known as "municipal solid waste" (MSW) (Rajkumar et al., 2010). Poverty, bad governance, urbanization, population expansion, low standards of living, low levels of environmental consciousness (Rachel et al., 2009; Ogu, 2000), and inadequate management of environmental knowledge are all linked to the rapid increase in the careless dumping of municipal solid garbage (Yatoo et al., 2024). Materials that can be broken down biologically include paper, textiles, food waste, straw, and yard waste, while materials that can only be broken down partially include wood, disposable napkins, and sludge. Non-biodegradable materials include leather, plastics, rubbers, metals, glass, ash from fuel burning such as coal, briquettes, or woods, dust, and electronic waste (Jha et al., 2011; Herat, 2009; Tchobanoglous et al., 1993). There has been a massive shift from rural to urban areas in India as a result of the country's rapid industrialization and population boom, and as a result, cities now produce hundreds of tonnes of MSW every day. The widespread presence of MSW is the result of insufficient collection and transportation (Bundela et al., 2010; Gidde et al., 2008). The lack of adequate facilities to treat and dispose of the greater volumes of MSW generated every day in metropolitan cities has reached a crisis level in MSW management. Disposing of MSW scientifically harms all aspects of the environment and human health (Gupta et al., 2007; Rathi, 2006; Sharholy et al., 2005; Jha et al., 2003).

Many studies have found that cities in developing countries are poorly managing their solid waste (Mohanty et al., 2014; Das and Bhattacharya, 2013; Noorjahan et al., 2012; Jafari et al., 2010). As a result of sloppy trash disposal, solid waste has joined the ranks of the pollution sources responsible for a wide range of negative effects on the environment and people's health and safety (Shazwin and Nakagoshi, 2010).

Even though India has the world's second-largest urban population, it ranks among the least urbanized countries. More than 42% of India's urban population lives in India's four largest metropolitan areas (Delhi, Mumbai, Kolkata, and Chennai) and these cities are also among the country's top MSW producers due to their high occupancy rates. According to the Central Pollution Control Board (CPCB), the total quantity of Solid waste generated in the country is ~160,000 metric tonnes per day (TPD). About 153,000 TPD of waste is collected at a collection efficiency of ~96%, 80,000 TPD (50%) of waste is treated, and ~30,000 (18.4%) TPD is landfilled. To assess MSWM's current state and highlight its shortcomings, this study aims to provide a full evaluation of the topic for Indian cities. Institutional/financial concerns are highlighted as the most critical ones impeding advancements in SWM in a study of the literature on SWM in India (Parthan and Milke, 2009). To establish and implement realistic investment estimates, local and state governments require better medium-term planning, as noted by Hanrahan et al. (2006).

11.3 TREATMENT AND DISPOSAL OF SOLID WASTE

11.3.1 COMPOSTING

Various odour and smell issues might arise from the decomposition of the organic materials included in MSW. The MSW disposal process should be as pollution-free as possible, hence it is recommended that the waste be processed in several ways. When organic materials are composted, they undergo microbial decomposition in a combination of aerobic and anaerobic conditions at high temperatures and high moisture. To treat the organic component of MSW, composting is the simplest and least expensive process available. It works well with the organic, biodegradable byproducts of MSW, yard (or garden) trash/waste containing a large proportion of lignocelluloses components (which do not rapidly decompose under anaerobic conditions), waste from slaughterhouses, and waste from dairies. There are two types of composting methods: aerobic and anaerobic. Aerobic microorganisms in compost piles turn organic matter into carbon dioxide, nitrite, and nitrate during the aerobic composting process. Anaerobic microorganisms perform a process of reduction on organic molecules during their metabolism of nutrition during the anaerobic phase. When applied to land, the final product of an anaerobic process undergoes some little oxidation. Research into the benefits of employing compost made from MSW has led to its promotion in many places across the world (Gautam et al., 2010; Pokhrel and Viraraghavan, 2005; Abigail, 1998; John, 1997).

11.3.2 VERMICOMPOSTING

The process of using earthworms to convert organic waste into nutrient-rich soil is called vermicomposting (Yatoo et al., 2020). Earthworms break down any kind of organic waste (like kitchen waste, aquatic waste, and municipal waste) and excrete the processed material into humus-like material rich in organic material and stable plant-available nutrients (Yatoo et al., 2022). Vermicompost typically has a lot more nutrients than regular garden compost (Yatoo et al., 2021). The bio-fertilizer known as vermicompost is odourless and dark brown in colour (Manyuchi and Phiri, 2013; Manyuchi et al., 2013; Abbasi et al., 2009). Countries like Canada, the United States, Italy, and Japan are among the pioneers in the commercialization of vermicomposting operations. This is the perfect time for India to start considering how to implement vermitechnology on a large scale (Aalok et al., 2008).

11.3.3 LANDFILLING

The process of landfilling comprises covering rubbish with a variety of liners and, eventually, dirt. The cheapest alternative, especially in developing countries, is to throw garbage into a depression or an old mine (Daskalopoulous et al., 1998). A landfill is a safe place to discard garbage without worrying about the public. The bottom liner and the top cover of a landfill are, by and large, considered to be the two most crucial components. Seepage of leachate is a major problem for landfills (Alekhya et al., 2013). Landfills represent a risk to human health and the environment because of the toxic gases and leachate they release (Yatoo et al., 2024). Methane and carbon dioxide (up to 90%) are the most abundant components of landfill gases (LFGs), which are produced by methanogens during the breakdown of complex molecules (El-Fadel et al., 1997). Leachate can contaminate both surfacewater and groundwater (Smahi et al., 2013; Nagarajan et al., 2012; Aljaradin and Persson, 2012).

11.3.4 BIO-METHANATION

Bio-methanation, also known as anaerobic digestion, is the process by which microorganisms break down organic matter in wastes into methane and manure without the need for oxygen. Due to the high organic content of wastes generated by agro-based companies, bio-methanation is the most economically sound method of treating this waste, yielding valuable byproducts such as biogas and composted manure. Acidification and methanization are the two steps in the bio-methanation process (CPCB, 2007). Biogas, which is made up of methane and carbon dioxide, can be burned as fuel or transformed into energy using a generator (Annepu, 2012). In December 2003, Lucknow launched a high-rate bio-methanation-based 5-MW MSW-based electricity facility. Asian Bio-Energy Ltd. of Chennai carried it out on a BOOM (Build, Own, Operate, and Maintain) basis. The mill can process 500–600 tonnes of MSW per day from Lucknow (Axelsson and Kvarnstrom, 2010). Vegetable and fruit waste bio-methanation schemes have been proposed for several different urban areas (Prakash and Singh, 2013; Velmurugan and Ramanujam, 2011).

11.3.5 INCINERATION

Raw or unprocessed garbage can be utilized as feedstock in the incineration process, which is a thermal waste treatment method (Zaman, 2010). When garbage is incinerated at 850°C in the presence of air, it produces carbon dioxide, water, and non-combustible elements with a solid residue, known as bottom ash (Zaman, 2009; DEFRA, 2007). Incinerators can be classified as either moving or fixed grates, rotational or fluidized beds (Moustakas et al., 2010). In most urban areas of India, incineration is restricted to hospitals and other biological garbage. The low calorific value content (800–1,100 kcal/kg) and significant organic material (40–60%) in solid waste may be to blame (Rajput et al., 2009; Kansal, 2002; Joardar, 2000). Studies on how to best deal with the ash left over from MSW incinerators have been conducted on multiple occasions (Sabbas et al., 2003; Sakai and Hiraoka, 2000). In 1987, an incinerator with a capacity of 3.75 MW was built in Timarpur, Delhi, to convert 300 TPD of MSW into electricity. The low net calorific value of MSW rendered its operations impossible. The factory is idle, and the money is wasted.

Claims that burning combustible solid waste is an absolute treatment method for eradicating pathogenic components found in such wastes are not without dispute, but the claims are not without merit either. However, incineration is also linked to the production and release of carcinogenic and toxic compounds, and the widespread consumption of potential hazards stemming from incineration is overemphasized relative to any hazards that are perceived to occur from non-thermal treatment or disposal methods, especially in countries where the performance of waste management and treatment facilities has failed to join public confidence. Even though incineration has declined as a result of ever-decreasing detection limits in analytical chemistry, the denial of hazardous releases has been effective due to the failure to establish the non-effective concentration thresholds from either individual or mixtures of gas phase combustion products.

11.3.6 Pyrolysis

When a solid fuel is heated without an oxygen source, a thermochemical change known as pyrolysis occurs (in an inert atmosphere). There are three byproducts of the pyrolysis process: (i) a gaseous mixture, (ii) a liquid (bio-oil/tar), and (iii) a solid residue (char). Fast pyrolysis, used for making bio-oil, and slow pyrolysis, used for making charcoal, are two technologies that exist but use different heat transfer processes (Becidan, 2007). Because of the high CO concentrations in the produced gases, they cannot be vented directly into the cabin without additional treatment, which is one of the main drawbacks of pyrolysis processing (Serio et al., 2000). The pyrolysis of solid waste to produce energy has been the subject of numerous high-quality studies (Serio et al., 2001, 2003; Serio and Kroo, 2002).

11.4 TREATMENT AND DISPOSAL OF LIQUID WASTE

Wastewater or sewage that is generated from a home or community including toilet, bath, There are three broad categories into which the wastewater and sewage produced in a home or community (such as those from a bathtub, sink, washing machine, shower, or lavatory, and surface runoff) might be placed.

a. Sanitary sewage from private residences, public facilities, and commercial and industrial facilities is known as "domestic sewage."
b. The term "industrial sewage" refers to wastewater that has been utilized in industrial operations and typically contains a wide range of chemical compounds.

c. Rainfall's surface runoff, known as storm sewage or stormwater, collects organics, suspended and dissolved particles, and other contaminants as it flows over the land.

11.4.1 SANITARY SEWAGE DISPOSAL

This technique is used to dispose of certain liquid wastes; in addition to preventing leaks, it lowers the overall disposal cost (NIUA, 2021). Before being thrown out, it undergoes a chemical process of disinfection to remove any remaining bacteria or microbial burden. For chemical treatment, you can use a sodium hypochlorite solution of 1% with a minimum contact time of 30 minutes, or you can use any of the other standard disinfectants available, such as 4% formaldehyde, 70% ethanol, 70% isopropyl alcohol, 6% hydrogen peroxide, or 2.5% povidone–iodine. Culture media disinfection is very different from regular disinfection due to the high microbial load and rich protein content of culture media, which necessitates rigorous disinfection, where inactivation will be done by 5.23% sodium hypochlorite in a 1:10 dilution and should be left for a minimum of 8 hours before disposal into the sanitary sewer, followed by flushing with large volumes of cold water for a minimum of 10 minutes. Bleach, or sodium hypochlorite, is an excellent disinfectant against a wide variety of microorganisms, including non-enveloped viruses like *Candida albicans* and *Mycobacterium avium* complex, as well as vegetative bacteria like *Pseudomonas*, *Staphylococcus aureus*, and *Salmonella*. Depending on the brand, household undiluted bleach can sit on a shelf for anywhere from six months to a year before it completely breaks down into salt and water.

11.4.2 PLACING DIRECTLY IN THE BIOHAZARDOUS WASTE BIN

In this case, the liquid waste is disposed of immediately into a red or yellow biohazard container, depending on the intended treatment at the medical institution. For liquid waste, a yellow biohazard bin is used in hospitals with incinerators, whereas a red biohazard bin is used in hospitals without incinerators, and the contents of the red bin are then autoclaved. Sterile liquid waste sterilization typically takes 45 to 90 minutes at 250°F, and the autoclave pressure should be 15 psi.

11.4.3 LIQUID WASTE SOLIDIFICATION

When a powdered solidifying agent is added to liquid waste containers, the contents harden within 5 to 10 minutes, rendering the fluids no longer hazardous and removing the need to carry them in a liquid state. When the solidification is complete, it may be thrown away in a red garbage bag. When it comes to turning liquid waste into solid trash, this application of microencapsulation technology utilizes dry granular highly absorbent polymers that can hold enormous volumes of liquids. Before the solidification process may begin, some solidifiers add sanitizing agents like chlorine or glutaraldehyde (Spill Control Technologies Ltd., 2023). Since these polymers can soak up liquids 300 times over with only a 1% increase in volume, the amount of water-based spills and the associated transportation costs were cut by around half.

There are several drawbacks associated with this technique, such as the fact that the canisters get too heavy after solidification and the hospital has to monitor for the landfill operators to deposit the solidified waste. Since these powders haven't been put to the test in actual bodily fluids, we can't be sure of their efficacy, and they also increase pollution in our landfills.

11.4.4 Closed Disposal Systems

Most healthcare institutions either flush treated or untreated liquid waste down the drain, dispose of full or partially filled containers of liquid waste, intact as red bag trash, or solidify the waste, which is then disposed of in either a red or yellow bag, depending on the state's regulations. Fluid waste systems of today are built on a closed disposal system, flushing wastewater directly into the sewer with minimum human intervention. As a result of many of them being connected to the vacuum system, the canisters are emptied directly into the sewer, lowering the danger of exposure.

The closed disposal system has numerous downsides, such as the need for a sizeable initial investment and the need for a lot of human labour to complete tasks including collection, processing, log book maintenance, cleaning, disinfecting collection, and redistribution. When these containers require servicing, that work must be done as well (Hoffmann, 2023).

11.4.5 Chemically Hazardous Liquid Waste Disposal

Formaldehyde and other chemicals obtained from pathology labs and embalming techniques are among the most common types of hazardous waste. These chemicals include solvents like xylene, acetonitrile, acetone, ethanol, isopropanol, Toluene, methanol, and ethyl acetate. The mercury released by damaged thermometers poses a similar risk to public health. When there is a need to dispose of huge amounts of formalin, the chemical is burned (University of California, 2023). The majority of hospitals' radioactive waste comes from the nuclear medicine department, and it may be safely stored there until it decays to the point where it can be safely transferred and disposed of.

11.4.6 Pharmaceutical Liquid Waste Disposal

Hospitals generate a lot of trash, the majority of which is liquid pharmaceutical waste that, if detected in tiny volumes, may be readily diluted with water and sent to the sewers. It can be returned to the provider, sent to a secure landfill, or even burned if the volume of garbage is small enough. An estimated 250 million pounds of medicines are lost annually in the United States sewage system.

11.4.7 Disposal Procedures for Photographic Liquid Waste

When it comes to getting rid of liquid waste from photography, 25 L containers are commonly utilized for temporary storage. Liquid photogenic waste requires special attention during the phases of storage, management, and transportation. The Resource Conservation and Recovery Act classifies silver as a dangerous substance when its concentration is more than 5 parts per million. Most of the silver is extracted by silver recovery equipment before being discarded.

11.4.8 WATER TREATMENT PLANT

Medical facilities such as operating rooms, labour and delivery rooms, laboratories, laundries, kitchens, and restrooms all create liquid waste, which must first be collected, disinfected, and then released into a central drainage system. Dangerous illnesses might spread if this garbage is dumped in water sources like rivers and lakes. Hospitals need also to build their effluent treatment plants to recycle their wastewater and comply with the Biomedical Waste Management Act of 1998. Because no effluent treatment plant is available, hospitals must first chemically treat their wastewater before discharging it into a shared sewage system that is piped into municipal water purification plants. Whether the wastewater contains organic material, inorganic material, or microorganisms may be determined with the help of BOD and COD tests. Both biodegradable and oxidizable contaminants may be measured by their oxygen requirements; BOD and COD, respectively. A high BOD in wastewater suggests that there is an abundance of organic carbon, which is very polluting. For proper disposal, wastewater (Ministry of Environment, Forest and Climate Change, 2021) must be treated to get rid of most of the contaminants and then be discharged into the environment safely. When it comes to cleaning their wastewater, hospitals with their effluent treatment facility use a three-stage process, starting with primary treatment and progressing through secondary and tertiary stages (Kadam et al., 2008).

11.4.8.1 Primary Treatment

This system involves collecting sewage in a basin, removing the solids that have settled and those that have floated to the surface, and then treating the residual liquid. The treatment reduces BOD and COD concentrations to 25% of their original levels.

11.4.8.2 Secondary Treatment

Local, water-borne microorganisms execute the process of biological waste removal. Microbes degrade organic molecules in either an aerobic or anaerobic secondary treatment. Using both primary and secondary therapy, we can reduce BOD by 80–90%. Sludge is the name given to the slurry that forms when materials in suspension sink to the bottom. The final stage of purification removes the last of the contaminants from the water, often around 95%.

11.4.8.3 Tertiary Treatment

In this process, chemicals are employed to get rid of the inorganic substances and the harmful bacteria. Phase three of therapy is complete. Here, the effluent is treated with sodium hypochlorite before being filtered using sand, anthracite, and activated carbon in a dual media filter and an activated carbon filter, respectively (Nemade et al., 2009).

11.5 PROSPECTS OF WASTE MANAGEMENT

Rapid garbage generation is a global problem, and appropriate waste management of the resulting mountains of trash is still a mystery. Almost everyone will agree that the current methods of garbage collection relying on human labour are inefficient. The proper management of trash includes all activities connected to waste, from its inception to its final disposal, and is given by municipalities or various firms dealing with garbage and preserving a circular economy to promote the health and happiness of its residents. The proliferation of hazardous waste, which needs careful management to mitigate the risks it poses, is a major factor driving the demand for improved waste administration. To be sure, this is just one of the fundamental truths. We are now hearing even more about the environmental and economic benefits of effective waste management. However, what does the future hold for waste management?

Improvements in technology are the foundation and the future of waste management. The waste management sector, like many others, will benefit from digitization and a data-driven approach if it is to continue developing. Smart and competitive times lie ahead. Businesses need to stay ahead of the curve. Applying waste management systems that utilize big data over time allows for the collection of useful information. With this information at hand, sensors can determine fill trends, improve driver schedules and routes, and cut expenses. The price of these sensors continues to drop, making smart bins an increasingly appealing option for businesses and municipal governments.

When we talk about how smart the future is, we're also referring to how useful it will be in everyday life. Because of careful container selection, garbage collection teams are utilized as little as possible. Minimizing the time and effort put into collecting procedures maximizes returns on investment. Utilizing user-friendly, comprehensive, and compact platforms and mobile applications at both ends of the waste management process helps cut down on both hardware costs and management time.

Wastewaters have more than ten times the energy content of most fluids, making them a prime candidate for energy processing. That's why this water-with-energy nexus is no longer just a theoretical idea; it's already helping to make energy generation utilizing wastewater a highly efficient and lucrative system. More energy, less pollution, and clean water that can be put to many uses are just a few of the many benefits (Rubbish Begone, 2020).

For solid waste in particular, efficient ways of handling solid waste improve sanitation and hygiene. In turn, this boosts public health and quality of life by decreasing the prevalence of disease vectors that thrive in open garbage. Hygiene is now of paramount significance and should not be dealt with manually. Smart solutions can help with solid waste in a few different ways: by cutting down on the cost of unnecessary fuel consumption, ensuring that pickups are never missed, analysing waste generation based on geographical factors, and lowering the amount of carbon dioxide released into the atmosphere.

11.5.1 THE FUTURE OF E-WASTE MANAGEMENT

The corporation Electronic Recyclers International is accountable for the deployment of this cutting-edge technology that can recycle up to 30,000 pounds per hour of electronic waste. This is an enormous increase from the typical quantity of electronic garbage processed before this future machine became available. It took electronic recyclers international one billion pounds of recycled electronic garbage to reach this point, and now the organization is doing an even bigger and better job of preparing for the future of waste management and research. Think about how far the firm will come in the next ten years. One day, maybe, the world will find a way to deal with the unprecedented amount of electronic garbage, which has risen to more than 93 million tonnes annually around the globe.

11.5.2 Biodegradable Plastics of the Future

Even though we've seen plenty of similar items before, the market and demand for them are expanding rapidly. Recent estimates put the annual growth rate of sales of biodegradable plastics at 19%, demonstrating the world's incapacity to recycle or compost the vast quantities of plastic bags that have already entered the environment.

Plastics are highly effective; however, they cannot be recycled and hence cannot biodegrade when thrown away. However, this cannot be done without the development of suitable recycling technology. In the next several years, we might expect to see plastics that break down in as little as three hours.

A team of researchers at the University of North Dakota has developed a plastic with built-in photo triggers, allowing it to change colour in response to sunshine. This technique has the potential to end the centuries-long problem of plastic trash contaminating the planet if it is adopted by the majority of the world's population.

11.5.3 Automated Waste Segregation

The automated waste management system prototype created by Texas Instruments, a multinational semiconductor firm present in 35 countries, has already been documented in online films.

While the device's findings may seem straightforward at first appearance, they reflect the use of sophisticated robotics technology capable of distinguishing between different sorts of trash and responding appropriately.

One of the most important procedures for a more optimized and successful waste management strategy is rubbish segregation, which requires a fast and dependable system to be effective. With the rate at which technology is advancing, very soon there will be a state-of-the-art, error-free system that can sort trash with unrivalled efficiency.

11.5.4 Pneumatic Refuse Collection System

This cutting-edge piece of equipment also called an automated vacuum collection system, is responsible for the prompt disposal of garbage via pneumatic tubes buried beneath the surface. After entering the system, trash is transported to a collecting depot, where it is placed in containers and further sealed in preparation for its eventual disposal.

Montreal attempted to implement this system and spent \$3 million on it before giving up. Pneumatic garbage collection is under consideration for a new project in Carmel, Indiana (Foth Infrastructure & Environment, LLC, 2013).

The idea's prospective potential lies in the fact that, with enough development, it may eventually be employed on a global scale, with a single centralized system governing waste collection and disposal via a network of subterranean tubes stretching across the whole planet. At the end of the collecting procedure, an innovative automated segregation system may perform the necessary processing before the recycling process completes the full cycle.

11.5.5 Water–Energy Nexus: Waste Waters Used for Energy Production

In most cases, the energy contained in wastewater is ten times greater than the energy needed to treat them. Because of this, the water-energy nexus is a notion that has made energy generation from wastewater a very economical and efficient way of utilizing waste as a source of energy. The benefits are too many to list, but they include less pollution, increased energy production, and cleaner water that may be put to a variety of different uses.

A team of researchers at the University of Barcelona identified a bacterium that might help remove the organic materials from wastewater. The method converts the organic waste present in the wastewater into hydrogen, which can then be easily converted into clean energy. This not only reduces the cost of treating wastewater, but also generates a byproduct that may be utilized, sold, or saved for later (Lim, 2019).

Given that any contact with waste matter shortens the life of all living organisms, the future of sustainable waste management and the reduction of electrical waste will undoubtedly revolve around processes like those described earlier, all of which have one overarching goal: to achieve maximum efficiency in terms of productivity, the minimum amount of resources being used, and the implementation of technology instead of actual people who do the hard work.

11.6 CONCLUSION

Waste creation, management, and disposal are complicated, involving public service provision, municipal and state budgets, private industry, ecological and human health, land use, resident satisfaction with service, facility siting, and operations. Since WWII, waste management services and programmes have improved, but the public no longer cares about solid waste issues, and when there's no crisis, they don't invest or participate. Global garbage management has had many crises. Early 20th-century trash pickup stories describe horse-drawn carts. There have been improvements, but the need to respond quickly has often postponed or impeded long-term planning for more effective and sustainable waste management systems. Well-designed materials management programmes (solid waste processing and disposal, recycling, organics, source reduction, reuse, etc.), citizen awareness, financial investments, and innovative institutions can assist achieve the waste hierarchy and reduce waste. Properly managed landfills should dispose of the last, considerably smaller garbage. India dumps 70% of urban waste in overcrowded landfills. India produces almost 3,500 tonnes of plastic rubbish daily, making it the biggest polluter. Waste management requires source segmentation, recycling, and resource recovery.

Rapid rubbish generation and waste management are global issues. Most agree that human-powered rubbish collection is inefficient. To promote health and happiness, towns or garbage companies manage rubbish from its inception to its final disposal. The rise of hazardous waste, which must be managed carefully, is a primary driver of waste management reform. This is one of the basic realities. Waste management relies on technological advancements. To grow, the waste management sector, like many others, needs digitization and data-driven approaches. Smart, competitive times ahead. Businesses must innovate. Big data waste management solutions collect relevant data over time. Sensors may use this data to identify fill trends, optimize driver schedules, and reduce costs. Businesses and municipalities are growing interested in smart bins as sensor prices drop. We mean the future's practicality when we say it's smart. Careful container selection minimizes waste collection crew use. Collecting data quickly maximizes return on investment. User-friendly, comprehensive, and compact platforms and mobile apps at both ends of waste management reduce hardware costs and management time.

REFERENCES

Aalok, A., Tripathi, A. K., & Soni, P. (2008). Vermicomposting: A better option for organic solid waste management. *Journal of Human Ecology*, *24*(1), 59–64. https://doi.org/10.1080/09709274.2008.11906100

Abbasi, T., Gajalakshmi, S., & Abbasi, S. A. (2009). Towards modeling and design of vermicomposting systems: Mechanisms of composting/vermicomposting and their implications. *Indian Journal of Biotechnology*, *8*, 177–182.

Abigail, A. M. (1998). Using municipal solid waste compost in nursery stock production. *BioCycle*, *39*(5), 63–65.

Alekhya, M., Divya, N., Jyothirmai, G., & Reddy, K. R. (2013). Secured landfills for disposal of municipal solid waste. *International journal of Engineering Research and General Science*, *1*(1).

Aljaradin, M., & Persson, K. M. (2012). Environmental impact of municipal solid waste landfills in semi-arid climates—Case study—Jordan. *Open Waste Management Journal*, *5*(1), 28–39. https://doi.org/10.2174/1876400201205010028

Annepu, R. K. (2012). *Sustainable SolidWaste management in India [MSc Thesis] Earth Engineering Center*. Columbia University.

Axelsson, C., & Kvarnstrom, T. (2010). *Energy from municipal solid waste in Chennai, India—A feasibility study*. Committee of Tropical Ecology, Minor field study. Uppsala University.

Barles, S. (2005). *L'invention des déchets urbains, France, 1790–1970*. Champ Vallon.

Becidan, M. (2007). Experimental studies on municipal solid waste and biomass [Pyrolysis PhD Thesis]. Norwegian University of Science and Technology.

Belgrand, E. (1887). *Les travaux souterrains de Paris, 5. Les égouts, les vidanges*. Dunod, 14.

Boudriot, P.-D. (1990). Les égouts de Paris aux XVIIe et XVIIIe siècles: Les humeurs de la ville préindustrielle. *Histoire, Économie et Société*, *9*(2), 197–211. https://doi.org/10.3406/hes.1990.2380

Bundela, P. S., Gautam, S. P., Pandey, A. K., Awasthi, M. K., & Sarsaiya, S. (2010). Municipal solid waste management in Indian cities. *International Journal of Environmental Sciences*, *1*(4), 591–605.

Central Pollution Control Board (2007). *Bio-methanation potential of solid wastes from agro-based industries.* Ministry of Environment and Forests.

Chadwick, E. (1842). *Chadwick's report on sanitary conditions. excerpt from report.* From the Poor Law commissioners on an Inquiry into the Sanitary Conditions of the Labouring Population of Great Britain (pp. 369–372) (online source). Added by Laura Del Col: To The Victorian Web. Accessed on 8 November 2009.

Chevallier, A. (1849). Notice historique sur le nettoiement de la ville de Paris. *Annales d'Hygiène Publique et de Médecine Légale, 42*, 264–321.

Das, S., & Bhattacharya, B. K. (2013). Municipal solid waste characteristics and management in Kolkata, India. *International Journal of Emerging Technology and Advanced Engineering, 3*(2), 147–152.

Daskalopoulos, E., Badr, O., & Probert, S. D. (1998). An integrated approach to municipal solid waste management. *Resources, Conservation and Recycling, 24*(1), 33–50. https://doi.org/10.1016/S0921-3449(98)00031-7

DEFRA—Department for Environment, Food and Rural Affairs (2007). *Incineration of municipal solid waste. Waste management technology brief, the new technologies work stream of the Defra waste implementation programme.*

El-Fadel, M., Findikakis, A. N., & Leckie, J. O. (1997). Environmental 497 impacts of solid waste landfilling. *Journal of Environment Management, 50*(1), 1–25.

Foth Infrastructure & Environment, LLC. (2013). *Alternative technologies for municipal solid waste.*

Gandy, M. (1994). *Recycling and the politics of urban waste.* Earthscan Publications. ISBN 9781853831683. Accessed on 7 March 2013.

Gautam, S. P., Bundela, P. S., Pandey, A. K., Awasthi, M. K., & Sarsaiya, S. (2010). Composting of municipal solid waste of Jabalpur City. *Global Journal of Environmental Research, 4*(1), 43–46.

Gidde, M. R., Todkar, V. V., & Kokate, K. K. (2008). *Municipal solid waste management in emerging mega cities: A case study of Pune city.* In Proceedings of the Indo Italian Conference on Green and Clean Environment, Pune, India (March 20–21).

Gille, Z. (2007). *From the cult of waste to the trash heap: The politics of waste in socialist and postsocialist Hungary.* Indianapolis, Bloomington.

Giuntini, A. (2006). *Cinquant'anni puliti puliti. [I rifiuti a Firenze dall'] Ottocento alla Società Quadrifoglio.* Franco Angeli.

Guillerme, A. (1983). *Les temps de l'eau. La cité, l'eau et les techniques. Nord de la France, fin IIIedébut XIXe siècles.* Champ Vallon.

Guillerme, A. (1988). *The age of water: The urban environment in the North of France.* Texas A & M University Press, College Station.

Gupta, S., Choudhary, N., & Alappat, B. J. (2007). Bioreactor landfill for MSW disposal in Delhi. Proceedings of the International Conference on Sustainable Solid Waste Management, Chennai, India (September 5–7).

Hanrahan, D., Srivastava, S., & Ramakrishna, A. S. (2006). *Improving management of municipal solid waste in India.* World Bank.

Harpet, C. (1998). *Du déchet. Philosophie des immondices.* Corps, ville, industrie. L'Harmattan.

Herat, S. (2009). Electronic waste: An emerging issue in solid waste management in Australia. *International Journal of Environment and Waste Management, 3*(1/2), 120–134. https://doi.org/10.1504/IJEWM.2009.024704

Herbert, L. (2007). *Centenary history of waste and waste managers in London and South East England (PDF).* Chartered Institution of Wastes Management.

Herbert, L. (2013). *Centenary history of waste and waste managers in London and south East England.*

Hoffmann Baumaschinen, M. (2023). *Industrie service.* www.baumaschinen-hoffmann.de/

Jafari, A., Godini, H., & Mirhousaini, S. H. J. (2010). Municipal solid waste management in KhoramAbad city and experiences. *World Academy of Science, Engineering and Technology, 4,* 163–168.

Jha, A. K., Singh, S. K., Singh, G. P., & Gupta, P. K. (2011). Sustainable municipal solid waste management in low income group of cities: A review. *Int. Soc. Tropical Ecology, 52*(1), 123–131.

Jha, M. K., Sondhi, O. A. K., & Pansare, M. (2003). Solid waste management—A case study. *Indian Journal of Environmental Protection, 23*(10), 1153–1160.

Joardar, S. D. (2000). Urban residential solid waste management in India. *Public Works Management and Policy, 4*(4), 319–330. https://doi.org/10.1177/1087724X0044006

John, H. (1997). Apply compost and mulches to control erosion. *BioCycle, 38*(5), 63–65.

Kadam, A., Ozaa, G., Nemadea, P., & Dutta, S. (2008). Shankar H. Municipal wastewater treatment using novel constructed soil filter system. *Chemosphere, 71*(5), 1975–1981.

Kansal, A. (2002). Solid waste management strategies for India. *Indian Journal of Environmental Protection, 22*(4), 444–448.

Leguay, J.-P. (1999). *La pollution au moyen age. Dans le royaume de France et dans les grands fiefs.* Editions Gisserot.

Lim, X. Z. (2019). *Turning organic waste into hydrogen.* A version of this story appeared in volume 97, Issue 14.

Maier, D. (1979). Nineteenth-century Asante medical practices. *Comparative Studies in Society and History, 21*(1), 63–81. https://doi.org/10.1017/S0010417500012652. JSTOR 178452.

Manyuchi, M. M., & Phiri, A. (2013). Vermicomposting in solid waste management: A review. *International Journal of Scientific Engineering and Technology, 2*(12), 1234–1242.

Manyuchi, M. M., Phiri, A., Muredzi, P., & Chirinda, N. (2013). Effect of drying on vermi-compost macronutrient composition. *International Journal of Inventive Engineering and Sciences, 1*(10), 1–3.

Melosi, M. V. (2005). *Garbage in the cities: Refuse, reform and the environment* (2nd ed.). University of Pittsburgh Press.

Ministry of the Environment (2021). *Forest and climate change.* https://moef.gov.in/en/division/environment-divisions/hazardous-substances-management-hsm/introduction/

Mohanty, C. R., Mishra, U., & Beuria, P. R. (2014). Municipal solid waste management in Bhubaneswar, India—A review. *International Journal of Latest Trends in Engineering and Technology (IJLTET), 3*(3), 303–312.

Moustakas, K., Loizidou, M., & Kumar, Er. S. (2010). *Solid waste management through the application of thermal methods.* Waste manage. ISBN: 978-953-7619-84-8. InTech. www.intechopen.com/books/waste-management/solid-waste-management-through-the-application-of-thermal-methods.

Nagarajan, R., Thirumalaisaumalaisamy, S., Lakshumanan, & Eakshumanan, E. (2012). Impact of leachate on groundwater pollution due to non-engineeredmunicipal solid waste landfill sites of erode city, Tamil Nadu, India. *Iranian Journal of Environmental Health Science and Engineering, 9*(1), 1–12.

Nemade, P. D., Kadam, A. M., & Shankar, H. S. (2009). Wastewater renovation using Constructed Soil Filter (CSF): A novel approach. *Journal of Hazardous Materials, 170*(2–3), 657–665. https://doi.org/10.1016/j.jhazmat.2009.05.015

NIUA. (2021). *Urban scenario of India.* www.niua.org/urban-scenario-of-india

Noorjahan, A., Dhanakumar, S., Mohanraj, R., & Ravichandran, M. (2012). Status of heavy metals distribution in municipal solid waste in Tiruchirappalli City, India. *International Journal of Applied Biology and Pharmaceutical Technology, 3*(3), 252–258.

Oatman-Stanford, H. (2013). A filthy history: When New Yorkers lived knee-deep in trash. *Collectors Weekly.* Accessed on 8 January 2021.

Ogu, V. I. (2000). Private sector participation and municipal waste management in Benin City, Nigeria. *Environment and Urbanization, 12*(2), 103–117. https://doi.org/10.1177/095624780001200209

Parthan, S., & Milke, M. W. (2009). Cost analysis of municipal solid waste management in India. Proceeding of waste MINZ, 21st Annual Conference, Christchurch, New Zealand (October 14–16).

Pokhrel, D., & Viraraghavan, T. (2005). Municipal solid waste management in Nepal: Practices and challenges. *Waste Management, 25*(5), 555–562. https://doi.org/10.1016/j.wasman.2005.01.020

Prakash, E. V., & Singh, L. P. (2013). Biomethanation of vegetable and fruit waste in co-digestion process. *International Journal of Emerging Technology and Advanced Engineering, 3*(6), 493–495.

Rachel, O. A., Komine, H., Yasuhara, K., & Murakami, S. (2009). *Municipal solid waste management in developed and developing countries—Japan and Nigeria as case studies.* http://wwwgeo.civil.ibaraki.ac.jp/komine/640mypapers/JGSPaper/2009/JGS2009(973)Rachel.pdf

Rajkumar, N., Subramani, T., & Elango, L. (2010). Groundwater contamination due to municipal solid waste diposal—A GIS based study in Erode City. *International Journal of Environmental Sciences, 1*(1), 39–55.

Rajput, R., Prasad, G., & Chopra, A. K. (2009). Scenario of solid waste management in present Indian context. Casp. *Journal of Environmental Sciences, 7*(1), 45–53.

Rathi, S. (2006). Alternative approaches for better municipal solid waste management in Mumbai, India. *Waste Management, 26*(10), 1192–1200. https://doi.org/10.1016/j.wasman.2005.09.006

Rathi, S. (2007). Optimization model for integrated municipal solid waste management in Mumbai, India. *Environment and Development Economics, 12*(1), 105–121. https://doi.org/10.1017/S1355770X0600341X

Rubbish Begone (2020). *What is the future of waste management? What is the future of waste management.* rubbishbegone.co.uk

Sabbas, T., Polettini, A., Pomi, R., Astrup, T., Hjelmar, O., Mostbauer, P., Cappai, G., Magel, G., Salhofer, S., Speiser, C., Heuss-Assbichler, S. H., Klein, R., Lechner, P., & pHOENIX Working Group on Management of MSWI Residues. (2003). Management of municipal solid waste incineration residues. *Waste Management, 23*(1), 61–88. https://doi.org/10.1016/S0956-053X(02)00161-7

Sakai, S., & Hiraoka, M. (2000). Municipal solid waste incinerator residue recycling by thermal processes. *Waste Management, 20*(2–3), 249–258. https://doi.org/10.1016/S0956-053X(99)00315-3

Saxena, S., Srivastava, R. K., & Samaddar, A. B. (2010). Sustainablewaste 669 management issues in India. *The IUP Journal of Soil & Water Sciences, 3*(1), 72–90.

Serio, M. A., Chen, Y., Wójtowicz, M. A., & Suuberg, E. M. (10–30 July 2000). Pyrolysis processing for solid waste resource recovery in space. SAE Technical Paper Series. Proceedings of the 30th International Conference on Environmental Systems, Toulouse, France. https://doi.org/10.4271/2000-01-2286

Serio, M. A., & Kroo, E. (2002). Wójtowicz, Suuberg, E, & Filburn, T. Animproved pyrolyzer for solid waste resource recovery in space. *Proceedings of the 32nd International Conference on 681 Environmental Systems,* San Antonio, Texas (July 15–18), 682.

Serio, M. A., Kroo, E., Bassilakis, R., Wójtowicz, M. A., & Suuberg, E. M. (2001). A prototype pyrolyzer for solid waste resource recovery in space. SAE Technical Paper Series. Proceedings of the 31st International Conference on Environmental Systems, Orlando, Florida, 677(July 9–12), 678. https://doi.org/10.4271/2001-01-2349

Serio, M. A., Kroo, E., & Wojtowicz, M. A. (2003). Biomass pyrolysis for distributed energy generation. *Preprints of Papers: American Chemical Society, Division of Fuel Chemistry, 48*(2), 584–589.

Sharholy, M., Ahmad, K., Mahmood, G., & Trivedi, R. C. (2005). Analysis of municipal solid waste management systems in Delhi—A review. Book of Proceedings for the second International Congress of Chemistry and Environment, Indore, India (December 24).

Shazwin, T. M., & Nakagoshi, N. (2010). Sustainable waste management through international cooperation: Review of comprehensive waste management technique and training course. *Journal of International Development Cooperation, 16*(1), 23–33.

Smahi, D., Hammoumi, O. E., & Fekri, A. (2013). Assessment of the impact of the landfill on groundwater quality: A case study of Mediouna Site, Casablanca, Morocco. *Journal of Water Resource and Protection, 5*, 440–445.

Spill Control Technologies Ltd. (2023). *Liquid waste solidifier.* https://store-f2b01.mybigcommerce.com/liquid-waste-solidifier-lws-2lbs/

Strasser, S. (1999). *Waste and want: A social history of trash.* Metropolitan Books.

Syed, S. (2006). Solid and liquid waste management. *Emirates Journal for Engineering Research, 11*(2), 19–36.

Tarr, J. A. (1971, October). URBAN POLLUTION-Many long years ago. *American Heritage Magazine.* Ban horse drawn carriages. Accessed on 13 November 2016.

Tchobanoglous, G., Theisen, H., & Vigil, S. A. (1993). *Integrated solid waste management-engineering principles and management issues* (international Ed.). Tata McGraw Hill.

University of California (2023). *Blink. How to identify, Label, Package and Dispose of Biohazardous and Medical Waste.* https://blink.ucsd.edu/safety/research-lab/hazardous-waste/disposal-guidance/medical/dispose.html

Velis, C. A., Wilson, D. C., & Cheeseman, C. R. (2009). 19th century London dust-yards: A case study in closed-loop resource efficiency. *Waste Management, 29*(4), 1282–1290. https://doi.org/10.1016/j.wasman.2008.10.018

Velmurugan, B., & Ramanujam, R. A. (2011). Anaerobic digestion of vegetable wastes for biogas production in a fed-batch reactor. *International Journal of Emerging Sciences, 1*(3), 478–486.

Yatoo, A. M., Ali, M. N., Baba, Z. A., & Hassan, B. (2021). Sustainable management of diseases and pests in crops by vermicompost and vermicompost tea. A review. *Agronomy for Sustainable Development, 41*, 1–26.

Yatoo, A. M., Bhat, S. A., Ali, M. N., Baba, Z. A., & Zaheen, Z. (2022). Production of nutrient-enriched vermicompost from aquatic macrophytes supplemented with kitchen waste: Assessment of nutrient changes, phytotoxicity, and earthworm biodynamics. *Agronomy, 12*(6), 1303.

Yatoo, A. M., Rasool, S., Ali, S., Majid, S., Rehman, M. U., Ali, M. N., . . . & Farooq, S. (2020). Vermicomposting: An eco-friendly approach for recycling/management of organic wastes. *Bioremediation and Biotechnology: Sustainable Approaches to Pollution Degradation*, 167–187.

Yatoo, A. M, Hamid, B., Sheikh, T. A., Ali, S., Bhat, S. A., Ramola, S., Baba, Z. A., & Kumar, S. (2024). Global perspective of municipal solid waste and landfill leachate: Generation, composition, ecotoxicity, and sustainable management strategies. *Environmental Science and Pollution Research, 31*(16), 23363–23392.

Zaman, A. U. (2009). Life cycle environmental assessment of municipal solid waste to energy technologies. *Global Journal of Environmental Research, 3*(3), 155–163.

Zaman, A. U. (2010). Comparative study of municipal solid waste treatment technologies using life cycle assessment method. *International Journal of Environmental Science and Technology, 7*(2), 225–234. https://doi.org/10.1007/BF03326132

Zhu, D., Asnani, P. U., Zurbrugg, C., Anapolsky, S., & Mani, S. (2008). *Improving solid waste management in India: A sourcebook for policy makers and practitioners.* World Bank.

12 Groundwater Vulnerability Mapping Using DRASTIC Model in GIS Environment of Imphal East District, Manipur, India

Haobam Bidyapati and Thiyam Tamphasana Devi

12.1 INTRODUCTION

Groundwater is available on the surface of the earth, which is mostly used by living beings and cultivation of plants. The presence of groundwater affects the growth of the plantation. Groundwater makes up around 30% of the world's readily usable freshwater. Around 50% of the world's municipal water extraction comes through irrigation, with 40% coming from groundwater (volume is 982 km³/year), or almost one-fourth of total water extraction is the estimated global groundwater withdrawal. Of these groundwater amounts, 21% are for domestic use, 70% are for agricultural, and the remainder is mostly for industrial use. The aforementioned statements prove that how groundwater is the primary need of agriculture, and so if the groundwater is at high risk, it will affect the agricultural land as well as its practices (crop yield). DRASTIC model (first time introduced by Aller et al., 1987) is one of the techniques to help evaluate the vulnerability of groundwater using geographical information system (GIS) techniques and is the first time applied with GIS by Merchant et al. (1987). Additionally, understanding of the GIS scenario in terms of potential can be modelled using other models (AHP, analytic hierarchy process; and MIF, multi-influencing factors) (Singh and Devi, 2022) apart from the DRASTIC model. Parameters used in the DRASTIC model are represented by each letter such as depth to water, net recharge, aquifer media, soil media, topography, impact of vadose zone, and conductivity and these are the significant influencing factors to groundwater vulnerability. Several researchers effectively applied the DRASTIC model (Malik and Shukla, 2019) for the assessment of groundwater vulnerability risk using these parameters.

Aller et al. (1987) study used the DRASTIC model and aimed at providing a mechanism that will enable us to thoroughly evaluate any hydrogeological setting's

DOI: 10.1201/9781003359326-12

risk for groundwater contamination using data already available in the United States. In their study, only hydrogeological elements that affect contamination potential are included in the DRASTIC technique. By determining the hydrogeology of the location and its hydrogeologic setting condition, the seven factors are assessed. The scale (ranges) assigned were derived using a hydrogeologic setting. But net recharge, which is one of the parameters, was not available, and so the value considered was precipitation in their study. As the specific area's pollution potential is a result of a mix of hydrogeological elements, human impacts, and pollution sources, they conducted their study in two different locations and compared the results. To understand the relative contamination potential of each area associated with pesticide application, pesticide–DRASTIC indices of two areas were assessed, and it was observed that this specific index should vary from 88 to 251 in any study conducted at any geographical location. It is concluded from their study that the higher the DRASTIC index, higher is the vulnerability. Arzu Firat and Fatma (2013) also used the DRASTIC model to evaluate groundwater vulnerability to pollution in a catchment situated in Turkey for a shallow aquifer. The depth of the groundwater in the Merzifon aquifer is the shallowest in the catchment, and its intensity ranges from 9 to 40 metres below the surface. The river of Gumussuyu (major river in the study region) and its tributaries are directly recharged into from local small tributaries in the area which is the most utilized and hence vulnerable to contamination. According to drilling data, the clay layer that lies on top of the permeable sediment devoid of cement is 2–10 metres thick. Seven layers of hydrogeological data were used to construct the vulnerability map in the model and found that the DRASTIC scores varied from 58 to 177. The susceptibility of the research region was categorized as low (100), medium (100–140), and high (>140) based on the hydrogeological survey data. They concluded that the central part of the basin is at high risk of groundwater contamination. Bera et al.'s (2021) study aims to evaluate the vulnerabilities of the Totko River in West Bengal, and an overlay approach for assessing vulnerabilities, DRASTIC, and AHP was employed in this study. The accuracy of the map was checked using nitrate concentration. It comes to the conclusion that pollutants from point and non-point sources are now contaminating groundwater aquifers as a result of numerous anthropogenic activities. The socioeconomic and health situations of the population are negatively impacted both directly and indirectly by this. Sarkar and Pal (2021) applied agricultural DRASTICA and modified DRASTIC to assess the groundwater vulnerability of the Malda area in West Bengal, India. Standard weights (scale 1–5) are used to indicate the use of pesticides in agricultural activities (Aller et al., 1985). Pesticide applications were said to occur in agricultural DRASTICA and are utilized in agricultural regions, while the original DRASTIC is employed in industrial and urban locations. Agricultural DRASICA was used because most of the study area is agricultural. In this model, 43.16% of areas are of low vulnerability (range 95 to 143), and 20.51% of areas are of high vulnerability (range 170 to 197).

Borah and Deka (2023) analysed the capacity of groundwater in the watershed of the Jamuna located at Assam using the MIF technique. The influencing factors (geomorphology, lithology, lineament density, land use sample, slope, rainfall, and distinct soil) scored are given 0–1, according to relatively minor and major effects, then both major and minor were added, and the summation of each factor is represented as proposed score for each influencing factor. Using the overlay weighted, the entire influencing factor maps were overlaid, and potential zones of the study area were produced. The map results in four levels from lowest to best index of the study area. The validation of this technique is done using the well data of the study area. Then groundwater yield and groundwater potential value were plotted using a linear graph resulting in direct linear to each other. Finally, it was concluded that 70% of the study is good and potential zones are mostly in the northern part of the study area. It was revealed that this information will help the survey or any query about the Jamuna watershed.

In this chapter, the groundwater vulnerability (in terms of contamination) mapping of the Imphal East district, Manipur, is conducted using the DRASTIC model with GIS techniques.

12.2 STUDY AREA

Imphal East district (Figure 12.1) is one of the districts of Manipur (northeastern state of India), which has a total population of 4,52,661 people (Census, 2011) and a

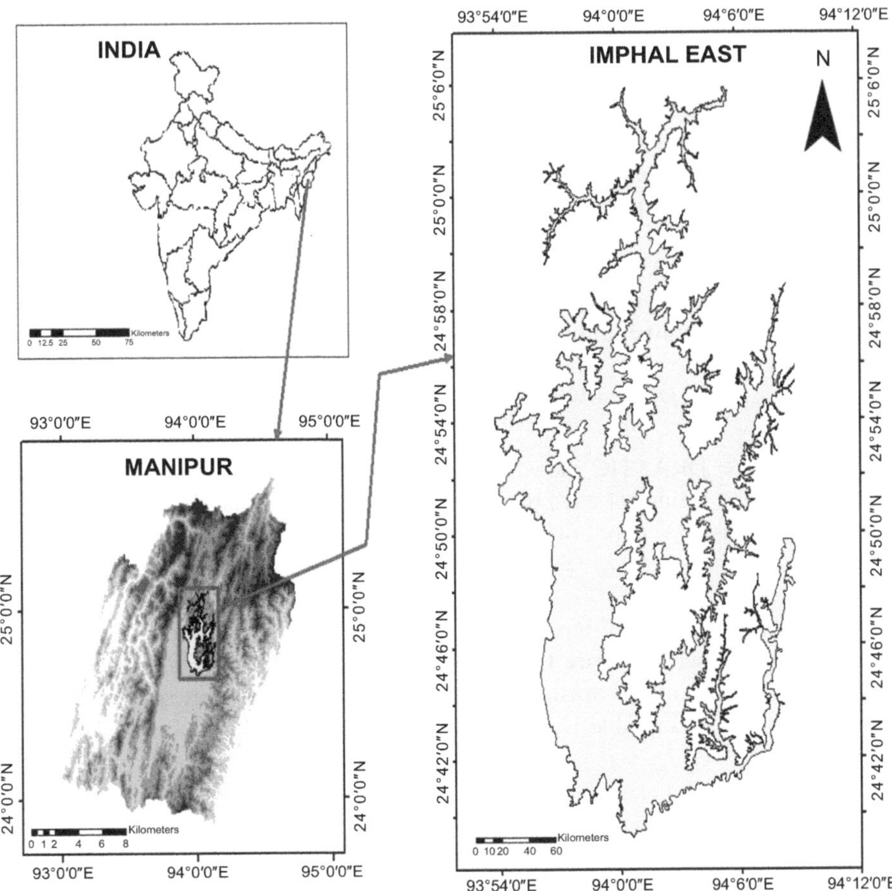

FIGURE 12.1 Imphal East district map (MARSAC, Government of Manipur)

geographical area of around 709 km². It is located at an altitude of 790 m above sea level, at latitude 24.780654 N, and longitude 93.967437 E. A subtropical to temperate climate may be found there, and the temperature is between 0°C and 40°C. The south-west tropical monsoon's effect is a phenomenon that the region encounters. The monsoon season's heaviest rain falls between May and August of every year. The region resembles a flat, long, narrow valley with solitary hills that rise up towards the south.

12.3 METHODOLOGY

The data used and equations of the method in the DRASTIC model with the theoretical background to find out the groundwater pollution vulnerability in this study are presented in this section.

12.3.1 DRASTIC Model

The DRASTIC model is used to evaluate vertical vulnerability mapping, and each letter in this model represents the required input parameters to get the final output. Therefore, D represents depth to water, R as net recharge, A as aquifer media, S as soil media, T as topography, I as the impact of the vadose zone, and C as hydraulic conductivity. The source of these parameters was given in Table 12.1. It is presumed that all of these input parameters significantly affect the health of groundwater (pollution or contamination of groundwater) and play an important role in the overall results. In this model, each component (input parameters) is given a weighting multiplier to balance and enhance its importance. Thus, a scale of 1 to 5 as weight and a scale of 1 to 10 ratings are normally assigned to each input parameter, which is provided in Table 12.2. After weights and rank are assigned to these input parameters, weighted overlay analysis is performed, and the weighted total of the seven elements results in the final vulnerability index, as the DRASTIC index (D_i) is given as:

$$D_i = D_r D_w + R_r R_w + A_r A_w + S_r S_w + T_r T_w + I_r I_w + C_r C_w \qquad (12.1)$$

where D_i = DRASTIC index for a mapping unit w weighting factor for each parameter r rating for each parameter D, R, A, S, T, I, and C, the seven parameters, that is, D = depth of water, R = net recharge, A = aquifer media, S = soil media, T = topography, I = vadose zone impact, and C = hydraulic conductivity.

The sequential structure of input parameters of the DRASTIC model in the form of a flowchart is shown in Figure 12.2. The groundwater vulnerability criterion, which is based on the calculated drastic vulnerability index (DVI) value of the DRASTIC model, is presented in Table 12.1.

12.3.2 Data Used

The data used (input parameters) with its sources and mode of derivation and extraction are provided in Table 12.2. The description of these input parameters provides insight meaning of the possessed characteristics. The collected data retrieved from different sources were again processed through the GIS tool (ArcGIS).

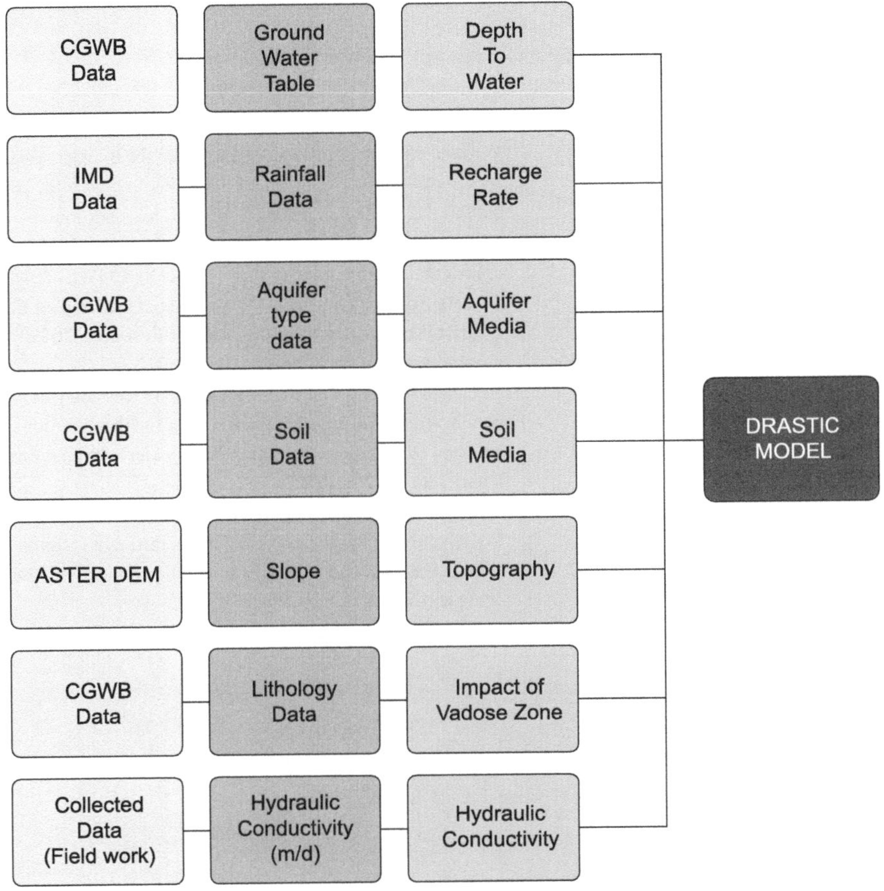

FIGURE 12.2 Structure of data used in the DRASTIC model

TABLE 12.1
Vulnerability criteria based on drastic vulnerability index value of the DRASTIC model

	Calculated index value	
Class vulnerability	**Aller et al. (1987)**	**Engel et al. (1996)**
Low	47–92	<101
Moderate	93–136	101–140
Very high	137–184	141–200
High	>184	>200

Field work was conducted to calculate the hydraulic conductivity. Mini disc infiltrometer (MDI) is used for the measurement (Figure 12.3). It has a suction tube on top (7 cm) of the suction regulation tube (10.2 cm) followed by a mariotte tube (28 cm) and a stainless steel porous disc (4.5 cm diameter, 3 mm thick). Water should be

TABLE 12.2
Sources of data used

Parameters (data)	Mode	Sources
Depth to water	Interpolation	Collected data (depth to water table) from the booklet issued by Central Groundwater Board (CGWB), Imphal East, 2018
Net recharge	Integration	Rainfall data collected from CGWB, 2018, as per Indian Meteorological Department (IMD) from 2013 to 2017
Aquifer media	Digitization	Aquifer data from booklet—East District by CGWB, 2018
Soil	Digitization	Soil data from groundwater information booklet Imphal East District, Manipur by Ministry of Water Resource, 2013, Groundwater Information Booklet, 2013
Topography (slope)	Spatial Analyst tool	Developed from Shuttle Radar Topography Mission (SRTM) data in ArcGIS using spatial analyst tool with resolution 30
Impact of vadose zone	Digitization	Collected the lithology data from the booklet—Imphal East District by CGWB, 2018
Hydraulic conductivity	Digitization	Collected the data by using the mini disc infiltrometer (MDI), performing the experiment in a different location of Imphal East and calculated the hydraulic conductivity value using a graph in an Excel sheet

FIGURE 12.3   MDI at the site

poured into the regulation tube up to 10 cm, and the suction tube is inserted according to the soil porosity nature. The volume of the mariotte tube in the initial time is noted and thereafter for every 30 s. It can be continued until the difference between the initial and the final volume is less than 30 ml. Then the cumulative value is calculated from the difference between the initial and final volumes divided by the area of the infiltration. Thereafter, hydraulic conductivity is calculated by cumulative infiltration divided by the area (from the Van Genuchten parameters according to the suction type and soil type). Different locations were selected for the measurement of hydraulic conductivity in the study region, and Figure 12.3 shows the MDI at the measurement site (Napet Pali).

12.4 RESULTS AND DISCUSSION

In this section, the derived input parameters in the form of raster and vector from the collected ground data will be provided along with the final weighted overlay result of the groundwater vulnerability map. All the input parameters were derived using inverse distance weighting (IDW) method in ArcGIS.

12.4.1 Input Parameters

12.4.1.1 Depth of Water

For mapping of parameter, D, depths (m) of certain locations of the study area are collected from CGWB (2018). Depth ranges from 0 to 9 m. Then using the coordinates of the locations, interpolation is done by IDW, shown in Figure 12.4, and is classified per Table 12.2.

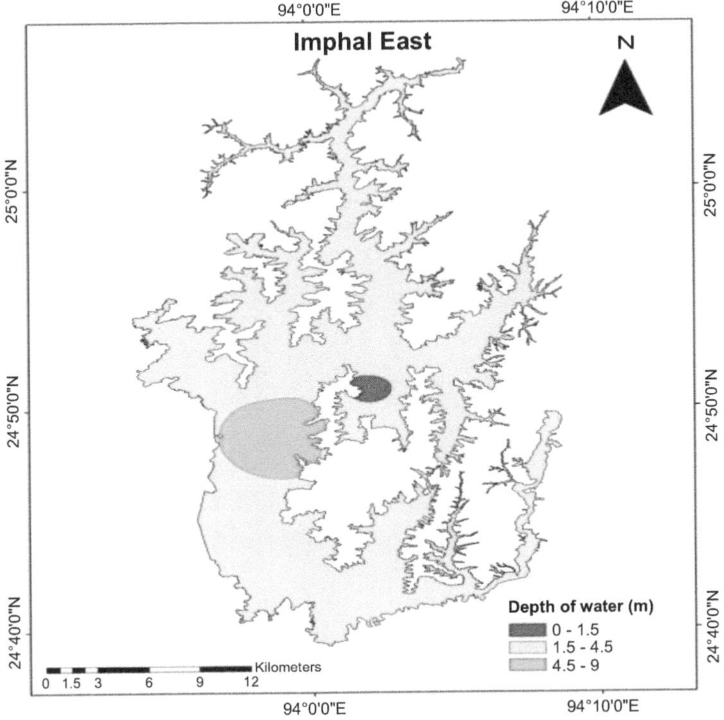

FIGURE 12.4 Depth of water (m)

12.4.1.2 Net Recharge

The next parameter, R, is calculated by the summation of rainfall for the year 2021. After the summation, the data is added to the tool, and then net recharge mapping is done through IDW using the coordinates of the study area, providing the map in Figure 12.5. The sum of annual rainfall is about 1,099 mm in the study area.

12.4.1.3 Aquifer Media

The aquifer medium is the medium that is beneath the water table of the ground, and Imphal East is mostly covered by sand and gravel (CGWB, 2018). Like the other parameters, mapping is done through IDW, shown in Figure 12.6 for aquifer media.

12.4.1.4 Soil Media

Soil media data are collected from CGWB (2018) and interpolated using IDW. The soil type of the study area is mostly covered by loamy sand. The map of the soil type is shown in Figure 12.7.

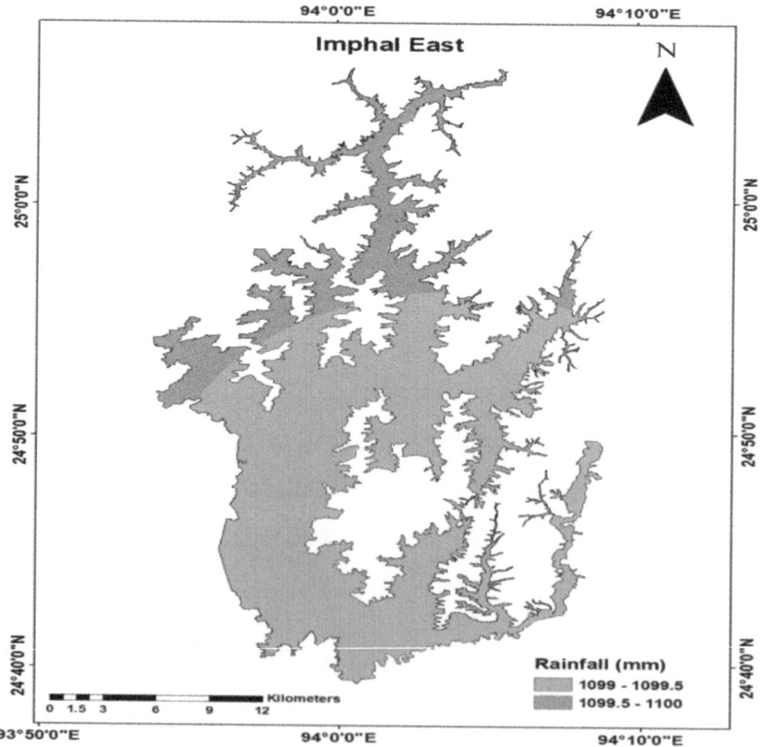

FIGURE 12.5 Rainfall (mm)

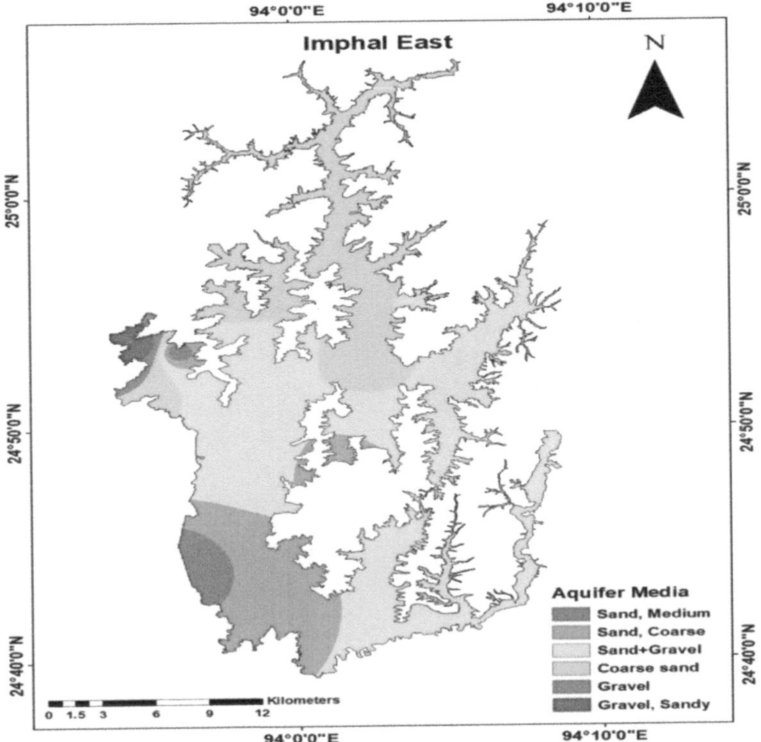

FIGURE 12.6   Aquifer media

12.4.1.5   Topography

Topography is the percentage slope. The digital elevation model is downloaded from Earth Explorer website.

Using the slope tool in ArcGIS, the percentage has been calculated and further classified per Table 12.2. The map in Figure 12.8 represents the slope in the percentage of the area, ranging from $0 \rightarrow 18$.

12.4.1.6   Impact of Vadose Zone

Impact of vadose zone is the zone just beneath the soil surface, which is followed by the aquifer media. For Imphal East, sand silt and clay as well as shale and siltstone constitute the vadose zone, which is shown in Figure 12.9.

12.4.1.7   Hydraulic Conductivity

Finally, hydraulic conductivity is calculated using the mini-MDI at different locations of the study area. In the field work, the volume change in the tube of the MDI is noted in the time interval of 30 s. Then, the collected data are analysed in excel

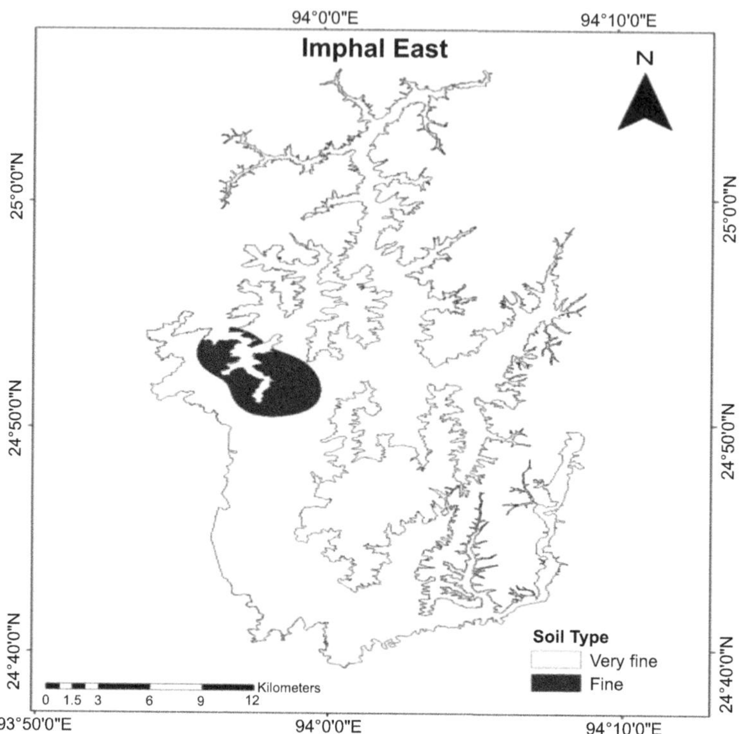

FIGURE 12.7 Soil type

sheet, and the value of hydraulic conductivity is provided in m/d. Interpolating the values of it, a map is generated, which is provided in Figure 12.10.

12.4.2 Drastic Vulnerability Index

Using the seven maps given in Figures 12.4 to 12.10, an overlay weighted analysis is performed and a final map is generated (Figure 12.11). After reclassifying the seven maps according to their relative rates from 1 to 10, which was given in Table 12.3, the maps are layered over to the raster calculated in Spatial Analyst tool.

As per the assigned weights in Table 12.2, the weights are multiplied with their rates, then the DVI is calculated, and is given in Figure 12.11. It shows different levels of vulnerable area in the study region, which is classified into three levels as low vulnerability (8%), moderate vulnerability (12%), and high vulnerability (80%), presented in Figure 12.11. It is observed that the index ranges from 111 to 142, and the northern part of the district is comparatively high from other parts of the district.

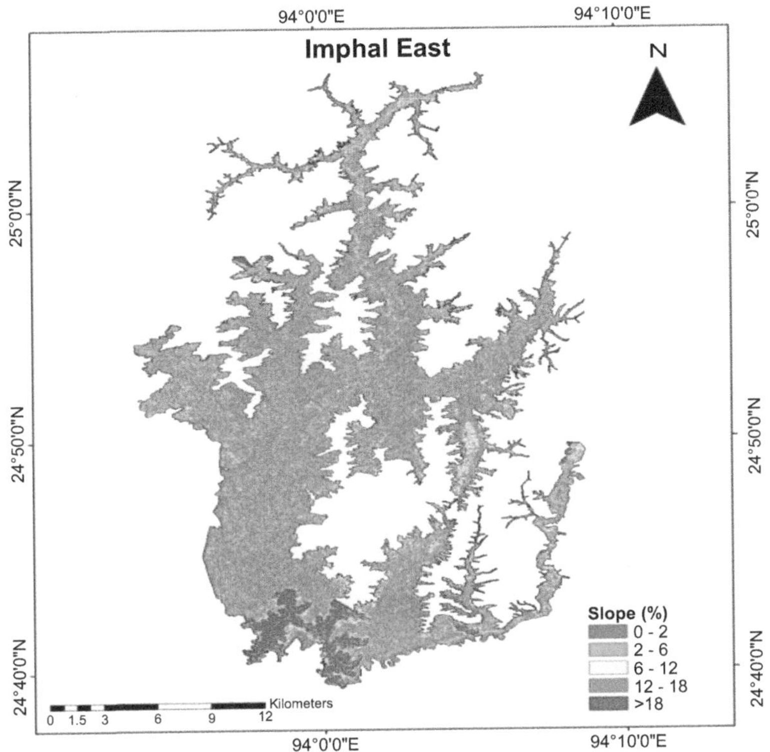

FIGURE 12.8   Slope (%)

12.4.3 VALIDATION

Total dissolved solids (TDSs) is the presence of different types of solids, which include salts, minerals, metals, and other dissolved substances in a given volume of water. It is considered to be one of the parameters for the health of water. To validate the vulnerability of the study area, the presence of the chemical in the groundwater like nitrate, pH, and TDS can prove the pollution of groundwater if the DVI and chemical value is linearly correlated. In this study, the predicted DVI is plotted with TDS data collected from Central Ground Water Board, Government of India, for the year 2018, which is shown in Figure 12.11. It is depicted from Figure 12.12 that the predicted DVI (it directly indicates the quality of groundwater) is linearly correlated ($R^2 = 0.6$) with the collected ground data of TDS (mg/l) and is considered reasonable for modelling and simulation studies.

12.5 CONCLUSION

DRASTIC model of the study region for groundwater vulnerability mapping is classified as low- (8%), moderate- (12%), and high-vulnerability (80%) areas. The

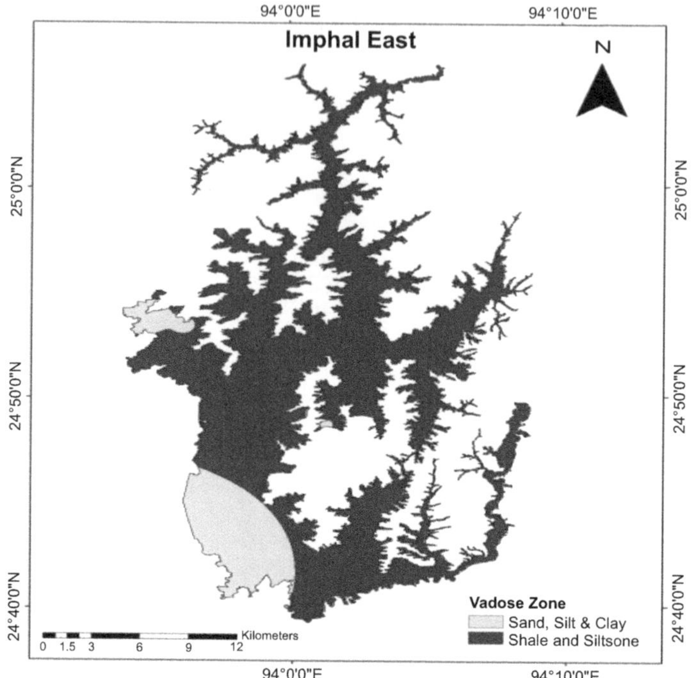

FIGURE 12.9 Vadose zone

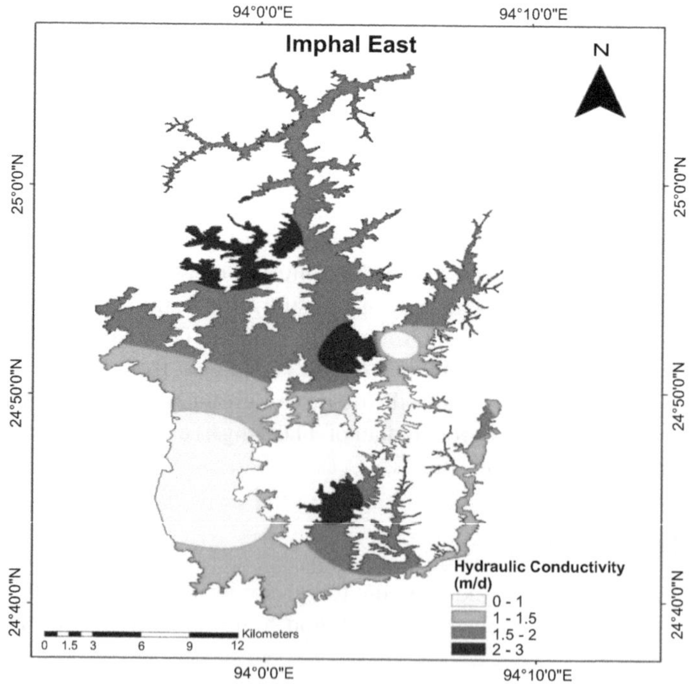

FIGURE 12.10 Hydraulic conductivity (m/d) of the study area

TABLE 12.3
Weight and rank assigned to the input parameters of the DRASTIC model (Aller et al., 1987)

Parameters	Range	Weight	Rating
Depth of water (m)	0–1.5	5	10
	1.5–4.5		9
	4.5–9		7
Net recharge (mm)	10–99	4	9
Aquifer media	Sand, medium	3	10
	Sand, coarse		9
	Sand, gravel		8
	Gravel		7
Soil media	Loamy sand	2	6
Topography (% slope)	0–2	1	10
	2–6		9
	6–12		5
	12–18		3
	>18		1
Impact of vadose zone	Shale and siltsone	5	6
	Sand, silt, and clay		3
Hydraulic conductivity (m/d)	0–4.1	3	1

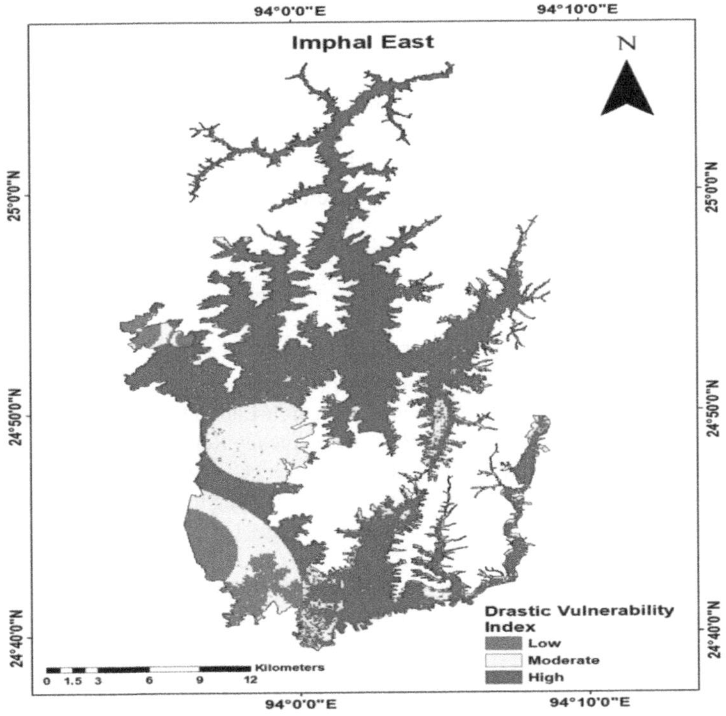

FIGURE 12.11   Drastic vulnerability index (DVI) of the study area

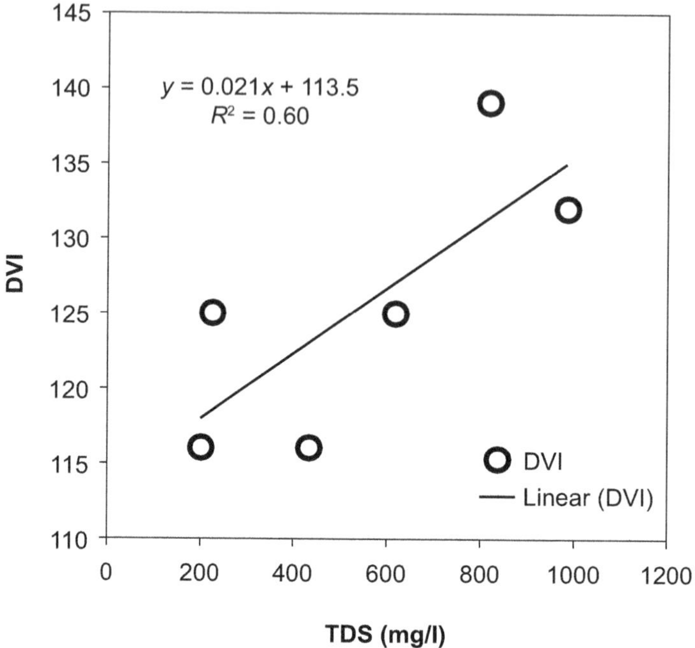

FIGURE 12.12   Comparison of predicted DVI with TDS value

calculated DVI range from 111 to 142 indicates that groundwater contamination is high and may affect the overall ecosystem in the region, including agricultural practices and yield. As a suggestion, cultivation can be done in those areas where the vulnerability is low (though it is a very small area). And, for the high areas, an understanding of what pollutes the groundwater and how to conserve the groundwater is needed, and, therefore, such studies are encouraged to be conducted regularly and widely.

REFERENCES

Aller, L., Bennet, T., Lehr, J. H., Petty, R. J., & Hachet, G. (1987). DRASTIC: A standardized system for evaluating groundwater pollution potential using hydrogeologic settings (EPA 600/2–87). Environmental Research Laboratory, Office of Research and Development, US Environmental Protection Agency Report, Tucson, 622.

Aller, L., Bennet, T., Lehr, J. H., Petty, R. J., Hacket, G. (1985). DRASTIC: A standardized system for evaluating groundwater pollution using hydrological settings. Prepared by the National water Well Association for the US EPA Office of Research and Development, Ada.

Arzu Firat, E., & Fatma, G. (2013). DRASTIC-based methodology for assessing groundwater vulnerability in the Gümüshaciköy and Merzifon basin (Amasya, Turkey). *Earth Sciences Research Journal, 17*(1), 33–40.

Bera, A., Mukhopadhyay, B. P., Chowdhury, P., Ghosh, A., & Biswas, S. (2021). Groundwater vulnerability assessment using GIS-based DRASTIC model in Nangasai River

Basin, India with special emphasis on agricultural contamination. *Ecotoxicology and Environmental Safety, 214*, 112085.

Borah, H., & Deka, S. (2023). Exploration of potential zones of groundwater in Jamuna watershed, Assam, by applying multi-influencing factor technique. *Journal of the Indian Society of Remote Sensing, 51*(1), 75–91.

Census. (2011). *District population report.* Government of Manipur.

Central Ground Water Board (CGWB). (2018). *Ministry of Jal Shakti: Aquifer Mapping and Management Plan Imphal East District, Manipur.*

Engel, B. A., Navulur, K. C. S., Cooper, B. S., Hahn, L. (1996). Estimating groundwater vulnerability to nonpoint source pollution from nitrates and pesticides on a regional scale, *International Association of Hydrological Sciences Publication, 235*, 521–526.

Ground Water Information Booklet Imphal East District, Manipur by Ministry of Water Resources (2013).

Malik, M. S., Shukla, J. P. (2019). GIS modeling approach for assessment of groundwater vulnerability in parts of Tawa river catchment area, Hoshangabad, Madhya Pradesh, India, *Groundwater for Sustainable Development, 9*, 100249.

Merchant, J. W., Whittemore, D. O., Whistler, J. L. McElwee C. D., & Woods, J. J. (1987). Groundwater pollution hazard assessment: A GIS approach. In: *Proceedings of the International Geographic Information Systems (IGIS) Symposium, Volume 3* (pp. 103–115). Association of American Geographers, Washington, DC.

Sarkar, M., & Pal, S. C. (2021). Application of DRASTIC and modified DRASTIC models for modeling groundwater vulnerability of Malda District in West Bengal. *Journal of the Indian Society of Remote Sensing, 49*, 1201–1219.

Singh, N. M., & Devi, T. T. (2022, October). Assessment and identification of drought prone zone in a low laying area by AHP and MIF method: A GIS based study. In: *IOP Conference Series: Earth and Environmental Science* (Vol. 1084, No. 1, p. 012047). IOP Publishing.

13 Constructed Wetlands as a Nature-Based Solution for Wastewater Management
A Sustainable Approach

Manoj Kumar

13.1 INTRODUCTION

With the rapid economic expansion and industrialization (Mudakkar et al., 2013), water resources are severely strained due to the production of huge amounts of industrial effluents and other factors (Olvera et al., 2017; Xiaoming et al., 2018). The water cycle is anticipated to be significantly impacted by climate change (Bates et al., 2008; Finger et al., 2012), causing problems in cities, including heat waves, droughts, and floods of water resources (Snep et al., 2020; Stefanakis et al., 2021). More than 80% of the wastewater produced globally is not collected or treated, making the city's parts the key source of contamination. In metropolitan areas, a lot of wastewater is generated and used. Large cities with centralized treatment plants have quite advanced wastewater management systems. They employ conventional engineering techniques that were expensive to create and maintain. Decentralizing minor treatment facilities is one approach that could provide rural communities with access to clean water and lessen environmental pollution. Each decentralized wastewater treatment plant (d-WWTP) can be specifically designed to accommodate the environment, aesthetic standards, water quality, and usage of the area. Constructed wetlands (CWs) have been widely employed in sanitation and pollution management as an example of d-WWTPs. They offer an inexpensive, low-maintenance substitute for conventional wastewater treatment. Another 2.1 billion people have insufficient sanitation, and 2.4 billion lack improved sanitation. Despite advancements made over the years to address this issue, public health, the environment, and the economy continue to be seriously endangered by the improper handling of faeces and wastewater. Growing interest is being shown in inexpensive sanitation options that utilize natural processes. Yet, understanding the circumstances under which such nature-based solutions (NbSs) might be appropriate and how to most effectively combine conventional infrastructure with natural solutions can be challenging for

DOI: 10.1201/9781003359326-13

wastewater utility management and other interested parties. For human life, water is a natural and necessary resource (Nika et al., 2020). Due to the increase in global population over the past century, water shortage has become one of the most urgent problems of the twenty-first century (Barbosa et al., 2020; Mansoori et al., 2020; McMillan et al., 2016).

While this is going on, there is currently no cycle treatment for water in urban ecologies, comprising wastewater, drinking water, stormwater, groundwater, surface waters, and a variety of urban environments where water is necessary. For instance, only countries with severe water shortages or hydric stress are now prioritizing the renewal and reuse of wastewater; however, in some cases, this regeneration can take place without additional water treatment (Guerra-Rodríguez et al., 2020). According to the Chinese Ministry of Ecology and Environment, 73 billion cubic meters of wastewater were produced in 2015. However, only 2.9% of this wastewater was treated for reuse. In contrast, in Europe, 964 million cubic meters of water were reused annually, which represented approximately 2.4% of the treated urban wastewater effluents (Zhu and Lam, 2009). Achieving the convergence of nutrient and material recovery, freshwater generation, and energy production requires a circular economy approach to wastewater management and sanitation. This is essential for water conservation and sustainability (Neczaj et al., 2018).

Hence, cities are increasingly incorporating NbS into their water management strategies. According to the EC, NbS refers to "keys that are cost-effective, maintained by wildlife, strategies that encourage resilience while also advancing the environment, society, and economy". These methods expand and diversify the amount of nature and natural features and processes in cities, landscapes, and seascapes through locally suitable, resource-effective, and systemic interventions. Within urban ecosystems, NbS addresses social issues and facilitates reserve retrieval, climate control and reworking difficulties, human well-being, and ecosystem restoration (Langergraber et al., 2020). NbS can also successfully reduce urban floods brought on by heavy rain occurrences (WWAP, 2018; Kolokotsa et al., 2013).

NbS for managing water resources includes the deliberate and planned use of ecology facilities to boost water availability, excellence, and resilience to climate change. "Grey infrastructure" is complemented by natural solutions. Although regulatory requirements are typically more focused on grey than on green solutions, each has a role to play in achieving sustainable management of water resources. For services where there is no ecosystem equivalent, such as the distribution of water to homes, seasonal water storage, and so forth, traditional engineering is still required. Although built-for-purpose infrastructure, such as levees, canals, reservoirs, and dams, may not always provide services at a predictable rate, NbSs can supplement and possibly even improve the advantages of such infrastructure, ensuring a return on investment. Using natural solutions can be extremely effective in reducing water-related risks and their negative effects, especially in light of climate change, which will increase already-existing pressures. Sustainable water management solutions can be provided through infrastructure portfolios that include developed infrastructure with nature-based alternatives.

Actions that help communities address various environmental, social, and economic issues in sustainable ways are known as "Nature-based Solutions" (NbSs).

These solutions are often inspired by and maintained by natural processes (DG Research and Innovation, 2015). Some meanings emphasize the role that diversified healthy ecosystems contribute to sustainable development and increased human resilience, emphasizing ecological facilities. For example, Maes and Jacobs (2017) define "Nature-based Solutions" (NbS) as any change in the utilization of ecosystem services that involves using fewer non-renewable resources and adopting more environmentally friendly methods. Eggermont et al. (2015) add that NbS aims to enhance the provision of a variety of ecosystem services through these sustainable practices. Three main types of NbSs have been identified, solutions that do not or only minimally intervene in ecologies intervene in landscapes and ecosystems that are managed for sustainability and a variety of purposes, and build new ecosystems or dramatically alter existing ones, generally to build green or blue infrastructure. Investing in environmentally friendly keys can have significant socioeconomic benefits and create jobs. Several of the naturally designed keys previously utilized in water management and urban development (e.g., vegetation in street canyons, bio-infiltration rain gardens, and green roofs) have shown to be more effective than the supposed "grey substructure" substitutes in terms of efficiency, cost, adaptability, multipurpose use, and durability (e.g., Gill et al., 2007; Pugh et al., 2012; Ellis, 2013; Flynn and Traver, 2013; Raje et al., 2013). Environmentally responsible technology called CWs treatment systems imitates the function of natural wetlands to enhance water quality in a practical way, with straightforward instructions and a more maintainable approach (Vymazal, 2013; Wu et al., 2013; Chen et al., 2014; Kumar and Singh, 2017).

13.2 NBS TECHNOLOGIES FOR SUSTAINABLE WATER MANAGEMENT

With an emphasis on the potential co-benefits, the recently released NbS for the treatment of wastewater by an International Water Association (IWA) Publishing offers a starting point for identifying and comparing NbS solutions that can be implemented the home and municipal wastewater treatment processes. Peer-reviewed factsheets and case studies highlight a variety of NbS as part of the process of treating household wastewater while also delivering ecological and social benefits, building on the body of current research. Case studies demonstrate the practical application of these natural wastewater treatment methods and shed light on the range of potentially useful solutions. The objective is for a wide range of stakeholders to gain a basic grasp of the design criteria, removal efficiency, costs, co-benefits for people and the environment, and trade-offs to consider in their own local context.

13.3 NBS INNOVATIONS IN SUSTAINABLE WATER MANAGEMENT

The idea, or improved still, the idea, of NbS is founded on a process known as biomimicry, in which we observe, study, and imitate nature. To use ecosystems' power and sophistication as substructure to afford usual facilities that improve society and the environment, and to turn issues with the environment, society, and the economy into possibilities for innovation, prior knowledge of the natural world and its processes is crucial for NbS. According to this perspective, NbS varies from solutions obtained from nature, such as natural energy from the wind, waves, and sun, which, while not directly dependent on healthy ecosystems, help meet our low-carbon energy needs by using production methods that were inspired by natural sources. To achieve the hoped-for and anticipated results, NbS utilizes the traits of complicated natural system processes, such as the ability to regulate water flow and store carbon. NbS can, therefore, also play a significant role in promoting well-being, high living standards, plenty, biodiversity, green growth, social and economic sustainability, and of course, environmental sustainability (Kolokotsa et al., 2020; Spano et al., 2021; Cohen-Shacham et al., 2019).

A resolution was approved by the members of the International Union for Conservation of Nature (IUCN) (WCC-2016-Res-069-EN) at the 2016 World Conservation Congress and Members' Assembly, that described the usage of NbS as "activities to safeguard, sustainably manage" for the first time and reestablish natural or modified ecosystems that successfully and adapt to societal issues, promoting benefits for both people and biodiversity. Putting in place permeable pavements to reduce flooding risk and green roofs to boost biodiversity and minimize the need for cooling and heating in cities are straightforward NbS examples that will help you better comprehend their concept. NbS has been hailed as a powerful paradigm for addressing and reversing global environmental problems like biodiversity loss, ecological restoration, and the depletion of natural resources (Cohen-Shacham et al., 2016) based on a set of best practice guidelines for catastrophe risk, environmental degradation, water security, human health, and socioeconomic progress.

The European Commission offers a different definition, this time from the perspective of the social impact of NbS application: "NBS to social difficulties are keys that are supported by and inspired by nature, which are efficient, afford advantages for the environment, society, and the economy at the same time, and foster resilience". These methods expand and diversify the amount of nature and natural features and processes in cities, landscapes, and seascapes through locally customized, resource-effective, and systemic interventions. Biodiversity must be benefited by NbSs, and they must help supply a variety of ecosystem services (Bulkeley, 2020).

13.4 NBS TYPES

In the implementation experiences, many sorts of NbSs are offered as technical replies. Sustainable urban drainage systems (SUDS), green walls and roofs, riverside parks, agroforestry, parks, permeable paving, and wetland-related methods, including natural, artificial, and purpose-built wetlands, are examples of NbS. Phytorid sewage treatment involves purifying wastewater in artificial wetlands by employing a certain type of plant (Dhyani et al., 2018), in addition to bioswales and rain gardens. Solutions using SUDS specifically for their filtration or drainage capabilities include wetlands, detention and retention basins, bioswales, green roofs, permeable pavement, and rain gardens. NbSs are mutual within fusion methods of grey infrastructures rather than being deployed singly. When compared to conventional infrastructure projects, NbS as a place-based intervention changes how landscape management is approached. This is accomplished by highlighting the relationship between the blue (water) and green (vegetation) areas, as well as the impact of changes in land uses and land covers. Using flood risk reduction as an illustration, according to several publications, the hybrid technique is the most common; the blue and green approaches are then used (Sahani et al., 2019). Some, however, contend that the green strategy is more vulnerable than the blue plan because it is affected and transformed by land use and cover changes (Gunnell et al., 2019). NbSs advocate a change from the design and, despite its vulnerability, use of grey flood control infrastructure or NbS planning for water management (La Rosa and Pappalardo, 2020).

13.5 SELECTING AN NBS FOR WASTEWATER TREATMENT

Choosing the right NbS for wastewater treatment can be difficult. It is frequently challenging for communities to learn more and choose whether NbS could be appropriate for treating wastewater. To help with this, a web-based platform that incorporates various NbS technologies and case studies from across the world is currently being developed. These tools can increase awareness and enhance comprehension of the NbS requirements and practical options. However, to choose the optimum NbS or combination of NbS, wastewater operators should seek out additional technical advice and experience. The context in which NbS is applied must be taken into consideration when designing and implementing the system, as well as any potential trade-offs.

13.6 NBS TECHNOLOGIES AND PROJECTS/ CASE STUDIES APPROACH

For wastewater utility managers and operators, local governments and municipalities, and regulators, easily available technical material and case studies for adopting NbS as part of wastewater treatment can serve as a platform for thinking about cost-effective NbS choices. These are some situations illustrating several well-known and more recently created NbS techniques. For more information, see the book *Nature-Based Solutions for Wastewater Treatment*. Also, the green roof system helps to reduce runoff from rainfall, which can lessen urban flooding and the strain on the drainage system. Using willow systems, which are treatment wetlands dominated by willows, is another more recent NbS strategy appropriate for peri-urban or rural areas. To screen the technological possibilities and choose the preferred alternative, a feasibility study was carried out. Also, it included the economic and financial analysis, determining the need for capacity building and organizing plant operator training, as well as the technical and financial evaluation of the chosen alternative. The results of the feasibility study point to CWs as wherever there is a lack of space for manmade wetlands, this is a good alternative that might be used for decentralized treatment. CWs are a type of nature-based reactor. The previous study, however, found that the decentralized strategy gives only significant economic, social, and environmental benefits when extremely reliable and low-maintenance technology is employed. Therefore, the basin-scale water strategy must take these factors into account (Cross et al., 2021).

13.7 URBAN WATER POLLUTION CONTROL: CWS

Artificial wetlands were investigated as a potential natural alternative with several benefits. It is affordable to build, operate, and maintain; effective; and simple to use. So, it is probable that other rural communities in the region may be interested in implementing it. About 1,100 residents of Kosovo's Kramovik hamlet will directly benefit from the project. This remedy was meant to cut pollutants by 90%, per the construction company's specs. It was constructed on municipal property, and the GEF Drin Project provided the capital investment. The activity's implementation, including the construction, was coordinated by UNDP and GWP-Med with the help of national and local organizations. To guarantee a high level of ownership by individuals who would be directly impacted by this project, stakeholders were contacted and kept informed on the technical specifics of construction throughout the process. The infrastructure included a screening chamber where effluent is pre-treated. A baffled anaerobic reactor (ABR), a type of septic tank, can reduce nutrients and BOD5 by up to 90%. This was a 60×20 m "filter" packed with sand and gravel layers and covered in *Phragmites australis* reeds. After passing through the ABR, the wastewater flows into the filter, where it serves as "food" for the reeds in the artificial wetland. Sewage distribution tubes go throughout the swamp. Impermeable geomembranes stop wastewater from permeating and penetrating the soil (UNSCR, 1999).

13.8 CONSTRUCTED WETLAND

To handle sewage, greywater, and stormwater runoff, an artificial wetland, also known as a constructed wetland (CW), was created. CWs are also used for treating industrial effluent. (Vymazal et al., 2021) and (Arden and Ma, 2018). Moreover, it might be used for the reclamation of the land after mining or in an effort to make up for the natural areas that land development has destroyed. Engineered structures known as CWs treat wastewater secondarily by utilizing the natural processes of vegetation, soil, and organisms. The type of wastewater that needs to be treated must be taken into consideration when designing the created wetland. CWs have been employed in both centralized and decentralized wastewater systems (Maiga et al., 2017).

13.8.1 CLASSIFICATION OF CWs

There are primarily two types of wetlands: (a) natural wetlands, and (b) CWs. Depending on how they operate, CWs might be of several sorts (Figure 13.1).

13.8.2 TYPES OF CWs

13.8.2.1 Free Water Surface CW

Free water surface-built wetlands (FWS CWs) (Figure 13.2), also known as created wetlands with surface flow (SF), are places with open water as well as floating, submerged, and emerging plants (Kadlec and Wallace, 2008). The presence of plant stalks and debris, combined with low flow velocity and shallow water depth, controls water flow, creating plug-flow conditions, particularly in long, narrow channels

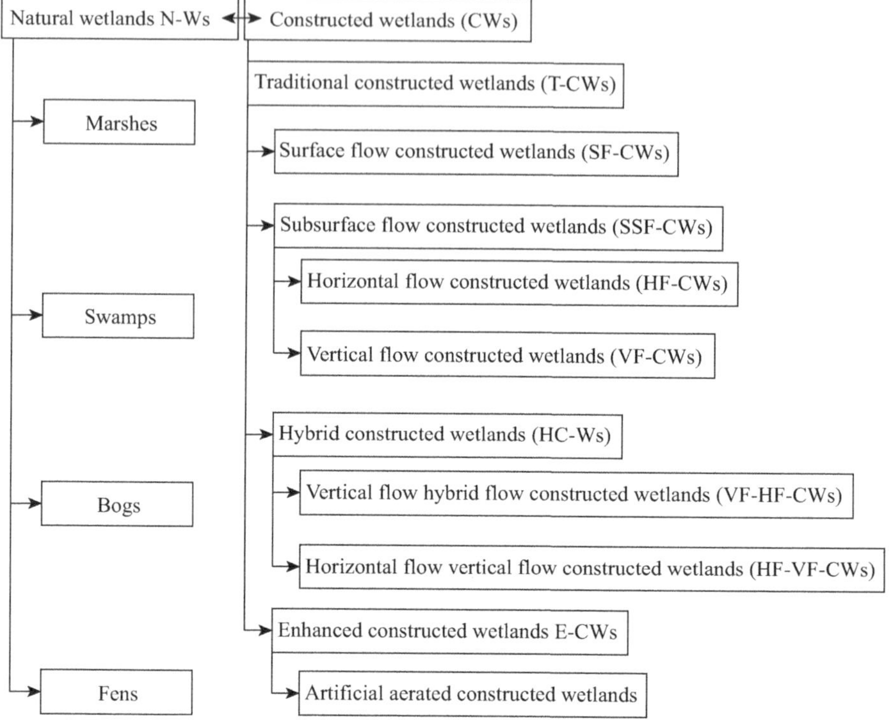

FIGURE 13.1 Classification of natural and artificial constructed wetlands (Adapted from Kumar et al., 2021)

(Crites et al., 2010). The surface flow of free water CWs may remove a lot of suspended particles while removing pathogens, nutrients, and other toxins like phosphate and heavy metals only moderately (Tilley et al., 2008).

13.8.2.2 Horizontal Subsurface Flow CWs

Wetlands with horizontal subsurface flow are one type of CW (Figure 13.3). The removal of the contaminant involves a number of methods, including organics, N, and PAs the primary biological components of these systems, macrophytes enhance various removal processes and directly absorb nitrogen (N), phosphorus (P), and other pollutants, thereby supporting depuration reactions (Liu et al., 2011; Ko et al., 2011). With a mean elimination effectiveness of 78.4%, suspended solids revealed substantial variances between intake and output. Biochemical oxygen demand had average elimination performance of 57.9% and 68.7%, respectively. TKN, nitrates, and total phosphate removal percentages were lower than those for other parameters (25.7%, 47.8%, and 29.9%, respectively). In the final effluent, faecal coliform bacteria dropped by an order of magnitude. Results for *Escherichia coli* and *Pseudomonas aeruginosa* were inconsistent. By the end of the monitoring period, total biomass had multiplied by 4.6 (Schierano et al., 2020).

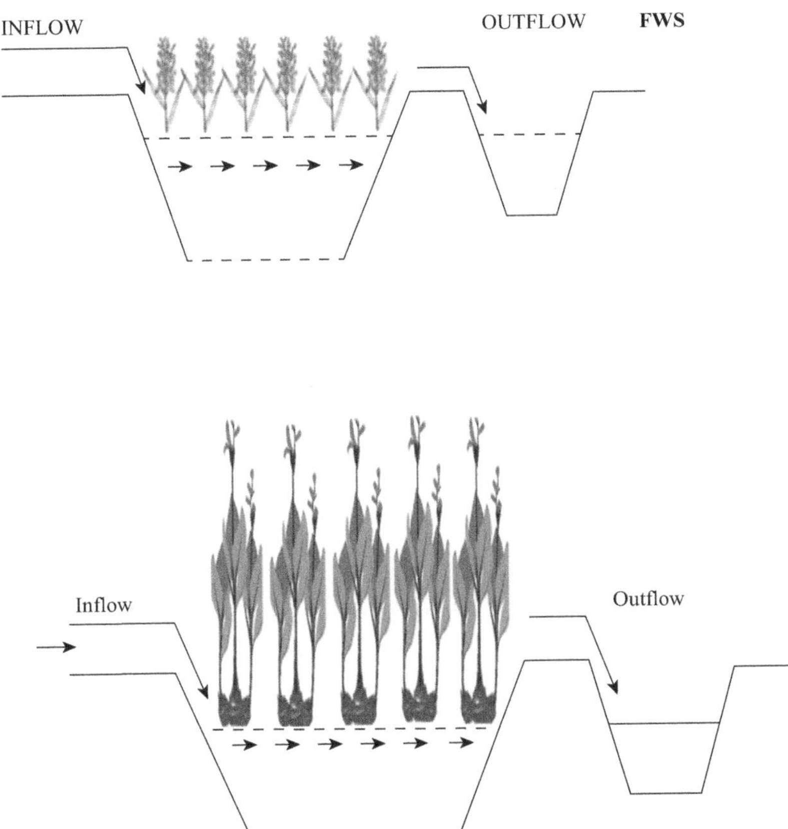

Free-water Surface Constructed Wetlands

FIGURE 13.2   Free water surface constructed wetland

13.8.2.3   Vertical Subsurface Flow CWs

Vertical Subsurface Flow CWs (Figure 13.4) were first used to oxygenize anaerobiotic septic tank effluents by Seidel (Seidel, 1965). However, Compared to HF-CWs, the VF-CWs did not spread as swiftly, most likely because of the greater operational and maintenance costs associated with the required systems to sporadically pump wastewater onto the wetland surface. Large batches of water are injected, and, after that, the liquid diffuses through the sand medium. Only after all the water has filtered through and the bed is dry is the new batch fed. This allows oxygen to diffuse into the bed from the surrounding air. Hence, CWs-VF is significantly additionally aerobic than CWs-HF and offers favourable nitrification circumstances. VF-CWs, in contrast, don't offer any denitrification. Moreover, VF-CWs are highly good at removing suspended particles and organics. Phosphorus removal is minimal unless mediums with large sorption capacities are utilized. CWs-VF systems require less

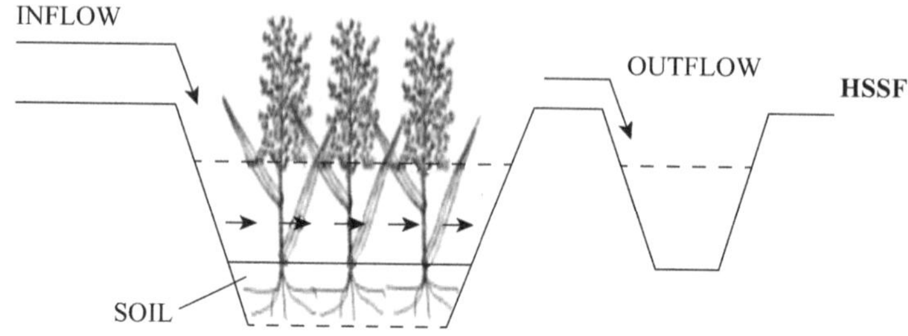

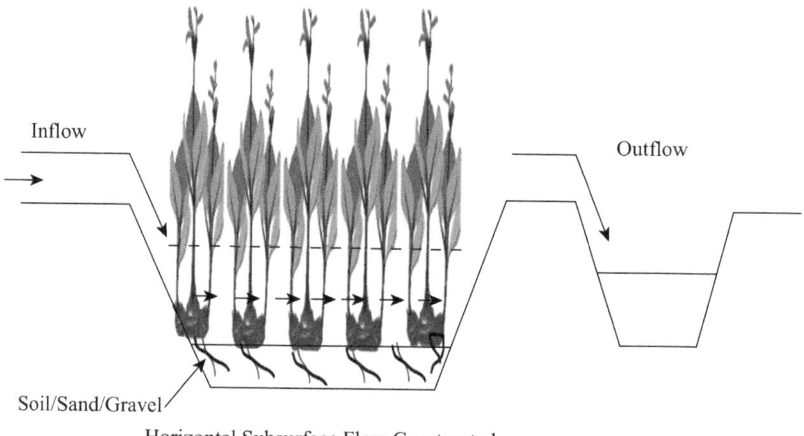

FIGURE 13.3   Horizontal subsurface flow constructed wetland

acreage than CWs-HF, typically 1–3 m² PE[1] (Brix and Arias, 2005; Cooper, 1999; Weedon, 2001; Molle et al., 2005). The early VF-CWs had multiple stages, with the first stage's beds being fed alternately. Currently, when CWs-VF are built, they are frequently only equipped with one bed; this type of system is known as "compact" VF-CWs. (Cooper, 1999).

13.8.2.4 Subsurface Flow CWs

A subsurface flow (SF) wetland is a type of wetland that is specifically created to treat or polish a particular kind of wastewater. It is normally built as a bed or channel that contains the necessary material. An example of a SF wetland is shown in Figure 13.5. There have been uses for coarse rock, gravel, sand, and several types of soils, although the most popular one in the United States and Europe is gravel. It is meant for the water surface to cling to the medium's top surface, and the average is frequently established with similar budding plants found in marshes. The main benefits

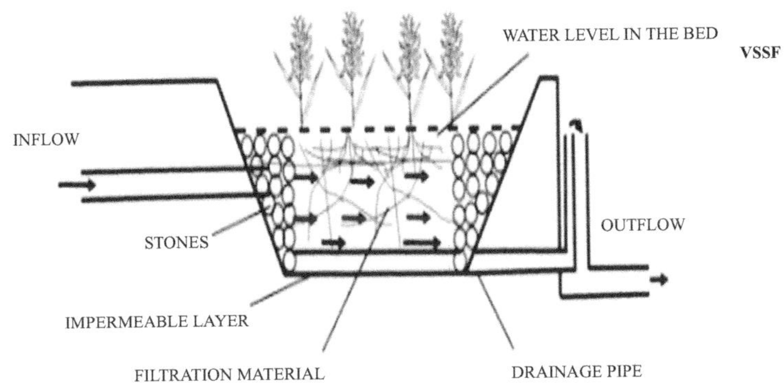

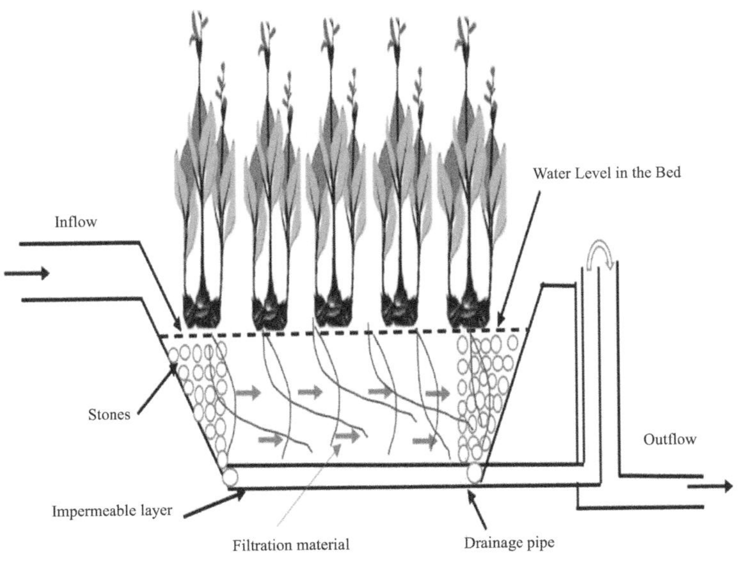

FIGURE 13.4 Vertica subsurface flow constructed wetlands

of this subterranean water level are the removal of the risk of community interaction with inadequate wastewater treatment and the control of odors and mosquitoes. As a result, there is a risk of mosquitoes and community contact when the water surface is exposed in naturally occurring marshes and wetlands with FWS. This widespread lack of oxygen availability restricts these SF wetlands' ability to biologically remove ammonia nitrogen (NH_3/NH_4^-N) through nitrification. Nonetheless, as biochemical oxygen demand, total suspended solids, metals, and several importance contaminant organics might be treated in either anoxic or aerobic environments; the system is still quite successful for removing these pollutants (USEPA, 2000).

13.9 INTEGRATING NBS INTO POLICY AND ACTION

As a tool to manage wastewater and water resources and take action on climate change, NbSs are becoming more and more incorporated into policy and practice. For instance, NbSs have been a part of the Horizon 2020 Project inside the European

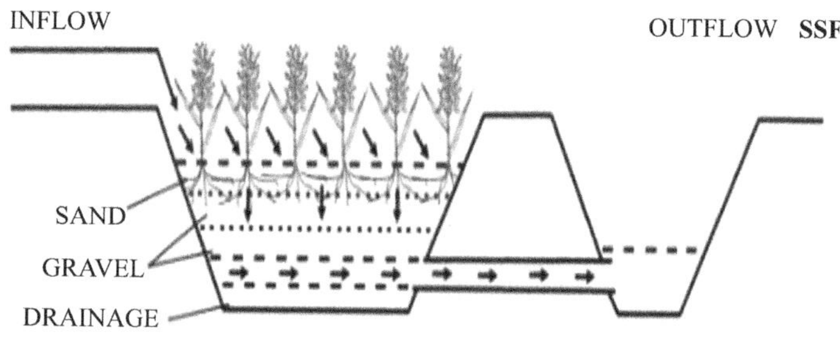

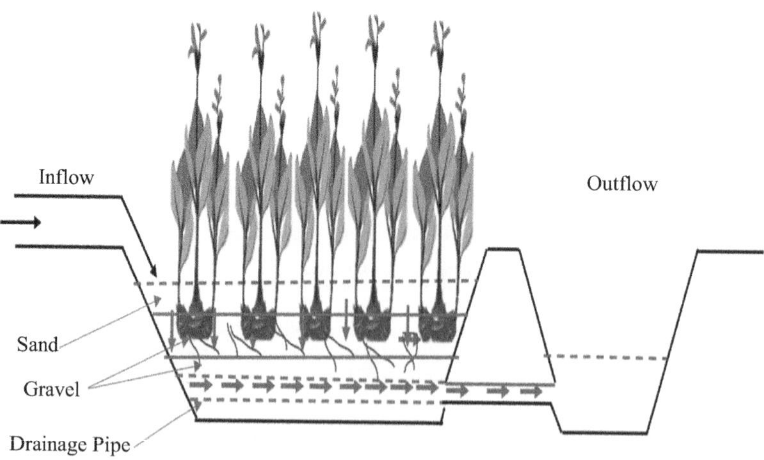

Subsurface Flow Constructed Wetland

FIGURE 13.5   Subsurface flow constructed wetlands

Commission, a financial instrument designed to guarantee that Europe produces top-tier science, reducing obstacles to innovation and simplifying public–private sector collaboration to deliver innovation.

To achieve its aims of innovation for growth and job creation, Horizon 2020 has pushed the alignment of biodiversity and environmental services. An additional illustration is the World Bank, which has included NbS into more than 100 projects in 60 countries and established its knowledge base to assist the use of NbS to manage disasters and water resources. The lack of infrastructure for three-quarters of 2050 presents a chance to investigate how to incorporate nature into its design. Recovery plans for COVID-19 present a chance to foster economic growth in a way that safeguards and restores the environment, addresses climate change, encourages the energy transition, and supports livelihoods. It is a good opportunity to rebuild better in a way that promotes, protects, and integrates nature in addition to the need for infrastructure. The Paris Agreement on Climate recognizes the connections between community resilience, livelihoods, and ecosystem resilience, and NbS can help those objectives. NbSs are also essential for achieving the Sustainable Development Goals, particularly SDG2 (zero hunger), SDG6 (water), SDG13 (climate change), SDG14 (oceans), and SDG15 (ecosystems). There is increasing support for implementation as a result of the rising commitment and interest throughout the policy landscape, which can be built upon a strong evidence base from ongoing experience across the wastewater and water sectors. The various advantages of incorporating NbSs as assets into water and wastewater treatment processes can be included as a key component of a business case that managers can use to better convince regulators, municipal governments, and wastewater utilities. With a recently established working committee, IWA has supported such interaction in the previous and will continue to do so in the forthcoming.

13.10 CONCLUSION

For the treatment of wastewater, examples of NbSs include treatment wetlands and artificial wetlands. CWs offer natural treatment methods that efficiently handle an extensive range of wastewaters (e.g., homes, oil refineries, as well as industrial effluent from pulp and paper mills, textile mills, and seafood processors). CWs are suitable for primary, secondary, and, or tertiary treatment of raw wastewater and can treat it to various levels of cleanliness. Considering that CWs are designed systems that both enhance and replicate natural environmental processes, they are regarded as supportable, ecologically approachable techniques for treating wastewater. CWs are less vulnerable to inflow volatility, have steady performance, and require little operational and maintenance. CWs have demonstrated their capacity to treat a variety of wastewater kinds. The interest in CWs as a treatment technique has increased due to a number of advantages and facts, including their minimal energy consumption, low construction and operating expenses, and few operational requirements. With a focus on merging the technical and financial sustainability with a circular economic approach and climate change resistance, it is explained how long-lasting CWs are as a sanitary solution.

REFERENCES

Arden, S., & Ma, X. (2018). Constructed wetlands for greywater recycle and reuse: A review. *Science of the Total Environment*, *630*, 587–599.

Barbosa, R. F., Souza, A. G., Maltez, H. F., & Rosa, D. S. (2020). Chromium removal from contaminated wastewaters using biodegradable membranes containing cellulose nanostructures. *Chemical Engineering Journal*, *395*, 125055.

Bates, B. C., Kundzewicz, Z. W., Wu, S., & Palutikof, J. P. (2008). Climate change and water. Technical paper of the intergovernmental panel on climate change, IPCC secretariat: Geneva, Switzerland. *The American Midland Naturalist*, *168*(1).

Brix, H., & Arias, C. A. (2005). The use of vertical flow constructed wetlands for on-site treatment of domestic wastewater: New Danish guidelines. *Ecological Engineering*, *25*(5), 491–500.

Bulkeley, H. (2020). Nature-based Solutions: Towards sustainable communities—Analysis of EU-funded projects. Nature-Based Solutions—State of the Art of EU-Funded Projects; Directorate-General for Research & Innovation (European, Commission), 156–180.

Chen, Y., Wen, Y., Zhou, J., Tang, Z., Li, L., Zhou, Q., & Vymazal, J. (2014). Effects of cattail biomass on sulfate removal and carbon sources competition in subsurface-flow constructed wetlands treating secondary effluent. *Water Research*, *59*, 1–10.

Cohen-Shacham, E., Andrade, A., Dalton, J., Dudley, N., Jones, M., Kumar, C., Maginnis, S., Maynard, S., Nelson, C. R., Renaud, F. G., & Welling, R. (2019). Core principles for successfully implementing and upscaling Nature-based Solutions. *Environmental Science & Policy*, *98*, 20–29.

Cohen-Shacham, E., Walters, G., Janzen, C., & Maginnis, S. (2016). Nature-based solutions to address global societal challenges. *IUCN: Gland, Switzerland*, *97*, 2016–2036.

Cooper, P. (1999). A review of the design and performance of vertical-flow and hybrid reed bed treatment systems. *Water Science and Technology*, *40*(3), 1–9.

Crites, R. W., Middlebrooks, E. J., & Reed, S. C. (2010). *Natural wastewater treatment systems*. CRC Press.

Cross, K., Tondera, K., Rizzo, A., Andrews, L., Pucher, B., Istenič, D., Karres, N., & Mcdonald, R. (2021). *Nature-based solutions for wastewater treatment*. IWA Publishing.

DG Research and Innovation (2015). *Towards an EU research and innovation policy agenda for nature-based solutions & re-naturing cities Final Report of the Horizon 2020 Expert Group on 'Nature-Based Solutions and Re-Naturing Cities' (full version)*. Publications Office of the European Union, Luxembourg, 70.

Dhyani, S., Lahoti, S., Khare, S., Pujari, P., & Verma, P. (2018). Ecosystem based Disaster Risk Reduction approaches (EbDRR) as a prerequisite for inclusive urban transformation of Nagpur City, India. *International Journal of Disaster Risk Reduction*, *32*, 95–105.

Eggermont, H., Balian, E., Azevedo, J. M. N., Beumer, V., Brodin, T., Claudet, J., Fady, B., Grube, M., Keune, H., Lamarque, P., Reuter, K., Smith, M., van Ham, C., Weisser, W. W., & Le Roux, X. (2015). Nature-based solutions: New influence for environmental management and research in Europe. *Gaia*, *24*(4), 243–248.

Ellis, J. B. (2013). Sustainable surface water management and green infrastructure in UK urban catchment planning. *Journal of Environmental Planning and Management*, *56*(1), 24–41.

Flynn, K. M., & Traver, R. G. (2013). Green infrastructure life cycle assessment: A bio-infiltration case study. *Ecological Engineering*, *55*, 9–22.

Gill, S. E., Handley, J. F., Ennos, A. R., & Pauleit, S. (2007). Adapting cities for climate change: The role of the green infrastructure. *Built Environment*, *33*(1), 115–133.

Guerra-Rodríguez, S., Oulego, P., Rodríguez, E., Singh, D. N., & Rodríguez-Chueca, J. (2020). Towards the implementation of circular economy in the wastewater sector: Challenges and opportunities. *Water*, *12*(5), 1431.

Gunnell, K., Mulligan, M., Francis, R. A., & Hole, D. G. (2019). Evaluating natural infrastructure for flood management within the watersheds of selected global cities. *Science of the Total Environment, 670*, 411–424.

Kadlec, R. H., & Wallace, S. (2008). *Treatment wetlands.* CRC Press.

Ko, C. H., Lee, T. M., Chang, F. C., & Liao, S. P. (2011). The correlations between system treatment efficiencies and aboveground emergent macrophyte nutrient removal for the Hsin-Hai Bridge phase II constructed wetland. *Bioresource Technology, 102*(9), 5431–5437.

Kolokotsa, D., Lilli, A. A., Lilli, M. A., & Nikolaidis, N. P. (2020). On the impact of nature-based solutions on citizens' health & well being. *Energy Build, 229*, 110527.

Kolokotsa, D., Santamouris, M., & Zerefos, S. C. (2013). Green and cool roofs' urban heat island mitigation potential in European climates for office buildings under free floating conditions. *Solar Energy, 95*, 118–130.

Kumar, M., & Singh, R. (2017). Performance evaluation of semi continuous vertical flow constructed wetlands (SC-VF-CWs) for municipal wastewater treatment. *Bioresource Technology, 232*, 321–330.

Langergraber, G., Pucher, B., Simperler, L., Kisser, J., Katsou, E., Buehler, D., Mateo, M. C. G., & Atanasova, N. (2020). Implementing nature-based solutions for creating a resourceful circular city. *Blue-Green Systems, 2*(1), 173–185.

La Rosa, D., & Pappalardo, V. (2020). Planning for spatial equity-A performance-based approach for sustainable urban drainage systems. *Sustainable Cities and Society, 53*, 101885.

Liu, W. J., Zeng, F. X., Jiang, H., & Yu, H. Q. (2011). Total recovery of nitrogen and phosphorus from three wetland plants by fast pyrolysis technology. *Bioresource Technology, 102*(3), 3471–3479.

Maes, J., & Jacobs, S. (2017). Nature-based solutions for Europe's sustainable development. *Conservation Letters, 10*(1), 121–124.

Maiga, Y., Von Sperling, M., & Mihelcic, J. (2017). Constructed wetlands. In: *Global Water Pathogen Project,* C. Haas, J. Mihelcic, M. Verbyla (Eds.), Michigan State University, E. Lansing, UNESCO.

Mansoori, S., Davarnejad, R., Matsuura, T., & Ismail, A. F. (2020). Membranes based on non-synthetic (natural) polymers for wastewater treatment. *Polymer Testing, 84*, 106381.

Kumar, M., Singh, N. K., & Singh, R. (2021). Application of constructed wetlands in degradation and detoxification of industrial effluents: Challenges and prospects. *Bioremediation for Environmental Sustainability*, 43–82.

Molle, P., Liénard, A., Boutin, C., Merlin, G., & Iwema, A. (2005). How to treat raw sewage with constructed wetlands: An overview of the French systems. *Water Science and Technology, 51*(9), 11–21.

Mudakkar, S. R., Zaman, K., Khan, M. M., & Ahmad, M. (2013). Energy for economic growth, industrialization, environment and natural resources: Living with just enough. *Renewable and Sustainable Energy Reviews, 25*, 580–595.

Neczaj, E., & Grosser, A. (2018, July). Circular economy in wastewater treatment plant—challenges and barriers. In: *Proceedings* (Vol. 2, No. 11, p. 614). MDPI.

Nika, C. E., Gusmaroli, L., Ghafourian, M., Atanasova, N., Buttiglieri, G., & Katsou, E. (2020). Nature-based solutions as enablers of circularity in water systems: A review on assessment methodologies, tools and indicators. *Water Research, 183*, 115988.

Olvera, R. C., Silva, S. L., Robles-Belmont, E., & Lau, E. Z. (2017). Review of nanotechnology value chain for water treatment applications in Mexico. *Resource-Efficient Technologies, 3*(1), 1–11.

Pugh, T. A., MacKenzie, A. R., Whyatt, J. D., & Hewitt, C. N. (2012). Effectiveness of green infrastructure for improvement of air quality in urban street canyons. *Environmental Science & Technology, 46*(14), 7692–7699.

Raje, S., Kertesz, R., Maccarone, K., Seltzer, K., Siminari, M., Simms, P., Wood, B., & Sansalone, J. (2013). Green infrastructure design for pavement systems subject to rainfall—runoff loadings. *Transportation Research Record, 2358*(1), 79–87.

Sahani, J., Kumar, P., Debele, S., Spyrou, C., Loupis, M., Aragão, L., Porcù, F., Shah, M. A. R., & Di Sabatino, S. (2019). Hydro-meteorological risk assessment methods and management by nature-based solutions. *Science of the Total Environment, 696,* 133936.

Schierano, M. C., Panigatti, M. C., Maine, M. A., Griffa, C. A., & Boglione, R. (2020). Horizontal subsurface flow constructed wetland for tertiary treatment of dairy wastewater: Removal efficiencies and plant uptake. *Journal of Environmental Management, 272,* 111094.

Seidel, K. (1965). Neue Wege zur Grundwasseranreicherung in Krefeld, Vol. II. Hydrobotanische Reinigungsmethode. *GWF Wasser/Abwasser, 30,* 831–833.

Snep, R. P., Voeten, J. G., Mol, G., & Van Hattum, T. (2020). Nature based solutions for urban resilience: A distinction between no-tech, low-tech and high-tech solutions. *Frontiers in Environmental Science, 8,* 599060.

Spano, G., Dadvand, P., & Sanesi, G. (2021). The benefits of nature-based solutions to psychological health. *Frontiers in Psychology, 12,* 646627.

Stefanakis, A. I., Calheiros, C. S., & Nikolaou, I. (2021). Nature-based solutions as a tool in the new circular economic model for climate change adaptation. *Circular Economy and Sustainability, 1,* 303–318.

Tilley, E., Atwater, J., & Mavinic, D. (2008). Effects of storage on phosphorus recovery from urine. *Environmental Technology, 29*(7), 807–816.

United States Environmental Protection Agency (USEPA) Office of Water, Washington, DC. EPA 832-F-00–023 September 2000.

UNSCR. (1999). *This designation is without prejudice to positions on status, and is in line with 1244/1999 and the ICJ Opinion on the Kosovo declaration of independence.*

Vymazal, J. (2013). The use of hybrid constructed wetlands for wastewater treatment with special attention to nitrogen removal: A review of a recent development. *Water Research, 47*(14), 4795–4811.

Vymazal, J., Zhao, Y., & Mander, Ü. (2021). Recent research challenges in constructed wetlands for wastewater treatment: A review. *Ecological Engineering, 169,* 106318.

Weedon, C. M. (2001). Compact vertical flow reed beds: Design rationale and early performance. *IWA Macrophytes Newsletter, 23,* 12–20.

Wu, S., Kuschk, P., Wiessner, A., Müller, J., Saad, R. A., & Dong, R. (2013). Sulphur transformations in constructed wetlands for wastewater treatment: A review. *Ecological Engineering, 52,* 278–289.

WWAP (United Nations World Water Assessment Programme) (2018). *The United Nations world water development report 2018: Nature-based solutions.* UNESCO.

Xiaoming, F., Qinglong, L., Lichang, Y., Fu, B., & Yongzhe, C. (2018). Linking water research with the sustainability of the human—natural system. *Current Opinion in Environmental Sustainability, 33,* 99–103.

Zhu, T., & Lam, K. C., (2009). *Environmental impact assessment in China.* Research Centre of Strategic Environmental Assessment for China and Chinese University of Hong Kong.

14 Urban Mining of E-Waste
Inception of Secondary Resources of Valuable Metals

Rohit Jha, Munmun Agrawal, Pushpa Gautam, Mudila Dhanunjaya Rao, and Kamalesh K. Singh

14.1 E-WASTE: ORIGIN AND ITS IMPACT

In the current era of industrialized society, the electronic industry is among the most growing industries due to the rapid advancements in different types of gadgets. It has been expanding significantly in recent years and creates a lot of jobs, nurtures technological advancement, and increases a strong demand for raw resources that are regarded as rare or scarce (Agrawal et al., 2021). The rapid advancement also increases the hasty obsolescence of electronic gadgets, resulting in tonnes of electronic waste (e-waste). E-waste includes all objects of electrical and electronic apparatus and its constituents that have been rejected by its owner.

Reuse, remanufacturing, and recycling, along with burning and dumping are the foremost possibilities for the handling of e-waste at present. The hierarchy of e-waste treatment is shown in Figure 14.1. The incineration and landfilling are not environmentally friendly treatments because of the hazardous material content of the e-waste. Moreover, the incineration produces dioxin and other gases due to the existence of flame retardants (FRs), chlorofluorocarbons, and copper content (acts as a catalyst for dioxin formation) in constituents of e-waste (Rao et al., 2020; Rautela et al., 2021).

Recycling of electronic waste may reduce environmental hazards, and also it will maintain the supply and demand chain of raw materials by recovering them from the e-waste. Consequently, the recycling of e-waste is recommended by most of the countries, and separate legislation has also been imposed in contrast to e-waste management and recycling.

From the recycling point of view, e-waste can be measured as a source of valuable metals (Au, Ag, Cu, other platinum group metals, etc.), glass, plastics, and other non-metallic fractions. Most of the work on recycling e-waste is focused on the recovery of metallic values because of the economical values. This chapter basically deliberated the origin of e-waste, global situation of e-waste production, and its controlling, and further the possible recycling routes of e-waste have been discussed.

DOI: 10.1201/9781003359326-14

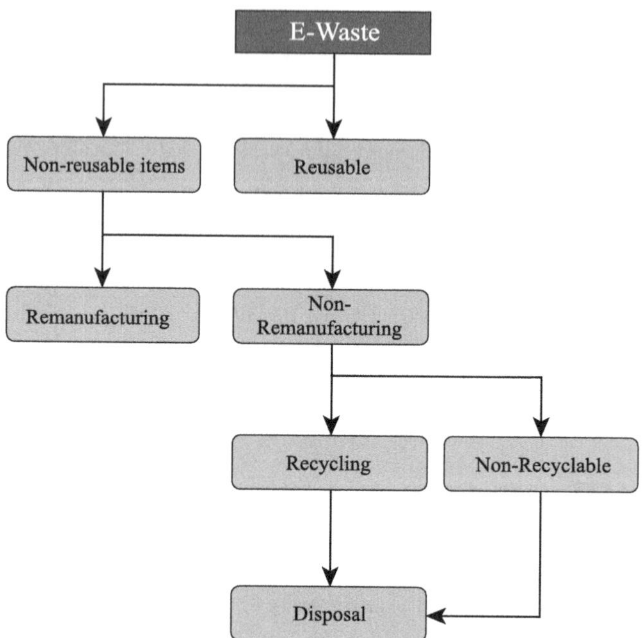

FIGURE 14.1   Hierarchy of e-waste treatment (Cui and Zhang, 2008)

14.2   GLOBAL SCENARIO

According to the Waste Electrical and Electronic Equipment (WEEE) directive (Directive 2002/96/EC), any gadget which depends on electrical currents or electromagnetic field to work is listed under the category of electrical and electronic equipment. WEEE directive, Annex 1A categorizes the following equipment under the list of electrical and electronic equipment (UNEP, 2007):

1. Bulky domestic appliances
2. Minor domestic appliances
3. IT and telecommunications gadgets
4. Consumer gadgets
5. Lighting apparatus
6. Electrical and electronic tools
7. Toys, leisure, and sports equipment
8. Medical devices
9. Monitoring and control instruments
10. Automatic dispensers

However, the definition of e-waste varies from country to country. Therefore, there is a mismatch in e-waste data collection worldwide. As per the report of Forti et al. (2020), nearly 53.6 million metric tonnes (Mt) of e-waste (without waste from solar panels) was produced in 2019, which is estimated to surpass 74 Mt by 2030 (Gautam et al., 2022; Jha et al., 2020). E-waste generation is continuously growing over the years at an alarming rate of approximately 2Mt/year. There are various factors which are responsible for this increasing trend in e-waste generation over the decades. The major factor is increasing consumer demand for technologically advanced electronic gadgets with updated designs, which causes a high obsolescence rate for these gadgets. Besides that, another important factor is the dissimilarity between production and recycling, which is mainly due to the poor legislation and regulations on e-waste collection and recycling or their strict implementation and absence of consciousness among people (buyers and producers) on the importance of e-waste production and recycling. Out of overall e-waste produced, only 17.4% is collected legally and recycled, whereas 82.6% was not systematically treated in an environmentally sound manner. Considering the presence of a high metallic value, this is a huge loss for the economy. E-waste generation data for each continent and their corresponding recycling rate (RR) is presented in Figure 14.2. Asia has remained as largest e-waste producer over the years, with a total contribution of 24.9 Mt (46.46%) in 2019, followed by Americas, Europe, Africa, and Oceania with a total contribution of 13.1 (24.44%), 12 (22.39%), 2.9 (5.41%), and 0.7 (1.31%), respectively (Forti et al., 2020). In Asia, China contributes to the largest e-waste generation (10 Mt). Despite the largest e-waste production in Asia, the amount which was formally collected and recycled was just 2.7 Mt. In terms of per capita, Europe leads with 16.2 kg/capita waste

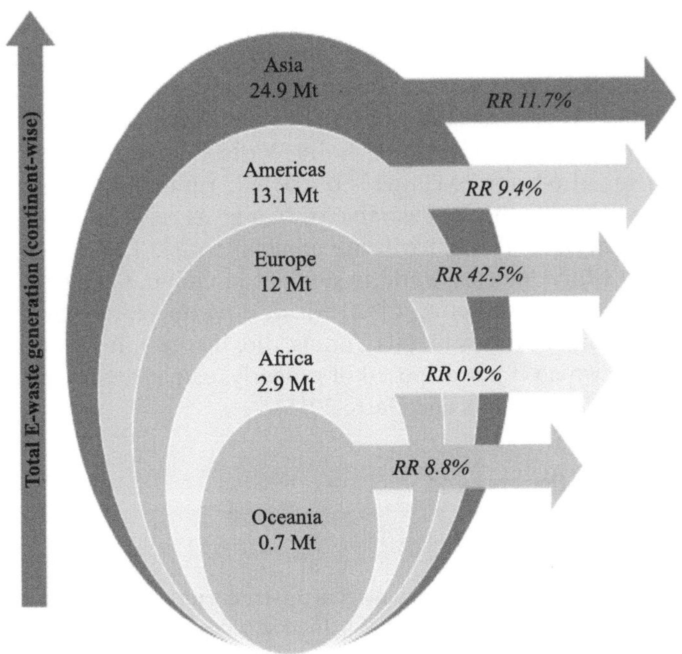

FIGURE 14.2 Total e-waste generation continent-wise and their corresponding recycling rate (RR) in 2019 (Forti et al., 2020)

generation. In recycling statistics, Europe is the continent that stands at the largest formal collection and RR (42.5%), followed by Asia (11.7%). The reason for such high collection and recycling in Europe is the existence of strict legislations/regulations on the collection and recycling of e-waste set by the WEEE directive (2012/19/EU), which set the aim for collection, reuse, recovery, and recycling. Developed countries (USA, Germany, United Kingdom, etc.) are causing relatively more e-waste generation than developing countries. This increasing amount of waste per year needs proper attention from each country for sustainable waste management, environmental protection, and resource utilization. Countries which do not have any existing legislation need to formulate it and implement, whereas the countries, where already some legislation/regulations exist, need to work on strict implementation of those rules to get maximum benefit from it.

14.3 PROCESSING OF E-WASTE

Developed countries export their waste electronics to developing countries like China, India, and Ghana (Wang et al., 2020). About 80% of the net e-waste created worldwide is transported illegitimately to developing nations despite many policies and regulations setup to monitor the export of e-waste (Lundgren, 2012). Several regulations have been enforced to keep a watch on the e-waste flow. Europe has formulated and enforced the EU Waste Shipment Regulation and WEEE directive to prevent unauthorized e-waste trade (Geeraerts et al., 2015). US legislation recognizes most e-waste as "special wastes", and not as "hazardous wastes", owing to the large quantity of e-waste from the US being traded to developing nations, which is inconsistent with the Basel Convention, according to which "A Party shall not authorize dangerous wastes or other wastes to be traded to a non-party or to be imported from a non-party". According to Article 25 of the Act on Prevention and Control of Solid Waste Pollution enforced in China, illegal importing of e-waste to China is restricted. Yet tonnes of electronic waste are imported to China either in the name of "Cu or Al recycling" or being mixed up with additional categories of hardware or metal scraps. The chief e-waste recycling station in China and the world is situated in Guiyu, Guangdong Province, which accounts for the recycling of 20 Mt of e-waste yearly (Geeraerts et al., 2015). E-waste contains heavy metals, furans, dioxins, brominated FRs, and other toxic chemicals, which if not disposed of properly, can adversely affect the ecology and human health (Davis and Garb, 2015).

14.3.1 INFORMAL PROCESSING

The informal solid waste sector is defined as

> individuals or companies who are involved in private sector recycling and waste management events which are not supported, funded, documented, sustained, planned or recognized by the formal solid waste experts, or which work in violation of or in rivalry with formal authorities.

> **(Tran and Salhofer, 2018)**

The informal sector gathers, repairs, and processes the e-waste, which might have been domestically generated or imported through international borders. E-waste generated domestically is generally collected by wandering hawkers who travel to various residential neighbourhoods or industrial estates with a loudspeaker on their vehicles (Wang et al., 2020). The illegal e-waste processing practices include manual disassembling of waste electronics; manual dismantling of printed circuit boards using hammers, chisels, and bare hands; recovering metals by opening cables; melting plastics; burning cables to recover copper; valuable metal recovery by treating e-waste in open acid baths (Tyagi et al., 2015). The resulting waste is then illegally land filled or disposed of in water bodies causing soil, air, and water pollution (Lepawsky, 2015; Shittu et al., 2021). In addition, such work practices put the health of the unskilled labours at risk. The startling environmental contaminations caused by the casual e-waste disposal sites cause China to be known as the universal "e-waste capital" or "e-waste heaven" in global media (Wang et al., 2020). Lack of consciousness of the possible toxicity and hazards of e-waste among the government and public spheres; deficiency of capital and infrastructure for profitable recycling; ineffective implementation of current regulations related to cross-border movement of e-waste; absence of mandatory voluntary take-back programmes (Extended Producer Responsibility, EPR) for obsolete electronics in developing countries is leading to more unmonitored primitive "backyard" recycling activities (Nivedha and Sutha, 2020).

14.3.2 Formalizing Informal Processing of E-Waste

According to the GTZ-MAIT survey conducted in 2007, about 95% of e-waste produced in India is informally recycled in urban slums where unregulated small-scale labour-intensive units have low-technology process waste (Annamalai, 2015). So a model has been proposed for the formalization of informal recycling where collection, segregation, and manual dismantling are performed by the informal sector for the optimization of the flow of resources to the formal sector, where mechanized, specialized, and full-scale recovery of source material from informal processors occurs (Figure 14.3). Thus the formal sector acts as a midway between the informal sector and large-scale smelting units (Niyati, 2014). One such ingenuity led to the formation of E-Parisaraa, a private, large-scale formal recycling unit which aims to recycle 99% of e-waste (Niyati, 2014). This "Best-of-2-Worlds" model has been implemented in many developing countries where initial collection and dismantling are carried out by established informal workers and the dismantled e-waste is transported to formal recyclers equipped with advanced end-processing facilities which cause the least environmental and health hazards. However, complete implementation of this model is challenging in the presence of a well-established informal e-waste management sector contesting with the formal recyclers. To reduce the gap between informal and formal recycling, the following measures can be taken:

a. Collecting e-waste by drop-off or buy-back process at collection centres setup in various parts of a country
b. Providing more storage space in optimal locations to facilitate easy and economical transportation of e-waste

c. Centralizing e-waste dismantling and recycling units
d. Creating industrial parks or e-waste hubs outside large cities
e. Implementing the concept of zero-waste creation by following industrial symbiosis
f. Certification for e-waste recycling activities.

In addition, the Swedish Ministry of Environmental and Natural Resources put forward the EPR to develop producers' responsibility based on the polluter-pays principle. EPR is "an environmental protection strategy to reduce the environmental issues from a product by creating the company of the product answerable for the whole life cycle of the product and particularly for the take-back, recycling and final discarding of the product" (Lindhqvist, 2000). In India, e-waste management regulations started in the form of "E-waste Management and Handling Rules, 2011", which proposed EPR to tackle the limitations associated with e-waste management (Dasgupta et al., 2022). In current scenario where new regulations are set up

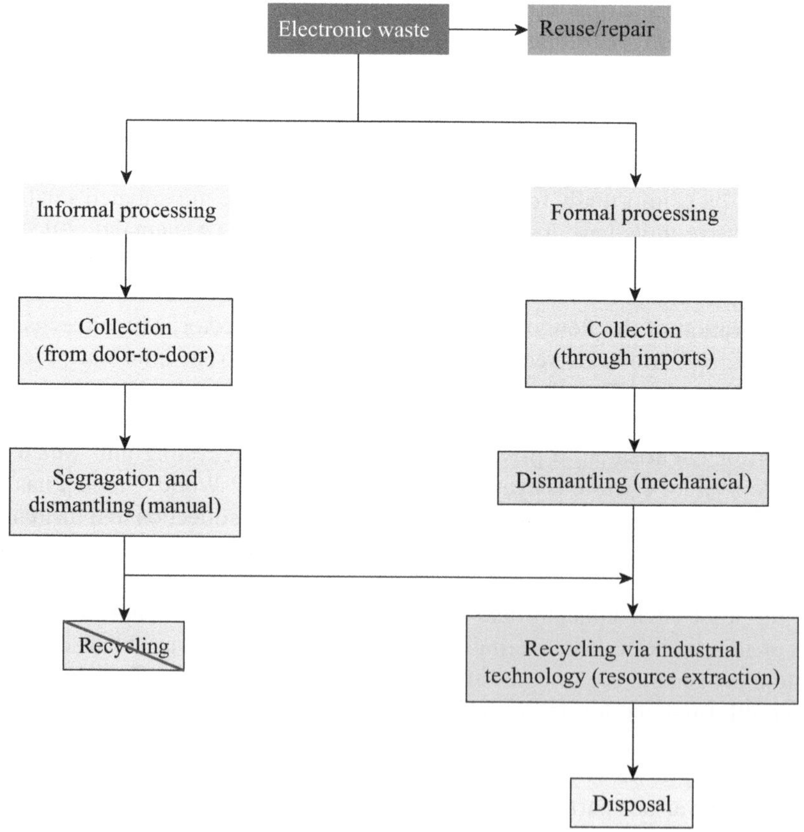

FIGURE 14.3   Formalizing informal processing of waste electronics

and formal entrepreneurs are developing, it is desirable, economically viable, and environmentally safe to incorporate informal processing units into formal units of e-waste management, to protect the livelihood of several informal sector workers and sustainable development. This would allow a drastic increase in e-waste collection and RRs by regulating existing informal collections, hence maximizing the yield of recovery and minimizing the loss of valuable metals to landfills.

14.4 METAL RECOVERY PROCESSING

14.4.1 Pyrometallurgy

Pyrometallurgical processes involve higher temperatures for the melting of the material and separation of the metallic layer from slag for further refining (Kaya, 2016; Syed, 2012). The output of this practice is generally copper bullion subjected to various purification processes for the separation of pure metal. The plastic content partially substitutes coke as a reducing agent in the smelter, and precious metals can be separated from the copper bullion. The existing traditional technology for the recovery of metallic values from mobile polychlorinated biphenyls (PCBs) is shown in Figure 14.4. Here, in this, WEEE is dismantled, downsized to about 5 mm, and put into the copper smelter for copper and other precious metals recovery (Lee et al., 2007). The pyrometallurgical practice compromises the benefit that a pretreatment is seldom required beyond the unit operations of recycling (Hoffmann, 1992; Lee et al., 2007). However, as e-waste components contain halogenated FRs, high-temperature processes lead to the formation of poisonous complexes such as dioxins, furans, and polyhalogenated organic pollutants, along with the generation of dust raises environmental concerns (Hagelüken, 2006; Rao et al., 2021a).

Several processes have been developed in the past decade for recycling e-waste through pyrometallurgical techniques, such as a top-blown rotary converter (TBRC)

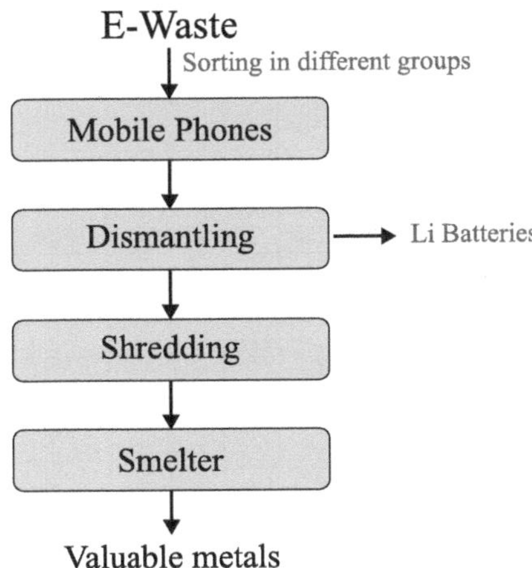

FIGURE 14.4   Process flow of pyrometallurgical treatment of waste mobile phones

smelter with a 360° rotating chamber system (Diaz et al., 2015) and Umicore's Hoboken plant, along with a waste gas utilization system (Hagelüken, 2006). TBRC is advanced smelting equipment for treating waste PCBs, non-ferrous metal smelting slag, and anode slime to recover gold, silver, and other rare and precious metals. This furnace is designated by a high compressive recovery rate, combining chemical oxidation and reduction in one furnace, and high adaptability of raw material. It also provides good performance in energy conservation for a circular economy (Metalcess, n.d.). TBRC-based recycling was introduced for the recycling of complicated metallic values by Diaz et al. (2015). This uses pyrolysis as a pre-processing technique to more effectively separate the non-condensable gas–liquid fraction from the solid residue. The resulting solid substance greatly facilitates the separation of the metals, glass fibre, and organic portions, increasing the viability of each fraction's recycling. A Moscow-based recycling company called Aurus teamed up with SMS Group in 2017 to provide an urban gold compact plant for recycling electronic waste (Group SMS, 2017). This plant is designed for an annual production of 6,000 tonnes of Waste printed circuit boards (WPCBs) every year for the production of copper, gold, silver, and platinum. This plant includes the facility for mechanical preparation of raw materials, TBRC, refining furnace, and automated finishing equipment (Greentech, 2017).

14.4.2 HYDROMETALLURGY

Severe environmental issues and serious health complications associated with the pyrometallurgical processes of WPCBs turn researchers into the eco-friendly hydrometallurgical route for recycling e-waste (Akcil et al., 2015). Hydrometallurgical processes are the potential techniques for the treatment of WPCBs with their relatively low capital cost, suitable for minor applications, reduction in environmental effect, and high metal recovery. This route involves a mechanical pre-treatment process for the facilitation of the extent of the leaching process. The pre-treated waste is

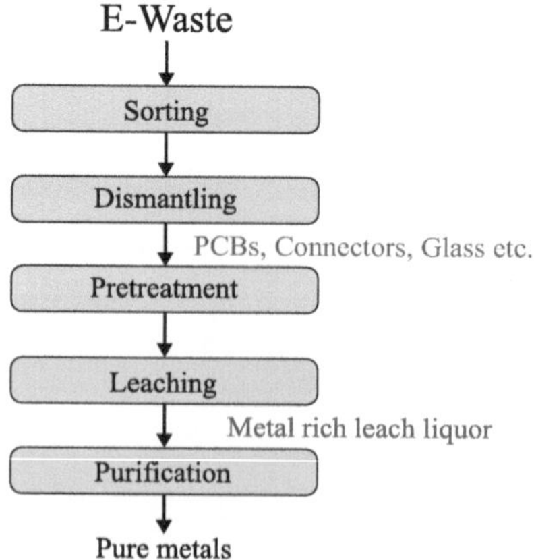

FIGURE 14.5 Flowsheet of the traditional hydrometallurgical process for the recovery of metal (Jha et al., 2022; Nigam et al., 2021; Rao et al., 2021b)

subjected to leaching by a suitable reagent, followed by purification of concentrated leach liquor and recovery of metals. The traditional hydrometallurgical method for the recovery of metal by treatment of WEEE is shown in Figure 14.5.

14.4.2.1 Leaching

Hydrometallurgy relies heavily on the selection or design of a leaching procedure. It must offer effective dissolving of metals and supply them in a form that is adequate for further separation process. The most common leaching methods are acidic, alkaline, and organic solvent leaching. It is significant to note that because metals like copper and gold are present in electronic waste in their elemental state, they must undergo oxidation to dissolve. In contrast, traditional mining uses ores that already contain metal cations as oxides or sulphides. The leaching process is enhanced by the presence of oxidative agents such as hydrogen peroxide (H_2O_2), as illustrated in Eqs. (14.1)–(14.4) (Tuncuk et al., 2012). Thermodynamic data of these reactions shows the effect of oxidizing agents under similar leaching conditions, which enhances the metal recovery. But the concentration of oxidant and temperature can affect the metal extraction from WPCBs (Deveci et al., 2010).

$$Cu^o + 2H^+ = Cu^{+2} + H_2 \qquad \Delta G^o = 65.50 \ Kj/mol \qquad (14.1)$$

$$Cu^o + \tfrac{1}{2} O_2 + 2H^+ = Cu^{+2} + H_2O \qquad \Delta G^o = -171.63 \ Kj/mol \qquad (14.2)$$

$$Au^o + 3H^+ = Au^{+3} + 3/2 \ H_2 \qquad \Delta G^o = 439.97 \ Kj/mol \qquad (14.3)$$

$$2Au^o + 3/2 \ O_2 + 6H^+ = 2Au^{+3} + 3 \ H_2O; \ \Delta G^o = -84.23 \ Kj/mol \qquad (14.4)$$

The precious metals content in e-waste are comparatively lower than the base metal content, which challenges their recovery (Fogarasi et al., 2014). To concentrate the precious metals in a solid residue for later recovery, the base metals are frequently selectively dissolved as the initial stage in the recovery of metals from WPCBs. Aqueous acids like sulphuric or hydrochloric acid are frequently used to leach base metals, but because WPCBs contain the metals in their elemental forms, they must be used in conjunction with an oxidant. H_2O_2, Cl_2, O_2, or bacteria are acceptable oxidants (Cui and Anderson, 2016). Nitric acid, however, is more efficient at leaching base metals due to its strong oxidizing behaviour (Kumar et al., 2014). Cyanide is a cheap and very effective reagent for gold leaching to form water-soluble cyanoaurate $[Au(CN)_2]^-$. However, cyanide is an extremely poisonous reagent. Many efforts have been made to create substitutes for cyanide leaching such as thiocyanate, thiosulphate, thiourea, and aqua-regia leaching (Aylmore, 2016; Gökelma et al., 2016; Syed, 2012). Halide leaching is another approach, in which powerful oxidizing agents such as Cl_2 or Br_2 are produced in situ via electrochemistry or through a reaction between H_2SO_4 and a halide salt (Kim et al., 2013; Torres et al., 2015). Many researchers have previously reported the leaching of metals from WPCBs of e-waste and consequent recovery of metallic values, shown in Table 14.1.

14.4.2.2 Purification of Leach Solution

Several methods, including adsorption, electrolysis, cementation, ion exchange, solvent extraction (SX), and polymer inclusion membranes (PIMs), can be used to

TABLE 14.1

Leaching reagent and the test conditions for recovery of various metals through hydrometallurgy route

Element	Conditions	% metal recovery	References
Copper	2 M sulphuric acid with 0.2 M H_2O_2	95	Oh et al. (2003)
Copper	Nitric acid	96	Kumar et al. (2014)
Copper	3.5 M nitric acid at 323 K in 1 h	99	Long Le et al. (2011)
Copper	≥ 2–3 M at 70°C		Bas et al. (2014)
Copper, nickel, and gold	1 M nitric acid for copper, 0.1 M nitric acid for nickel	97Cu	Kinoshita et al. (2003)
Gold	Thiourea	99	Jing-ying et al. (2012)
Copper	0.3 M HCl + 0.3 M $FeCl_3$	99.99	Imre-lucaci et al. (2017)
Copper, nickel	3 M HNO_3	99.9	Rao et al. (2021a)
Gold	3 M H_2SO_4 + 3 M NaBr	95	Rao et al. (2021b)
Copper	Acidic ferric chloride	99	Fogarasi et al. (2014)
Gold	Ammonium persulphate	98	Alzate et al. (2016)
Gold	Aqua regia	97	Park and Fray (2009)
Copper	H_2SO_4 + H_2O_2	96	Yang et al. (2011)
Copper	H_2SO_4 + H_2O_2	98	Kumar et al. (2014)
Gold	Ammonium thiosulphate	98	Ficeriová et al. (2011)

selectively recover a specific metal from multi-metallic solutions (Agrawal et al., 2022; Gurung et al., 2013; Mishra et al., 2021; Naumov et al., 2017).

I. Cementation and Adsorption

In cementation, a highly reactive metal or metal with higher oxidation potential replaces less reactive metals by allowing them into precipitation. However, the disadvantages are re-dissolution of sacrificial metal, time-consuming, and more critical at low pH. For example, zinc electrolytes are purified by the addition of zinc powder. While zinc is more electropositive (−0.76 V) than most of the common metals, it can replace them easily, and those impurities are precipitated as residue according to cementation, as shown in Eq. (14.5).

$$Zn + M^{2+} = M + Zn^{2+} \qquad (14.5)$$

where M = Cu, Pb, and so on.

The Merrill–Crowe method involves zinc cementation, whereby the gold cyanide solution obtained after leaching is reduced by zinc to precipitate gold (Muhtadi, 1988). After gold has precipitated, it is collected by filtering, blended with fluxes (such as borax and silica) to bind with impurities, and then smelted to create bars that are then sent for additional refining procedures. Another example of the selective separation of gold is the co-precipitation of the metal with α–cyclodextrin (Liu et al., 2013, 2016) and metal organic framework materials (Sun et al., 2018).

II. Ion-Exchange Process

Ion-exchange practice is the adsorption of metal from a leach liquor using an ion-exchange polymer resin for metal recovery from e-waste (Gomes et al., 2001). The basic process for the recovery of metal is the transfer of metal ions into a solid resin and back transfer (stripping). High temperatures are not necessary for the metal stripping stage, also known as elution, from the loaded resins. Recently amide-group scavengers have shown the selective adsorption of 78% of $[AuCl_4]-$ from aqua-regia leached PCBs. Back extraction has been seen to produce 99% gold recovery with repeated wash cycles using diluted nitric acid (Lahtinen et al., 2017). Another study has suggested the use of electro-generated chlorine gas as an oxidizing agent in the HCl leach system for the preferential gold recovery from e-waste [63]. Similar to prior investigations, 2 M HCl was used to separate the copper first, leaving a gold-rich residue, which was subsequently recovered (99.99%) using ion-exchange chromatography.

III. Solvent Extraction

SX is a predominantly adaptable method among purification processes due to its convenience of scale, and it finds application for the recovery of both trace and highly concentrated metals (Chauhan and Patel, 2014; Wilson et al., 2014). In SX, a mixed-metal aqueous liquor is contacted with an extractant or ligand to form a neutral complex that transports the preferred metal into a water-immiscible phase. Following phase separation, the aqueous solution (the raffinate) is recycled back into the leaching phase, while the water-immiscible organic phase is brought into contact with an aqueous stripping solution to release the metal ions for reduction to generate pure metal (Wilson et al., 2014). The whole process is depicted in Figure 14.6. Selecting an extremely selective organic carrier that will only transport the chosen metal from the metallic solution is a crucial step in this process. Coordination and supramolecular chemistry concepts are used to create selectivity by creating extractants that can distinguish amongst distinct metal ions according to size, charge, and shape (Carson et al., 2015; Turkington et al., 2012). When recycling used electronics, where base metal concentrations are much higher than concentrations of precious metals, the SX process plays a vital role. The effectiveness of this process resides mainly on the selectivity of the metal extractant and achieving adequate separation between the organic and aqueous phases.

IV. Polymer Inclusion Membrane

A PIM utilization is another technique for the separation and recovery of low-concentration metal ions from leach solution. The benefits of PIM include selective separations, concurrent extraction and stripping of metal ions, and being environmentally friendly. PIM also shows higher stability with minor loss of extracting agent and high efficiency of metal recovery. PIMs can be synthesized by dissolving the components such as a base polymer (PVC, CTA, etc.), plasticizer, and carrier or extracting agents in a suitable organic solvent like chloroform or tetrahydrofuran (Jha et al., 2022). The homogenous solution is then poured into a Petri dish of suitable size, or the casting of the membrane may done by using a casting knife. The solvent evaporates leisurely, and the remaining residue is the desired membrane for

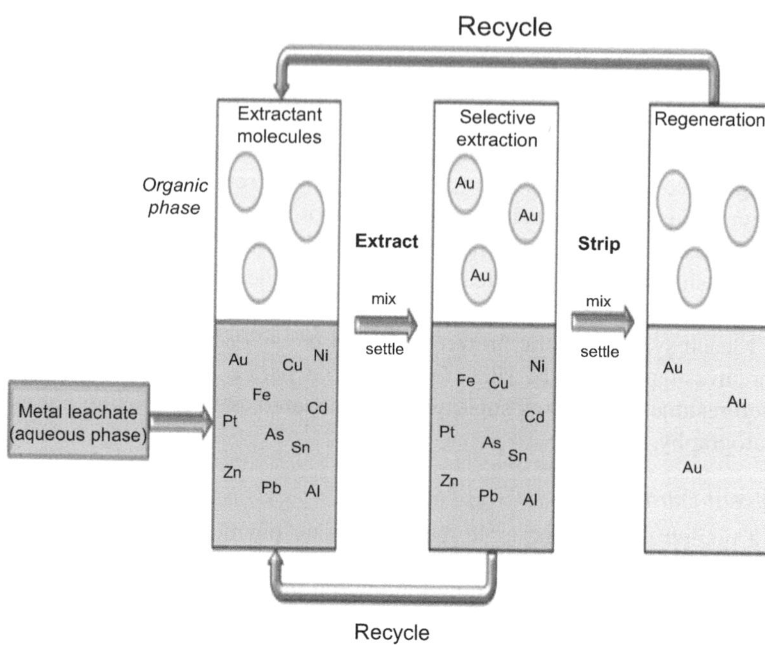

FIGURE 14.6   Solvent extraction route for gold recovery from e-waste solutions

metal extraction. In contrast, a carrier or extracting agent is used to transport metal ions, and a plasticizer provides elasticity to the membrane (Jha et al., 2020). However, the mechanical strength has been provided to the membrane with the base polymer. The separation of metal ions from the leach solution through PIMs generally involves two-compartment cells, where simultaneous extraction and back extraction processes occur. The carrier in the PIM phase actively participates by exchanging ions during the transportation of the membrane, and the separation of metal ions is not limited by the equilibrium conditions (Kavitha and Palanivelu, 2012). The PIM process has already been proven as a sufficient technique for metal removal from various aqueous streams and found effective for metal removal from WPCBs (Jha et al., 2023). The more research is required on the reduction of overall processing time to make it industrially feasible process.

14.4.3 Bio-Hydrometallurgy

The conventional methods for the recovery of various metals from electronic waste emit hazardous materials such as dioxins, furans, and acid vapours into the environment, while newly developed bio-metallurgical processes turn into a revolutionary practice (Erüst et al., 2013; Faramarzi et al., 2004). Mineral bio-oxidation also embodies 5% of the total gold production in the world. Industrial bodies use popular biohydrometallurgical reactors such as Autotech (BIOX) and Bactech-Mintek (Bacox) for the recovery of gold nowadays (Kaksonen et al., 2014). Bio-oxidation and bio-sorption are the two core methods for the recovery of gold through bio-hydrometallurgical processes. Bio-oxidation is applied for the successful recovery of gold from consumed electronic materials and metallic sulphides, which assist major gold-bearing minerals by the use of bacterial-aided reactions. The principle behind the static bio-oxidation systems is the circulation of water and air through the stacks of the consumed materials to initiate the evolution of microorganisms. This phenomenon enhances the rate of oxidation (Morin et al., 2006; Syed, 2012). In the case of bio-sorption, the contact between the ions in the solution and the charged surfaces of the microorganisms takes place. These microorganisms may include living or dead algae, yeast, fungi, and bacteria, which activate the bio-sorption of the gold (Mack et al., 2007). But the disadvantages of the bio-metallurgical processes are that sulphur by-products cannot be recovered and surplus acid generation (Morin et al., 2006).

14.5 CHALLENGES, OPPORTUNITIES, AND FUTURE PROSPECTIVE

In past decades, there have been some developments in the authorized, official, and infrastructural agenda for attaining rigorous management of e-waste in a few countries. However, precise e-waste laws on controlling e-waste are still deficient in most countries. The country-specific legislation must be imposed to stop informal recycling and to promote the formal recycling of e-waste for recovering valuable metals. The countries having e-waste management laws must emphasize the increase in the number of plants and for increasing the quantity of e-waste processed per day. Being a secondary resource of metallic and nonmetallic values, e-waste must be treated by the formal recycling routes to maintain the demand and stop environmental threats. The sorting of e-waste into different categories is still a challenge for most of the recycling plants. Moreover, the complex structure of electronic gadgets also requires specific and sophisticated pre-treatment or dismantling processes before applying the exact recycling routes, pyrometallurgy, hydrometallurgy, and so on. As a secondary resource, e-waste must be treated with advanced processes to extract high-purity metals from it. Therefore, the limitations associated with conventional recycling techniques may be overcome by using advanced processes like hydrometallurgy and bio-hydrometallurgy. Works on the different extracting agents used in hydrometallurgy should be explored for the regeneration and for enhancing the reusability of those regenerated extracting agents to reduce the cost of the overall process. Moreover, the extracting and stripping must be conducted as a single-step process to avoid the generation of large volumes of treated solutions. The reduction in large quantity of solutions may be done by utilizing bio-hydrometallurgy, and also it is a cost-effective process. Further research is required for the left-out categories of e-waste like medical equipment and monitors/LCDs and LEDs.

14.6 CONCLUSION

E-waste is considered a source of secondary materials including most of the precious metals and rare earth elements. Conventional pyrometallurgy processes are used for the recovery of valuable metals from electronic gadgets. However, this process shows a few severe limitations such as the route is costly and the process is energy-intensive and generates fumes and poisonous gases. Additionally, the gas scrubber and emission control systems have been attached to the furnaces; yet, the processes are found costlier in comparison to the other alternative routes, hydrometallurgy, and so on. Hydrometallurgy route involves the leaching of metals by using different acids and bases. Further SX or cementation is applied to recover selective metals for the e-waste solutions. Being an eco-friendly and low capital price process hydrometallurgy route gains much attention in the past few years. These processes are very effective for the recovery of lower-concentration metal ions from the aqueous phase.

In current years, the bioleaching of metals from e-waste has also gain attention as favourable technology due to its unique features like environmentally friendly, low operational costs, reduced quantity of chemical and/or biological waste to be moved and high efficiency in purifying wastes.

Further work should be done on enhancing the purity of recovered metals. The feasibility of combined routes of extraction should be assessed and implemented for making a industrially feasible low cost route for fully recovery of secondary materials from e-waste.

REFERENCES

Agrawal, M., Singh, R., Ranitović, M., Kamberovic, Z., Ekberg, C., & Singh, K. K. (2021). Global market trends of tantalum and recycling methods from waste tantalum capacitors: A review. *Sustainable Materials and Technologies*, *29*(August). https://doi.org/10.1016/j.susmat.2021.e00323

Agrawal, M., Singh, R., & Singh, K. K. (2022). Recovery of silica-free tantalum from epoxy-coated tantalum capacitors using hydrometallurgical routes. *Journal of Environmental Chemical Engineering*, *10*(4), 108182. https://doi.org/10.1016/j.jece.2022.108182

Akcil, A., Erust, C., Sekhar, C., Ozgun, M., Sahin, M., & Tuncuk, A. (2015). Precious metal recovery from waste printed circuit boards using cyanide and non-cyanide lixiviants— A review. *Waste Management, 45*, 258–271.

Alzate, A., López, M. E., & Serna, C. (2016). Recovery of gold from Waste Electrical and Electronic Equipment (WEEE) using ammonium persulfate. *Waste Management, 57*, 113–120. https://doi.org/10.1016/j.wasman.2016.01.043

Annamalai, J. (2015). Occupational health hazards related to informal recycling of E-waste in India: An overview. *Indian Journal of Occupational & Environmental Medicine, 19*(1), 61–65. https://doi.org/10.4103/0019-5278.157013

Aylmore, M. G. (2016). Alternative lixiviants to cyanide for leaching gold ores. In: Adams, M. D. (Ed.), *Gold Ore Processing* (pp. 447–484). Elsevier B.V. https://doi.org/10.1016/B978-0-444-63658-4.00027-X

Bas, A. D., Deveci, H., & Yazici, E. Y. (2014). Treatment of manufacturing scrap TV boards by nitric acid leaching. *Separation and Purification Technology, 130*(2), 151–159. https://doi.org/10.1016/j.seppur.2014.04.008

Carson, I., MacRuary, K. J., Doidge, E. D., Ellis, R. J., Grant, R. A., Gordon, R. J., Love, J. B., Morrison, C. A., Nichol, G. S., Tasker, P. A., & Wilson, A. M. (2015). Anion receptor design: Exploiting outer-sphere coordination chemistry to obtain high selectivity for chloridometalates over chloride. *Inorganic Chemistry, 54*(17), 8685–8692. https://doi.org/10.1021/acs.inorgchem.5b01317

Chauhan, S., & Patel, T. (2014). A review on solvent extraction of nickel. *International Journal of Engineering Research & Technology (IJERT)*, *3*(9), 1315–1322.

Cui, H., & Anderson, C. G. (2016). Literature review of hydrometallurgical recycling of Printed Circuit Boards (PCBs). *Journal of Advanced Chemical Engineering*, *6*(1), 1–11. https://doi.org/10.4172/2090-4568.1000142

Cui, J., & Zhang, L. (2008). Metallurgical recovery of metals from electronic waste : A review. *158*, 228–256. https://doi.org/10.1016/j.jhazmat.2008.02.001

Dasgupta, D., Majumder, S., Adhikari, J., Ghosh, P., Purchase, D., Garelick, H., Debsarkar, A., & Chatterjee, D. (2022). E-waste management in Indian microscale informal sectors— An environmental perspective towards policy regulation. *Research Square*, 1–28. https://doi.org/https://doi.org/10.21203/rs.3.rs-1089035/v1 License:

Deveci, H., Yazici, E. Y., Aydin, U., Yazici, R., & Akcil, A. U. (2010). Extraction of copper from scrap tv boards by sulphuric acid leaching under oxidising conditions. *Going Green Care Innovation*, January.

Davis, J. M., & Garb, Y. (2015). A model for partnering with the informal e-waste industry: Rationale, principles and a case study. *Resources, Conservation and Recycling*, *105*, 73–83. https://doi.org/10.1016/j.resconrec.2015.08.001

Diaz, F., Florez, S., & Friedrich, B. (2015). High recovery recycling route of WEEE : The potential of pyrolysis. *Proceedings of EMC 2015*, 1–18.

Erüst, C., Akcil, A., Gahan, C. S., Tuncuk, A., & Deveci, H. (2013). Biohydrometallurgy of secondary metal resources: A potential alternative approach for metal recovery. *Journal of Chemical Technology and Biotechnology*, *88*(12), 2115–2132. https://doi.org/10.1002/jctb.4164

Faramarzi, M. A., Stagars, M., Pensini, E., Krebs, W., & Brandl, H. (2004). Metal solubilization from metal-containing solid materials by cyanogenic Chromobacterium violaceum. *Journal of Biotechnology*, *113*(1–3), 321–326. https://doi.org/10.1016/j.jbiotec.2004.03.031

Ficeriová, J., Baláž, P., & Gock, E. (2011). Leaching of gold, silver and accompanying metals from circuit boards (PCBs) waste. *Acta Montanistica Slovaca*, *16*(2), 128–131.

Fogarasi, S., Imre-Lucaci, F., Imre-Lucaci, Á., & Ilea, P. (2014). Copper recovery and gold enrichment from waste printed circuit boards by mediated electrochemical oxidation. *Journal of Hazardous Materials*, *273*, 215–221. https://doi.org/10.1016/j.jhazmat.2014.03.043

Forti, V., Baldé, C. P., Kuehr, R., & Bel, G. (2020). *The Global E-waste Monitor 2020*.

Gautam, P., Behera, C. K., Sinha, I., Gicheva, G., & Singh, K. K. (2022). High added-value materials recovery using electronic scrap-transforming waste to valuable products. *Journal of Cleaner Production*, *330*(March 2021), 129836. https://doi.org/10.1016/j.jclepro.2021.129836

Geeraerts, K., Andrea, I., & Schweizer, J.-P. (2015). Illegal shipment of e-waste from the EU: A case study on illegal e-waste export from the EU to China. *A Study Compiled as Part of EFFACE Project*. IEEP, 320276.

Gökelma, M., Birich, A., Stopic, S., & Friedrich, B. (2016). A review on alternative gold recovery reagents to cyanide. *Journal of Materials Science and Chemical Engineering*, *4*, 8–17. https://doi.org/10.4236/msce.2016.48002

Gomes, C. P., Almeida, M. F., & Loureiro, J. M. (2001). Gold recovery with ion exchange used resins. *Separation and Purification Technology*, *24*(1–2), 35–57. https://doi.org/10.1016/S1383-5866(00)00211-2

Greentech. (2017). *New recycling process for circuit boards.*

Group SMS. (2017). *Aurus, russia, places order with sms group to supply urbangold compact plant for electronic scrap recycling.*

Gurung, M., Adhikari, B. B., Kawakita, H., Ohto, K., Inoue, K., & Alam, S. (2013). Recovery of gold and silver from spent mobile phones by means of acidothiourea leaching followed by adsorption using biosorbent prepared from persimmon tannin. *Hydrometallurgy*, *133*, 84–93. https://doi.org/10.1016/j.hydromet.2012.12.003

Hagelüken, C. (2006). Recycling of electronic scrap at Umicore precious metals refining. *Acta Metallurgica Slovaca*, *12*, 111–120.

Hoffmann, J. E. (1992). Recovering precious metals from electronic scrap. *The Journal of the Minerals, Metals & Materials Society*, *44*(7), 43–48. https://doi.org/10.1007/BF03222275

Imre-lucaci, Á., Nagy, M., Imre-lucaci, F., & Fogarasi, S. (2017). Technical and environmental assessment of gold recovery from secondary streams obtained in the processing of waste printed circuit boards. *Chemical Engineering Journal*, *309*, 655–662. https://doi.org/10.1016/j.cej.2016.10.045

Jha, R., Agrawal, M., & Singh, K. K. (2023). Synthesis and characterisation of PVC-based non-plasticised polymer inclusion membranes for selective metal extraction. *Canadian Metallurgical Quarterly*, 1–11. https://doi.org/10.1080/00084433.2023.2175302

Jha, R., Mishra, G., Agrawal, M., Rao, M. D., Meshram, A., & Singh, K. K. (2022). Opportunities for an en-route to polymer inclusion membrane approach from conventional hydrometallurgical recycling of WPCBs: A mini-review. *Canadian Metallurgical Quarterly*, *0*(0), 1–15. https://doi.org/10.1080/00084433.2022.2126576

Jha, R., Rao, M. D., Meshram, A., Verma, H. R., & Singh, K. K. (2020). Potential of polymer inclusion membrane process for selective recovery of metal values from waste printed circuit boards: A review. *Journal of Cleaner Production*, *265*, 121621. https://doi.org/10.1016/j.jclepro.2020.121621

Jing-ying, L., Xiu-li, X., & Wen-quan, L. (2012). Thiourea leaching gold and silver from the printed circuit boards of waste mobile phones. *Waste Management*, *32*(6), 1209–1212. https://doi.org/10.1016/j.wasman.2012.01.026

Kaksonen, A. H., Perrot, F., Morris, C., Rea, S., Benvie, B., Austin, P., & Hackl, R. (2014). Evaluation of submerged bio-oxidation concept for refractory gold ores. *Hydrometallurgy*, *141*, 117–125. https://doi.org/10.1016/j.hydromet.2013.10.012

Kavitha, N., & Palanivelu, K. (2012). Recovery of copper(II) through polymer inclusion membrane with di (2-ethylhexyl) phosphoric acid as carrier from e-waste. *Journal of Membrane Science*, *415–416*, 663–669. https://doi.org/10.1016/j.memsci.2012.05.047

Kaya, M. (2016). Recovery of metals and nonmetals from electronic waste by physical and chemical recycling processes. *Waste Management*, *57*, 64–90. https://doi.org/10.1016/j.wasman.2016.08.004

Kim, B., Lee, J., Jeong, J., Yang, D., Shin, D., & Lee, K. (2013). A novel process for extracting precious metals from spent mobile phone PCBs and automobile catalysts. *Materials Transactions*, *54*(6), 1045–1048. https://doi.org/10.2320/matertrans.M2013051

Kinoshita, T., Akita, S., Kobayashi, N., Nii, S., Kawaizumi, F., & Takahashi, K. (2003). Metal recovery from non-mounted printed wiring boards via hydrometallurgical processing. *Hydrometallurgy*, *69*(1–3), 73–79. https://doi.org/10.1016/S0304-386X(03)00031-8

Kumar, M., Lee, J. C., Kim, M. S., Jeong, J., & Yoo, K. (2014). Leaching of metals from Waste Printed Circuit Boards (WPCBs) using sulfuric and nitric acids. *Environmental Engineering and Management Journal*, *13*(10), 2601–2607.

Lahtinen, E., Kivijärvi, L., Tatikonda, R., Väisänen, A., Rissanen, K., & Haukka, M. (2017). Selective recovery of gold from electronic waste using 3D-printed scavenger. *ACS Omega*, *2*(10), 7299–7304. https://doi.org/10.1021/acsomega.7b01215

Lee, J. chun, Song, H. T., & Yoo, J. M. (2007). Present status of the recycling of waste electrical and electronic equipment in Korea. *Resources, Conservation and Recycling*, *50*, 380–397. https://doi.org/10.1016/j.resconrec.2007.01.010

Lepawsky, J. (2015). The changing geography of global trade in electronic discards: Time to rethink the e-waste problem. *Geographical Journal, 181*(2), 147–159. https://doi.org/10.1111/geoj.12077

Lindhqvist, T. (2000). Extended producer responsibility in cleaner production—policy principle to promote environmental improvements of product systems. *IIIEE, Lund University*, 196.

Liu, Z., Frasconi, M., Lei, J., Brown, Z. J., Zhu, Z., Cao, D., Iehl, J., Liu, G., Fahrenbach, A. C., Botros, Y. Y., Farha, O. K., Hupp, J. T., Mirkin, C. A., & Stoddart, J. F. (2013). Selective isolation of gold facilitated by second-sphere coordination with α-cyclodextrin. *Nature Communications, 4*(May). https://doi.org/10.1038/ncomms2891

Liu, Z., Samanta, A., Lei, J., Sun, J., Wang, Y., & Stoddart, J. F. (2016). Cation-dependent gold recovery with α-cyclodextrin facilitated by second-sphere coordination. *Journal of the American Chemical Society, 138*(36), 11643–11653. https://doi.org/10.1021/jacs.6b04986

Long Le, H., Jeong, J., Lee, J. C., Pandey, B. D., Yoo, J. M., & Huyunh, T. H. (2011). Hydrometallurgical process for copper recovery from waste Printed Circuit Boards (PCBs). *Mineral Processing and Extractive Metallurgy Review, 32*(2), 90–104. https://doi.org/10.1080/08827508.2010.530720

Lundgren, K. (2012). The global impact of e-waste: Addressing the challenge. In: *International labour office, programme on safety and health at work and the environment (SafeWork), Sectoral Activities Department (SECTOR)*.

Mack, C., Wilhelmi, B., Duncan, J. R., & Burgess, J. E. (2007). Biosorption of precious metals. *Biotechnology Advances, 25*(3), 264–271. https://doi.org/10.1016/j.biotechadv.2007.01.003

Metalcess. (n.d.). *Top blown rotary converter (TBRC) smelting*.

Mishra, G., Jha, R., Rao, M. D., Meshram, A., & Singh, K. K. (2021). Recovery of silver from Waste Printed Circuit Boards (WPCBs) through hydrometallurgical route: A review. *Environmental Challenges, 4*, 100073. https://doi.org/10.1016/j.envc.2021.100073

Morin, D., Lips, A., Pinches, T., Huisman, J., Frias, C., Norberg, A., & Forssberg, E. (2006). BioMinE—Integrated project for the development of biotechnology for metal-bearing materials in Europe. *83*, 69–76. https://doi.org/10.1016/j.hydromet.2006.03.047

Muhtadi, O. A. (1988). Metal extraction (recovery systems). In: *Introduction to evaluation, design and operation of precious metal heap leaching projects* (pp. 124–136). Society of Mining Engineers, Inc.

Naumov, K. D., Lobanov, V. G., & Zelyakh, Y. D. (2017). Gold electrowinning from cyanide solutions using three-dimensional cathodes. *Metallurgist, 61*(3–4), 249–253. https://doi.org/10.1007/s11015-017-0485-x

Nigam, S., Jha, R., & Prakash, R. (2021). Materials today : Proceedings a different approach to the electronic waste handling—a review. *Materials Today: Proceedings, xxxx*. https://doi.org/10.1016/j.matpr.2021.01.081

Nivedha, R., & Sutha, A. I. (2020). European journal of molecular & clinical medicine the challenges of Electronic Waste (E-Waste) management in India. *European Journal of Molecular & Clinical Medicine, 7*(3), 4583–4588.

Niyati, M. (2014). Role of informal sector in E-waste recycling. 社学研論集 , *23*(3), 279–286.

Oh, C. J., Lee, S. O., Yang, H. S., Ha, T. J., & Kim, M. J. (2003). Selective leaching of valuable metals from waste printed circuit boards. *Journal of the Air and Waste Management Association, 53*(7), 897–902. https://doi.org/10.1080/10473289.2003.10466230

Park, Y. J., & Fray, D. J. (2009). Recovery of high purity precious metals from printed circuit boards. *164*, 1152–1158. https://doi.org/10.1016/j.jhazmat.2008.09.043

Rao, M. D., Singh, K. K., Morrison, C. A., & Love, J. B. (2020). Challenges and opportunities in the recovery of gold from electronic waste. *RSC Advances, 10*(8), 4300–4309. https://doi.org/10.1039/c9ra07607g

Rao, M. D., Singh, K. K., Morrison, C. A., & Love, J. B. (2021a). Optimization of process parameters for the selective leaching of copper, nickel and isolation of gold from obsolete mobile phone PCBs. *Cleaner Engineering and Technology, 4*, 100180. https://doi.org/10.1016/j.clet.2021.100180

Rao, M. D., Singh, K. K., Morrison, C. A., & Love, J. B. (2021b). Recycling copper and gold from e-waste by a two-stage leaching and solvent extraction process. *Separation and Purification Technology, 263*(November 2020), 118400. https://doi.org/10.1016/j.seppur.2021.118400

Rautela, R., Arya, S., Vishwakarma, S., Lee, J., Kim, K. H., & Kumar, S. (2021). E-waste management and its effects on the environment and human health. *Science of the Total Environment, 773*, 145623. https://doi.org/10.1016/j.scitotenv.2021.145623

Shittu, O. S., Williams, I. D., & Shaw, P. J. (2021). Global E-waste management: Can WEEE make a difference? A review of e-waste trends, legislation, contemporary issues and future challenges. *Waste Management, 120*, 549–563.

Sun, D. T., Gasilova, N., Yang, S., Oveisi, E., & Queen, W. L. (2018). Rapid, selective extraction of trace amounts of gold from complex water mixtures with a Metal-Organic Framework (MOF)/polymer composite. *Journal of the American Chemical Society, 140*(48), 16697–16703. https://doi.org/10.1021/jacs.8b09555

Syed, S. (2012). Recovery of gold from secondary sources-A review. In: *Hydrometallurgy* (Vols. 115–116, pp. 30–51). https://doi.org/10.1016/j.hydromet.2011.12.012

Torres, C. M., Taboada, M. E., Graber, T. A., Herreros, O. O., Ghorbani, Y., & Watling, H. R. (2015). The effect of seawater based media on copper dissolution from low-grade copper ore. *Minerals Engineering, 71*, 139–145. https://doi.org/10.1016/j.mineng.2014.11.008

Tran, C. D., & Salhofer, S. P. (2018). Processes in informal end-processing of e-waste generated from personal computers in Vietnam. *Journal of Material Cycles and Waste Management, 20*(2), 1154–1178. https://doi.org/10.1007/s10163-017-0678-1

Tuncuk, A., Stazi, V., Akcil, A., Yazici, E. Y., & Deveci, H. (2012). Aqueous metal recovery techniques from e-scrap: Hydrometallurgy in recycling. *Minerals Engineering, 25*(1), 28–37. https://doi.org/10.1016/j.mineng.2011.09.019

Turkington, J. R., Cocalia, V., Kendall, K., Morrison, C. A., Richardson, P., Sassi, T., Tasker, P. A., Bailey, P. J., & Sole, K. C. (2012). Outer-sphere coordination chemistry: Amido-ammonium ligands as highly selective tetrachloridozinc(II)ate extractants. *Inorganic Chemistry, 51*(23), 12805–12819. https://doi.org/10.1021/ic301691d

Tyagi, N., Baberwal, S. K., & Passi, N. (2015). E-waste: Challenges and its management. *DU Journal of Undergraduate Research and Innovation, 1*(3), 108–114.

UNEP. (2007). E-waste volume I: Inventory assessment manual. *United Nations Environment Programme*, 127. www.unep.org

Wang, K., Qian, J., & Liu, L. (2020). Understanding environmental pollutions of informal E-waste clustering in global south via multi-scalar regulatory frameworks: A case study of Guiyu Town, China. *International Journal of Environmental Research and Public Health, 17*(8). https://doi.org/10.3390/ijerph17082802

Wilson, A. M., Bailey, P. J., Tasker, P. A., Turkington, J. R., Grant, R. A., & Love, J. B. (2014). Solvent extraction: The coordination chemistry behind extractive metallurgy. *Chemical Society Reviews, 43*(1), 123–134. https://doi.org/10.1039/c3cs60275c

Yang, H., Liu, J., & Yang, J. (2011). Leaching copper from shredded particles of waste printed circuit boards. *Journal of Hazardous Materials, 187*(1–3), 393–400. https://doi.org/10.1016/j.jhazmat.2011.01.051

15 Removal of Methylene Blue on Industrial Solid Waste Using a Novel Air-Agitated Tapered Bubble Column Adsorber

Soham Kundu, Modhurima Misra, Sandip K. Ghosh, and Amitava Bandyopadhyay

15.1 INTRODUCTION

The constant discharge of dye(s)-contaminated effluent in significant quantity by many industries like textile, tannery, dyeing and bleaching, and dye manufacturing is a matter of serious concern. The subsequent addition of these partially treated or untreated dye(s) wastewater into the environmental matrices is a major threat and adversely affects the health of different life forms (Chowdhury et al., 2009). Existing literature suggests different treatment methods for dye(s) removal from wastewater (Chowdhury et al., 2009), for example, coagulation–sedimentation, chemical oxidation, photo-oxidation, membrane separation, enzymatic decomposition, floatation, biological treatment, and adsorption (Ghosh and Bandyopadhyay, 2016). Adsorption is intrinsically an important step in the treatment of coloured effluent. Though the commercially available activated carbon has repeatedly shown promising results as a highly efficient adsorbent, it is costly.

Current research trend thus indicates the exploration of several alternatives in search of effective low-cost adsorbents such as the use of agricultural residues, industrial wastes like fly ash, and bio-adsorbents of plant and microbial origin with and without treatment. Literature survey of the past few decades pertaining to dye (MB, cationic dye) removal from synthetic wastewater onto various types of adsorbents include bamboo based activated carbon (Hameed et al., 2007), peach stone based carbon (Attia et al., 2008), steam activated bituminous carbon (El Qada et al., 2006), beech saw dust (Batzias and Sidiras, 2007), beer brewery wastes (Tsai et al., 2008), acid-treated bentonite (Banat et al., 2007), *Posidonia oceanic (L.)* fibres (Ncibi et al., 2007), coir pith carbon (Kavitha and Namasivayam, 2007), algal biomass (Vilar et al., 2007), neem (*Azadirachta indica*) leaf powder (Bhattacharyya and Sharma, 2005), peanut hull (Özer et al., 2007), rice husk (Han et al., 2007) cedar sawdust and crushed bricks (Hamdaoui, 2006), silica nanosheets (Zhao et al., 2008), and so on.

Some of these adsorbents are of low cost, and the rest are costly. However, the adsorptive removal of methylene blue (MB) in an air-agitated tapered bubble column adsorber (Ghosh and Bandyopadhyay, 2017a; Ghosh et al., 2017) onto an industrial solid waste such as biomass combustion residue (BCR) (Roy et al., 2015) as an adsorbent has not yet been reported. This study is performed under batch operating mode. The influence of salt on batch adsorption was investigated by utilizing a NaCl solution similar to that contained in effluents discharged from Indian textile industries. The experiments were orchestrated in a unique manner so as to exploit the advantages of the seasonal variation of temperature to observe its influence on the adsorption for analysing the thermodynamic variables adsorber (Ghosh and Bandyopadhyay, 2017a, 2017b). The temperature variations before and after the experimental runs were observed to be well within ± 1 K.

15.2 EXPERIMENTAL PROCEDURES

15.2.1 Adsorbent and Adsorbate

Biomass combustion residue was thoroughly washed in water without further treatment and then sieved to remove the coarse particles before experiments. Adsorbate used was powdered MB (molecular formula: $C_{16}H_{18}ClN_3S$; molecular weight of 319.851 g/mol; E.Merck Limited, Mumbai, India). Initially, a stock solution (1000 mg/L) was prepared, which was used to prepare desired concentrations by dilution for the batch investigations in the tapered adsorber.

15.2.2 Characterizing the Adsorbent

Standard procedures were adopted in measuring different densities (solid and bulk) of BCR. A particle size analyser (Malvern Mastersizer 2000E version. 5.60, UK) was used to determine the particle size. The zeta potential of different solutions containing 5 mg of adsorbent having varied pH was measured to determine the pH at the point of zero charge (pH_{PZC}). In brief, a series of 100-mL Erlenmeyer flasks with 50 mL of distilled water were taken, and the pH of each flask was maintained from 3 to 12 by adding either 0.1 N HCl or 0.1 N liquid NH_3. The initial solution pH was measured by a digital pH meter (Model #112, Electronics India, India). Adsorbent (5 mg) was added to the flasks, and the flasks were sonicated for 20 minutes to homogenize the mixture. The zeta potential of each mixture was finally measured by Zetasizer Ver. 6.01 (Malvern Zetasizer Nano ZS, UK) for determining the pH_{pzc}. A pore size analyser (Quantachrome Instruments, USA) and *Brunauer—Emmett—Teller* (BET) analyser (Model: Nova 1000e) were used to measure the adsorbent's mean pore size and surface area, respectively. The surface morphologies of BCR were analysed before and after the experiments using field emission scanning electron microscopy (FESEM) (JEOL JSM 7600F FESEM, Japan).

15.2.3 BATCH ADSORPTION EXPERIMENTS: SETUP AND PROCEDURE

The batch adsorption experiments were conducted in the tapered adsorber that was agitated with air, as shown in Figure 15.1. The detailed procedure has already been reported in the literature (Ghosh and Bandyopadhyay, 2017a, 2017b); however, for improved understanding it is summarized here. The tapered adsorber was made up of Perspex (ID at the bottom = 0.8 m, vertical height = 1.0 m; angle of divergence = 10.84⁰). To avoid splashing of slurry, the experimental height of the adsorber was less than 1.0 m. A Perspex-made sparger disc (number of holes = 93 holes; diameter of hole = 1 mm) was used for maintaining uniform air distribution into the adsorber. The airflow rate was measured with the help of a rotameter (R). All the experiments were carried out with a fixed volume of liquid (quiescent liquid volume) in the adsorber and at ambient temperature. The air supply from the compressor (COM) was maintained at a required flow rate for agitating the slurry. A few millilitres of samples were withdrawn from the adsorber column into a cuvette aseptically at specific time intervals, for analysing in a visible spectrophotometer (Perkin Elmer, UV–Vis Spectrophotometer, Lambda 25, USA) at a λ_{max} of 664 nm to measure the concentrations of MB. Under steady operating conditions, all the experimental observations were recorded. The adsorber was thoroughly washed in between two successive experimental runs. The conditions for the batch experiments were as follows: adsorbate (MB) concentration, C_o: 25, 50, 75, 100, 200 mg/L; adsorbent dose,

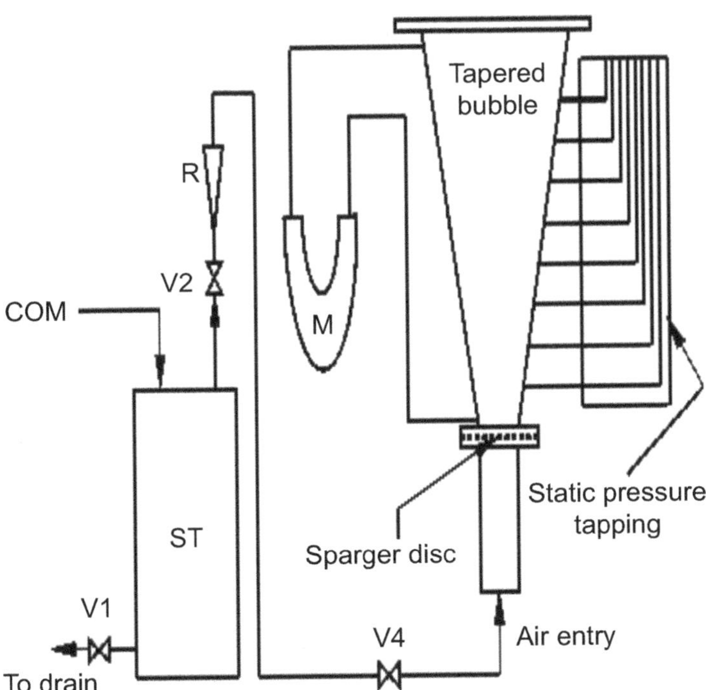

FIGURE 15.1 Diagram showing the different components of the experimental setup

m: 5, 10, 15, 20 g/L; contact time, t: 1, 2, 5, 10, 20, 30, 60, 90 and 120 min; air flow rate, Q_a: 1.17×10^{-4}, 1.67×10^{-4}, 5.00×10^{-4}, 8.33×10^{-4} m^3/s; batch quiescent liquid volume: 3.00×10^{-3} m^3 (3.0 L); salt (NaCl) concentration: 1, 5, 9 g/L; temperature, T: 295.5, 300, 306 K (for observing the temperature effect). All other experiments were performed at 306 K.

15.3 RESULTS AND DISCUSSION

15.3.1 ADSORBENT CHARACTERIZATION

The measured bulk and solid densities of BCR were 892.1 and 1709.1 kg/m^3, respectively. The average particle size measured was 34.02 mm. The distribution of the size of the particle of BCR as measured is shown in Figure 15.2.

The measured pH_{PZC} was 6.6. MB being a cationic dye needs a pH > pH_{PZC} for getting adsorbed. Thus the experiments were performed at a natural pH of 7.76 (>pH_{PZC}). The standard for pH for the discharge of dye effluent from textile industries in India is in the range of 5.5 to 9.0. Therefore, the pH for the current investigation was set at a natural pH to avoid additional burden, which is normally not desirable for plant operation. The influence of initial pH on the zeta potential for estimating pH_{PZC} is shown in Figure 15.3.

The observed total pore volume and surface area of the sample, from multipoint BET analysis, were 0.1102 cc/g and 34.27 m^2/g, respectively. The FESEM images showing the morphology of the surfaces of the adsorbent before and after batch adsorption experiments are shown in Figure 15.4(a) and (b), respectively. The heterogeneous

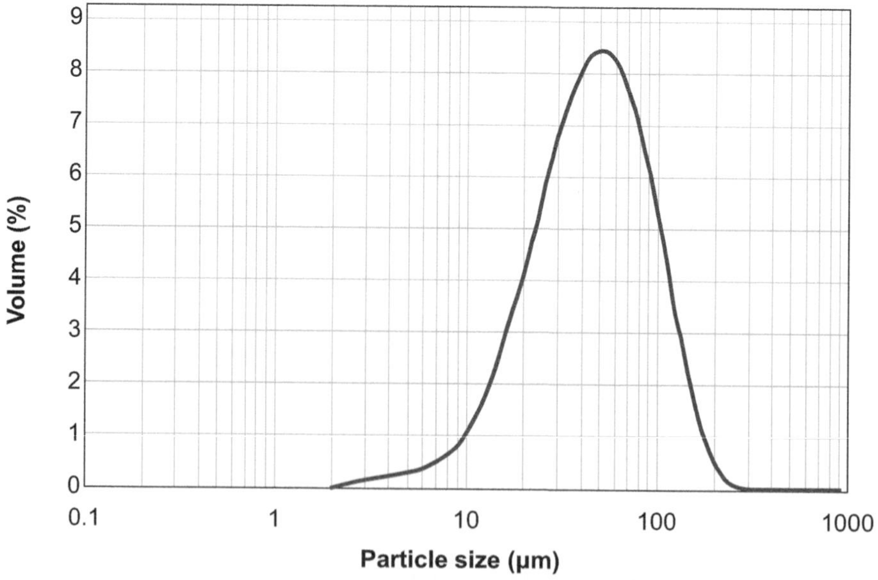

FIGURE 15.2 Distribution of the size of the particle of BCR using Mastersizer 2000

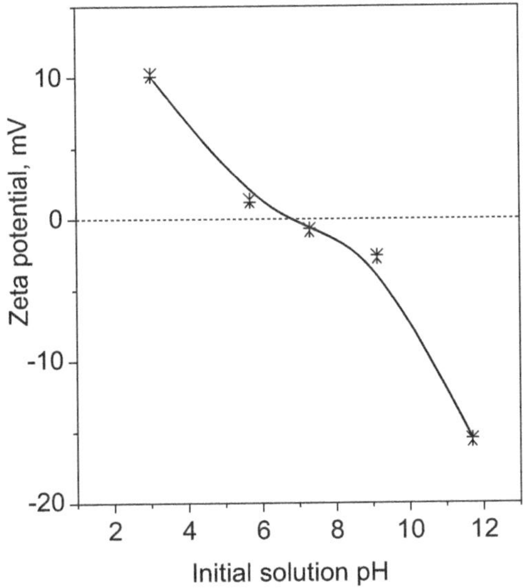

FIGURE 15.3 Determination of pH_{PZC} by measuring the zeta potential using Zetasizer (Ver. 6.01)

surfaces (Figure 15.4) and the occupancy of MB molecules (Figure 15.4(b)) could be seen.

15.3.2 EXPERIMENTS ON ADSORPTION UNDER BATCH MODE AND ADSORPTION ISOTHERM

The adsorption experiments under batch mode were performed under various operating conditions. The adsorption capacities BCR for MB at any time and at equilibrium were calculated using the relationships given as follows:

$$q_t = \frac{(C_o - C_t)}{W} V \tag{15.1}$$

$$\text{and } q_e = \frac{(C_o - C_e)V}{W} \tag{15.2}$$

The following relationship was used to calculate the removal of MB (%R)

$$\%R = \frac{C_o - C_f}{C_o} \times 100 \tag{15.3}$$

15.3.2.1 Influence of Initial Concentration of Dye and Contact Time

The influence of contact time on the removal of MB (%R) for all initial dye concentrations and other fixed operating conditions is depicted in Figure 15.5. It is evident

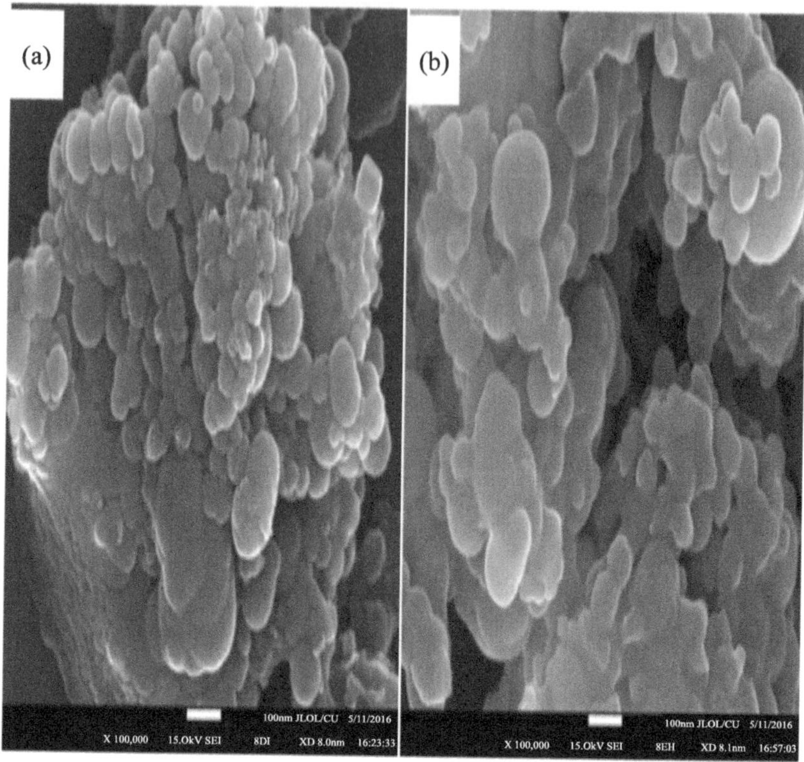

FIGURE 15.4 FESEM images showing the morphology of the surfaces of BCR (a) before batch experiments and (b) after batch experiments

that the removal of MB (%R) was increased at first for all initial dye concentrations, for a contact time of about 60 minutes, after that it became almost constant and ceased to improve further beyond the equilibrium concentration. It may be attributed to the fact that the uptake of dye molecules was enhanced with contact time; as a result, the final concentration decreased with time. The highest value of percentage of removal was found to be 99.2% for fixed operating conditions ($C_o = 50$ mg/L; $m = 15$ g/L; $Q_a = 5.00 \times 10{-4}$ m^3/s). The highest adsorption capacity was 11.44 mg/g for fixed operating conditions ($C_o = 200$ mg/L; $m = 15$ g/L; $Q_a = 5.00 \times 10^{-4}$ m^3/s).

15.3.2.2 Influence of Dose of the Adsorbent

Figure 15.6(a) depicts the influence of the dose of the adsorbent on the removal of MB. It is evident that the removal was increased with the adsorbent dose. The maximum removal obtained was 99.48% (adsorbent dose = 20 g/L). It is also evident that the removal reached almost a constant value (99.2%) at a dosage of 15 g/L. The most probable reason behind such behaviour might be the increased adsorption of MB onto the readily available addition sites with the initial increase in adsorbent

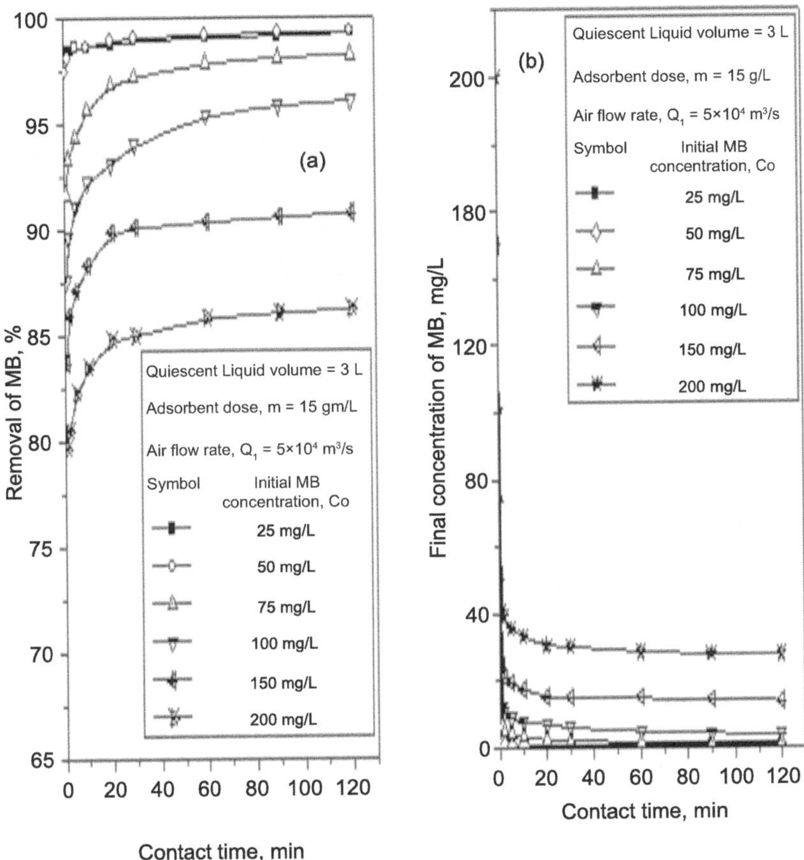

FIGURE 15.5 Influence of contact time on the (a) removal of MB and (b) final concentration of MB

dose; however, with the passage of time, as the sites reached equilibrium at a dose of 15 g/L, the percentage removal became constant beyond this value. Therefore, the optimum dose of BCR for this study was considered as 15 g/L. The influence of the adsorbent dosage on the final concentration of adsorbate is presented in Figure 15.6(b). It is evident that the final concentration declined sharply up to a dosage of 10 g/L, and thereafter it levels off due to reaching equilibrium as explained earlier. The final concentration was varied from 8.2 to 0.26 mg/L, and the variation of final concentration between the adsorbent dose of 15 and 20 g/L was marginal.

15.3.2.3 Influence of Flow Rate of Air

The influence of the flow rate of air on the percentage of removal of MB is presented in Figure 15.7(a). From the figure, it is evident that there was a decreasing trend in the percentage of removal with the airflow rate. A decrease in the percentage removal of MB indicates an increase in the final concentration of MB, as shown in Figure

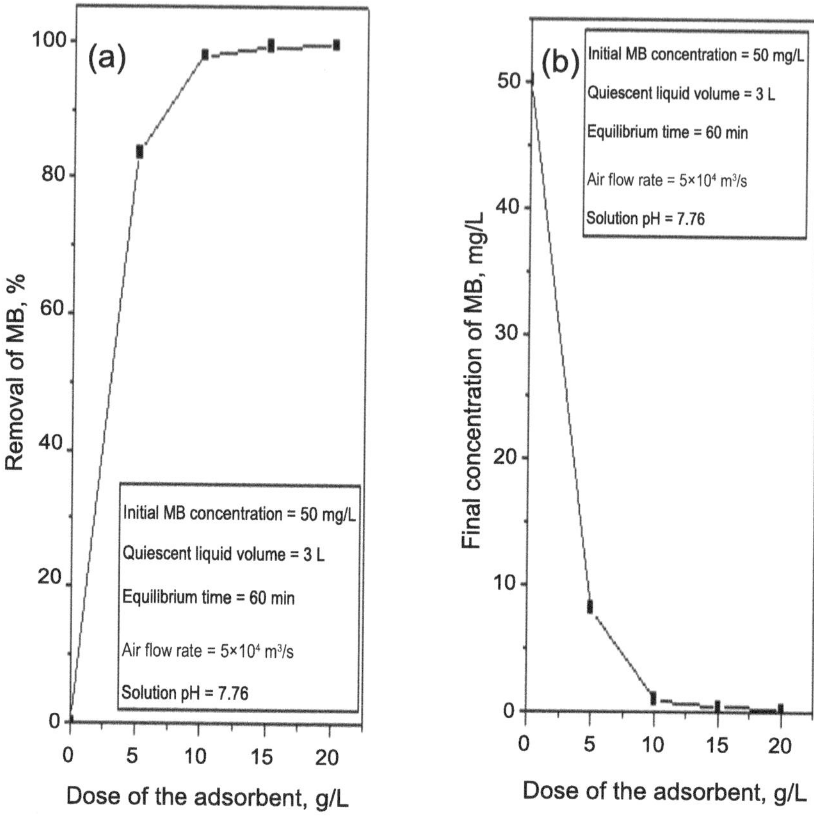

FIGURE 15.6 Influence of dose of the adsorbent on the (a) removal of MB and (b) final concentration of MB

15.7(b). Thus, there was an increasing trend of the final concentration of MB with the increase in the air flow rate. The maximum percentage removal (99.95%) was observed at the minimum air flow rate of 1.17×10^{-4} m³/s whence the final concentration was 0.025 mg/L in the present investigation. The decreasing trend of the percentage removal of MB with the rise in the air flow rate was attributed due to the enhanced splashing of adsorbent particles accumulated at the upper cross-section of the column which was unable to take part in the adsorption to the fullest extent.

15.3.2.4 Influence of Salt (NaCl)

Generally wastewater from textile industries as well as dyeing and bleaching units contains salts. This necessitates studying the effect of the presence of salt on the adsorptive removal of MB onto BCR. The NaCl was chosen as the salt (mostly present in real wastewater) in the present investigation the concentration of which was varied from 1.0 to 9.0 g/L. The variation in the percentage of removal of MB with contact time is depicted in Figure 15.8(a). From this figure, it is evident that the percentage removal of MB was increased initially significantly with the concentration of salt and

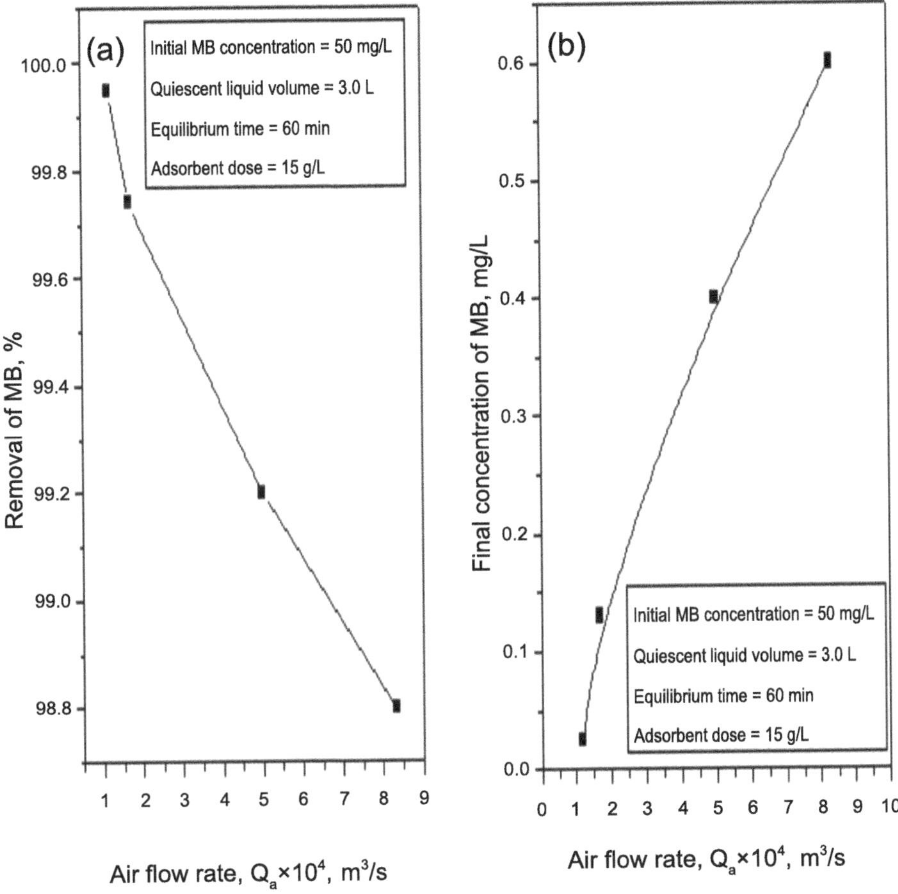

FIGURE 15.7   Influence of air flow rate on the (a) removal of MB and (b) final concentration of MB

then gradually levelled off. The solid–fluid interface may be modified in the presence of ions produced from NaCl, especially anion (chloride ions) which helps with the adsorption of MB, a cationic dye molecule. The final concentration of MB was gradually decreased with contact time, as shown in Figure 15.8(b). It is interesting to note that the equilibrium time is reduced with the increase in salt concentration. From Figure 15.8(a), it is further seen that the percentage of removal of MB was found to increase compared to that obtained without the presence of NaCl. The increase in the percentage removal of MB due to the presence of NaCl varied from 3% to 8% (Figure 15.8(a)). It could be perhaps owing to the molecular aggregation leading to the dimerization of MB molecules, which in turn increased the removal of MB increased NaCl concentration. As reported in the literature, several intermolecular forces like van der Waals forces, ion–dipole forces, and dipole–dipole forces, existing within MB molecule could play roles in molecular aggregation (Alberghina et al., 2000). This

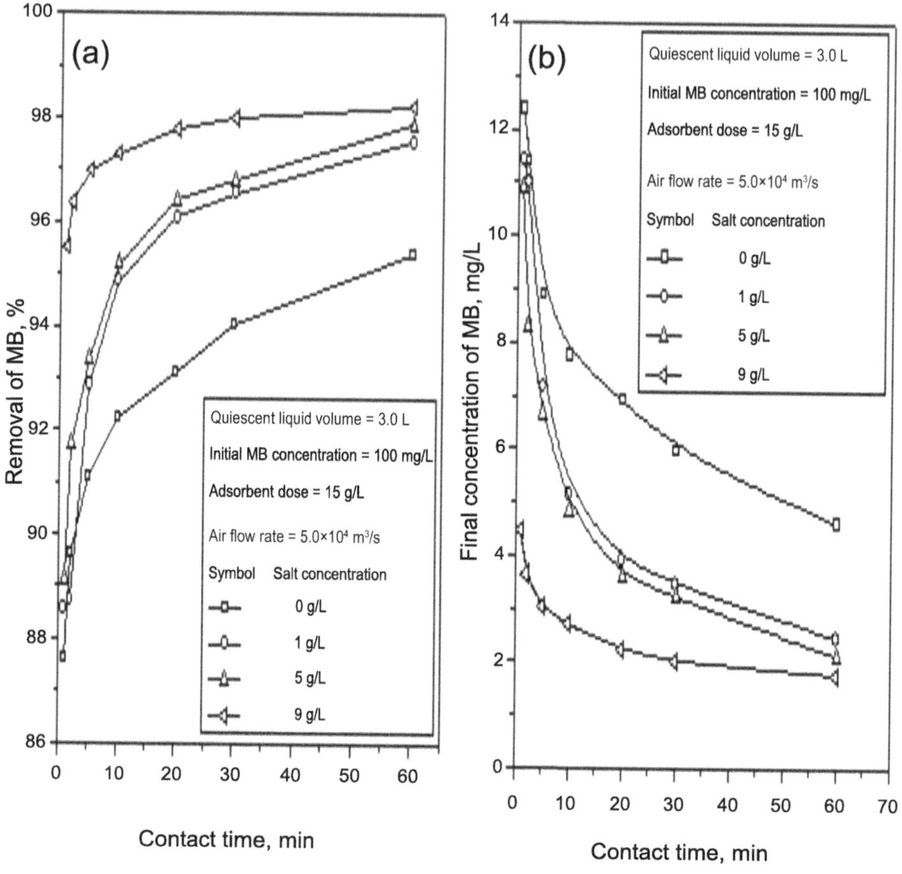

FIGURE 15.8   Influence of contact time on the (a) removal of MB and (b) final concentration of MB

observation is very much applicable in the practical field of the treatment of wastewater containing cationic dyes. Thus improved removal of MB could be achieved at higher salt concentration with shorter time of operation. The maximum percentage removal and maximum adsorption capacity observed were 98.24% and 6.549 mg/g, respectively, for fixed operating conditions (C_o = 100 mg/L; m = 9 g/L; Q_a = 5.00×10–⁴ m³/s).

15.3.2.5   Adsorption Isotherms

The adsorptive behaviour of MB onto BCR was studied utilizing the batch adsorption data of the initial concentration of MB against contact time. The batch experimental data were fitted into the non-linearized forms of several well-known isotherms models, namely Langmuir (1918), Freundlich (1906), Temkin (Temkin and Pyzhev, 1940), Halsey (1948), Dubinin–Radushkevich (D–R) (1947), Sips (1948), and Redlich–Peterson (R-P) (1959), as detailed in Table 15.1.

TABLE 15.1
Equations for isotherm models

Isotherm model	Non-linear isotherm equations
Langmuir	$q_e = \dfrac{q_m k_L C_e}{1 + k_L C_e}$
Freundlich	$q_e = k_F C_e^{1/n_F}$
Temkin	$q_e = \dfrac{RT}{b_T} ln(k_T C_e)$
Halsey	$q_e = \left(\dfrac{K_H}{C_e} \right)^{\frac{1}{n_H}}$
Dubinin–Radushkevich	$q_e = q_{mDR} \exp\left(-k_{DR}\varepsilon^2\right)$, where $\varepsilon = RT \ln\left[1 + \dfrac{1}{C_e}\right], E = \dfrac{1}{\sqrt{2k_{DR}}}$
Sips	$q_e = \dfrac{q_m k_S C_e^{1/n_s}}{1 + k_S C_e^{1/n_s}}$
Redlich–Peterson	$q_e = \dfrac{k_{RP} C_e}{\left(1 + a_{RP} C_e\right)^g}$

The values of the characteristic parameters for various isotherm models, calculated using their non-linearized forms, are depicted in Table 15.2.

Normally, the accuracy of data fitting is statistically determined from the correlation coefficient (r) and coefficient of determination (r^2) values. But, in this study, six different expressions for estimating statistical errors of each isotherm were further used for comparative assessment of the results of data fitting, obtained from the adsorption of MB onto BCR in the absence (Table 15.3a) and presence (Table 15.3b) of NaCl.

The correlation coefficient (r) and coefficient of determination (r^2) along with six types of error for each isotherm are analysed for comparative assessment of the isotherms, for adsorption of MB without NaCl (Table 15.3a) and for adsorption of MB with NaCl (Table 15.3b). It can be seen from Table 15.3a that the values of r and r^2 were highest for Freundlich and Halsey isotherm models. The values of errors for these isotherms were lowest amongst other isotherms. The overall errors of both models are more or less the same. Moreover, the experimental maximum adsorption capacity of 11.44 mg/g was closest to the calculated maximum adsorption capacity of 11.41 mg/g for both of these isotherms indicating that both Freundlich and Halsey models are fittest best for describing the behaviour of the adsorption of MB without NaCl onto BCR in the present study. The Freundlich isotherm is suitable mainly for multilayer adsorption with physisorption on heterogeneous surfaces while Halsey isotherm is suitable for multilayer adsorption. On the other hand, it can be seen from Table 15.3b that the highest numerical values of r and r^2 were for the Freundlich

TABLE 15.2
Parameters for various isotherm models

Isotherm models	Parameters, unit	Values of characteristic parameters	
		MB without salt	MB with salt
Langmuir	q_m, mg/g	11.0747	8.5827
	k_L, L/mg	0.4675	1.5
Freundlich	K_F, mg/g $(L/mg)^{1/n_F}$	3.7859	5.79475
	$1/n_F$	3.0313	6
Temkin	b_T, kJ/mol	1.9467	2.2162
	k_T, L/mg	8.4102	9
Halsey	k_H	0.0177	0.0012
	n_H	−3.03161	−4
Dubinin–Radushkevich	q_{mDR}, mg/g	10.2953	9.5986
	k_{DR}, mol²/kJ²	0.6725	0.5
	E, kJ/mol	0.8623	1.0
Sips	q_m, mg/g	39.4376	11
	k_S, L/mg	0.1057	0.9
	n_S	2.5105	1.6694
Redlich–Peterson	k_{RP}, L/g	16.1226	20
	a_{RP}, L/mg	6.4633	4
	g (lying between 0 and 1)	0.7102	0.8316

TABLE 15.3A
Comparison of r, r², and values of error of different isotherm models (without salt)

Isotherm models	r	r²	χ^2	SSE	SAE	ARE	Hybrid	MPSD	Exp, q_{max}	Cal, q_{max}
Langmuir	0.9708	0.9425	1.136	5.6497	4.8168	15.4228	28.4003	26.3813		11.0747
Freundlich	0.9932	0.9865	0.7698	1.3239	2.3337	13.5992	13.7988	20.9028		11.4109
Temkin	0.9875	0.9752	0.5519	2.4407	3.4665	15.2457	19.2444	32.525		10.6556
Halsey	0.9932	0.9865	0.7702	1.3239	2.3337	15.2447	13.7988	20.9028		11.4116
Dubinin–Radushkevich	0.962	0.9255	1.2541	7.3205	5.4473	15.9374	31.3529	25.8955	11.4438	10.2953
Redlich–Peterson	0.9919	0.9839	0.5534	1.5827	2.3914	13.6405	18.4479	30.3624		11.2502
Sips	0.9923	0.9846	0.6712	1.5136	2.4679	14.7302	22.3727	34.2547		39.4376

isotherm model, while the values of errors for this isotherm model were the lowest amongst other isotherms. Further, the experimental maximum adsorption capacity of 6.55 mg/g was closest to the calculated maximum adsorption capacity of 6.73 mg/g for this isotherm indicating that the Freundlich model is fittest best for describing the behaviour of the adsorption of MB without NaCl onto BCR in the present study showing multilayer adsorption with physisorption on heterogeneous surface. Thus, the presence of salt did not change the isotherm.

TABLE 15.3B

Comparison of r, r^2, and values of error of different isotherm models (with salt)

Isotherm models	r	r^2	χ^2	SSE	SAE	ARE	HYBRID	MPSD	Exp, q_{max}	Cal, q_{max}
Langmuir	0.9974	0.9948	0.0252	0.1647	0.5801	1.4808	0.6305	3.1067		8.5827
Freundlich	0.9987	0.9974	0.0129	0.0846	0.4377	1.1182	0.3245	2.2305		6.7282
Temkin	0.9952	0.9904	0.0469	0.3064	0.792	2.0219	1.1731	4.2383		6.8553
Halsey	0.9973	0.9946	0.0263	0.1713	0.5956	1.5209	0.6564	3.1712		6.7939
Dubinin–Radushkevich	0.9956	0.9912	0.0432	0.2822	0.7659	1.9549	1.0799	4.0656	6.5493	9.5986
Redlich–Peterson	0.9972	0.9943	0.0279	0.182	0.6108	1.5594	0.9289	3.7716		6.7729
Sips	0.9968	0.9935	0.0316	0.2065	0.6788	1.7312	1.0524	4.0108		11

TABLE 15.4

Equations for kinetic models

Kinetic models	Non-linear kinetic equations
Pseudo-first-order (PFO)	$q_t = q_e\left(1 - e^{-k_1 t}\right)$
Pseudo-second-order (PSO)	$q_t = \dfrac{q_e^2 k_2 t}{\left(1 + k_2 t q_e\right)}$
Film diffusion model	$F = 1 - e^{-k_{fd} t}$

15.3.2.6 Adsorption Kinetics

The kinetics of the adsorption of MB onto BCR was investigated by fitting the experimental data of batch adsorption into the pseudo-first-order (Lagergren, 1898) and pseudo-second-order (Ho and McKay, 1999) kinetic equations. The adsorption mechanism was predicted by fitting the experimental data of batch adsorption in the film diffusion model (Boyd et al., 1947). The batch experimental data was fitted using the non-linearized kinetic equations of the aforementioned models as presented in Table 15.4.

The rate constants, diffusion parameters, and statistical errors calculated for different kinetic and diffusion models based on the batch experimental data for the adsorption of MB without NaCl are presented in Table 15.5. From Table 15.5a, it is evident that the pseudo-first-order rate constant, k_1, varied between 2.81 and 6.12 min^{-1} (r = 0.9979 to 1.0, r^2 = 0.9908 to 1.0) (figure not shown). The pseudo-second-order rate constant, k_2, varied from 1.05 to 235.47 g/mg min (r = 0.9993 to 1.0, r^2 = 0.9986 to 1.0) (figure not shown) (Table 15.5b). The film diffusion constant (figure not shown) varied from 2.52 to 5.96 sec^{-1} (r = 0.9965 to 1.0, r^2 = 0.9931 to 1.0) (Table 15.5c). The batch experimental data, therefore, fitted best with the pseudo-second-order kinetic

TABLE 15.5
Rate constants and error calculations of different statistical parameters calculated for pseudo-first-order, pseudo-second-order, and film diffusion adsorption kinetics models (without salt)

(a) Pseudo-first-order (PFO)

Co (mg/L)	k_1 (min^{-1})	R	r^2	χ^2	SSE	SAE	ARE	Hybrid	MPSD
25	6.1169	1	1	0.000006	0.00001	0.0079	0.0533	0.00009	0.076
50	4.1879	0.9999	0.9999	0.0004	0.0013	0.0887	0.2996	0.0058	0.4188
75	2.9256	0.9987	0.9973	0.0115	0.0548	0.5955	1.3818	0.1642	1.8574
100	2.6807	0.9977	0.9955	0.0256	0.1582	1.0222	1.8343	0.366	2.4364
150	2.7306	0.9986	0.9972	0.0227	0.1988	1.1599	1.46	0.3239	1.9225
200	2.8107	0.9979	0.9958	0.0434	0.4779	1.7288	1.7278	0.6196	2.3723

(b) Pseudo-second-order (PSO)

Co (mg/L)	k_2 (g/mg min)	R	r^2	χ^2	SSE	SAE	ARE	Hybrid	MPSD
25	235.4747	1	1	0.000004	0.000007	0.0054	0.0364	0.00006	0.0585
50	16.3362	0.9999	0.9999	0.00008	0.0003	0.0427	0.1440	0.0012	0.1903
75	2.8911	0.9996	0.9992	0.0036	0.01697	0.3395	0.7896	0.0509	1.0331
100	1.7410	0.9992	0.9983	0.0095	0.0589	0.6770	1.2149	0.1357	1.4793
150	1.3354	0.9997	0.9993	0.0055	0.0479	0.5682	0.7170	0.078	0.9426
200	1.0472	0.9993	0.9986	0.0141	0.1554	1.0391	1.0422	0.201	1.3499

(c) Film diffusion

Co (mg/L)	k_{fd} (sec^{-1})	r	r^2	χ^2	SSE	SAE	ARE	Hybrid	MPSD
25	5.9608	1	0.9999	0.000005	0.000005	0.0041	0.0451	0.00007	0.0857
50	3.99925	0.9999	0.9998	0.0002	0.0002	0.0291	0.3261	0.0028	0.5318
75	2.67111	0.9979	0.9957	0.0039	0.0037	0.1291	1.4829	0.0554	2.4003
100	2.39357	0.996	0.992	0.0072	0.0068	0.1901	2.2044	0.1024	3.2784
150	2.55998	0.9981	0.9962	0.0035	0.0033	0.1172	1.3462	0.0493	2.2642
200	2.51852	0.9965	0.9931	0.0063	0.0059	0.1641	1.9034	0.0897	3.0719

Quiescent liquid volume = 3.0 L; initial MB concentration = 25 to 200 mg/L; adsorbent dose = 15 g/L; air flow rate = 5.00×10^{-4} m^3/s.

model since the values of r and r^2 were highest while that of other statistical errors was lowest for this model (Table 15.5a). The external film diffusion could have some influence on the rate. The overall statistical analysis followed the significance of the kinetics in the order as—pseudo-second-order > pseudo-first-order > liquid film diffusion.

The rate constants, diffusion parameters, and statistical errors calculated for the different kinetic and diffusion models based on the batch experimental data for the adsorption of MB in the presence of NaCl are presented in Table 15.6. It is evident from Table 15.6a that the pseudo-first-order rate constant, k_1, varied between 2.63 and 3.89 min^{-1} (r = 0.9969 to 0.9999, r^2 = 0.9939 to 0.9997) (figure not shown) The

TABLE 15.6
Calculated values of rate constants and error calculations of different statistical parameters of pseudo-first-order, pseudo-second-order, and film diffusion adsorption kinetics models (with salt)

(a) Pseudo-first-order (PFO)

Salt (g/L)	k_1 (min^{-1})	R	r^2	χ^2	SSE	SAE	ARE	Hybrid	MPSD
1	2.6255	0.9969	0.9939	0.0345	0.2099	0.9627	2.2253	0.6895	3.3674
5	2.6527	0.9985	0.9969	0.0169	0.1066	0.7216	1.6373	0.3396	2.3267
9	3.8947	0.9999	0.9997	0.0016	0.0105	0.2022	0.4455	0.0324	0.7073

(b) Pseudo-second-order (PSO)

Salt (g/L)	k_2 (g/mg min)	r	r^2	χ^2	SSE	SAE	ARE	Hybrid	MPSD
1	1.4513	0.9991	0.9982	0.0105	0.0633	0.5657	1.3184	0.209	1.8589
5	1.6608	0.9996	0.9993	0.0041	0.0256	0.3650	0.8298	0.0814	1.1366
9	5.7309	0.9999	0.9999	0.0004	0.0023	0.1152	0.2539	0.0071	0.3309

(c) Film diffusion

Salt (g/L)	k_{fd} (sec^{-1})	r	r^2	χ^2	SSE	SAE	ARE	Hybrid	MPSD
1	2.2158	0.9939	0.9879	0.0106	0.0098	0.1964	2.9926	0.2125	4.791
5	2.2968	0.9962	0.9924	0.0066	0.0062	0.1647	2.4683	0.1314	3.7208
9	3.5496	0.9996	0.9993	0.0006	0.0006	0.0487	0.7049	0.0125	1.1263

Quiescent liquid volume = 3.0 L; Initial MB concentration = 100 mg/L; Adsorbent dose = 15g/L; Air flow rate = 5.00×10^{-4} m^3/s.

pseudo-second-order rate constant, k_2, calculated (figure not shown) varied from 1.45 to 5.73 g/mg min (r = 0.9991 to 0.9999, r^2 = 0.9982 to 0.9999) (Table 15.6b). The film diffusion constant (figure not shown) varied from 2.22 to 3.55 sec^{-1} (r = 0.9939 to 0.9996, r^2 = 0.9879 to 0.9993) (Table 15.6c). The batch experimental data, therefore, fitted best with the pseudo-second-order kinetic model since the values of r and r^2 were highest while other statistical errors were lowest for this model (Table 15.6a). The external film diffusion could have some influence on the rate of adsorption in the presence of salt as well. The overall statistical analysis followed the significance of the kinetic models in the order as—pseudo-second-order > pseudo-first-order > liquid film diffusion. Therefore, the kinetics followed a similar trend as in the case of the adsorption without NaCl. However, a critical comparison between the various characteristic constants for both systems showed that higher values could be obtained by increasing the salt (NaCl) concentration.

15.4 EFFECT OF TEMPERATURE

The effect of operating temperature holds enormous significance in thermodynamic parameters estimation for adsorption at equilibrium. India being a tropical country, most of the cities experience a huge temperature difference between summer (about

40°C) and winter (about 10°C). Thus, investigating the effect of temperature on the adsorptive process in this work assumes considerable importance in determining its suitability and functioning as a part of a wastewater treatment system. The influence of temperature was studied in a novel way by utilizing the seasonal variation of temperatures as reported earlier in the literature (Ghosh and Bandyopadhyay, 2017a, 2017b). The percentage removals observed were 99.04% at 306 K, 99.08% at 300 K, and 99.16 at 295.5K (C_o = 50 mg/L; m = 15 g/L; Q_a = 5.00×10^{-4} m³/s; equilibrium time = 60 min). Since the percentage removal decreased with the rise in temperature, it can be concluded that for practical application the performance of the system would be better in winter than in summer.

In this study, the thermodynamic equilibrium constant, K_C was used to estimate the standard change in Gibbs free energy (ΔG^0), standard change in enthalpy (ΔH^0), and standard change in entropy (ΔS^0) for dilute solution with reasonable accuracy (Liu, 2009):

$$K_C = \frac{C_{ad}}{C_e} = \frac{q_e}{C_e}.m \tag{15.4}$$

The following standard equation was used for determining ΔG^0 at different temperatures:

$$\Delta G^0 = -RT \ln K_C \tag{15.5}$$

The Van't Hoff's linear relationship as given in Eq. (15.6) was used for determining the values of ΔH^0 and ΔS^0 from the slope and the intercept, respectively, of the linear plot of ln KC as a function of $\frac{1}{T}$

$$\ln K_C = -\frac{\Delta H^0}{RT} + \frac{\Delta S^0}{R} \tag{15.6}$$

The average values of ΔH^0 and ΔS^0 determined were 9.45 and 7.57 J/mol, respectively (figure not shown; r^2= 0.9243). The values of ΔG^0 obtained were ranging between −20 kJ/mol and 0.0 kJ/mol, which indicated physisorption, while chemisorption would predominate if the values of ΔG^0 were falling within the range of −80 to −20 kJ/mol (Ozcan et al., 2006; Machado, et al., 2011). In this study, it is observed that the values of ΔG^0 were well within the range of −20 to 0.0 kJ/mol indicating the predominance of physisorption in the adsorptive removal of MB on BCR. Furthermore, the activation energy (E, kJ/mol) estimated from the Dubinin–Radushkevich adsorption isotherm is often applied for the determination of the nature of the adsorption process. For instance, if E < 8 kJ/mol then the adsorption would be dominated by physisorption while the adsorption would be dominated by chemisorption for 8 < E < 16 kJ/mol (Chabani et al., 2006). In the current study, the activation energy estimated from the Dubinin–Radushkevich adsorption isotherm was 0.0863 kJ/mol (Table 15.2), for the adsorption of MB without NaCl. It further indicated towards the predominance of physisorption. This observation is in good agreement with that obtained from the thermodynamic analysis. The negative value of ΔG^0 indicated the spontaneity and feasibility of the adsorptive process, while the positive value of ΔH^0 and the greater value of $T\Delta S^0$ than ΔH^0 ensured the process was endothermic.

15.5 PRACTICAL SIGNIFICANCE

The tapered bubble column adsorber developed could be used for the treatment of wastewater generated from the extractive industries. Though the system performance has been characterized in terms of removal of MB by adsorption, similar adsorptive removal of pollutants could be performed in the device. The adsorbent has a crucial role to play in any adsorptive removal system. The adsorbent used in the present study was an industrial solid waste the loading of which was 15 g/L. At a higher air flow rate, the system performance was observed reduced due to the splashing of adsorbent particles in the upper cross-section of the adsorber column. Therefore, an adsorbent of higher adsorption capacity is needed for use to exploit the advantage of using a higher air flow rate in the system followed by its optimization for practical purposes. On the other hand, the settling of adsorbent particles followed by removal from the system is important for such batch adsorber. The tapered system could offer flexibility in achieving this target owing to its increased cross-sectional area in the upward direction. Since the air velocity is decreased in the flow direction, the settling velocity is increased opposite to the flow direction resulting in the faster settling of particles. The settling phenomenon, therefore, needs further exploitation. In addition, wastewater generated from extractive industries such as copper, gold, lead, nickel, silver, iron, bauxite, coal, limestone, magnesite, phosphate, oil shale, tin, uranium, and molybdenum requires adequate treatment before being discharged to prevent water and/or soil pollution from acidic/alkaline leaching of heavy metals. The present development could find potential application in extractive industries as well. The solid waste of the extractive industries could be explored as an adsorbent for treating the wastewater in situ and hence the simultaneous management of wastewater and solid waste could be achieved in tandem.

15.6 CONCLUSIONS

The performance of an air-agitated tapered bubble column adsorber was reported in this chapter for the removal of MB on BCR, an industrial solid waste. The adsorbent was characterized by measuring the bulk density, solid density, particle size, pH_{PZC}, BET-specific surface area, and SEM analysis. The bulk density, solid density, average particle size, pH_{PZC}, and BET surface area were 892.1 kg/m^3, 1709.1 kg/m^3, 34.2 μm, 6.6 m^2/g, and 34.27 m^2/g, respectively. The system performance was characterized based on various operating parameters [C_o = 25 to 200 mg/L; m= 5 to 20 g/L; t = 1 to 120 min; Q_a = 1.17×10–4 to 8.33×10–4 m^3/s; batch quiescent liquid volume = 3.00×10–3 m^3 (3.0 L); Salt (NaCl) concentration = 1 to 9 g/L; T = 295.5 to 306 K]. The equilibrium time observed was 60 min. The experimental results revealed a very high removal efficiency of 99.2% (at C_o = 50 mg/L, m = 15 g/L, and Q_a = 5.00×10–4 m^3/s, quiescent liquid volume = 3L). The observed highest adsorption capacity was 11.44 mg/g. The presence of NaCl increased the removal efficiency by about 8% which was attributed to the molecular association of MB molecules. Adsorption of MB without NaCl followed the Freundlich isotherm model and pseudo-second-order kinetic model. The effect of salt was also shown to offer similar adsorption isotherm and kinetic behaviour. The removal efficiency was increased at lower temperatures and hence the system would perform better in winter than in summer. In the present study, the values of ΔG^0 varied from -11.8 to -11.6 kJ/mol and were well within the range of—20 to 0.0 kJ/mol indicating the predominance of physisorption. The activation energy estimated from the Dubinin—Radushkevich adsorption isotherm also indicated the adsorption of MB without salt was dominated by physisorption (E = 0.0863 kJ/mol < 8 kJ/mol). The negative value of ΔG^0 indicated the feasibility and spontaneity of the adsorption while the positive value of ΔH^0 and the greater value of $T\Delta S^0$ than ΔH^0 ensured the adsorption process to be endothermic in nature. The present development could be potentially used for the adsorptive treatment of industrial wastewater including extractive industries as well.

ACKNOWLEDGEMENTS

Authors acknowledge the Indian Association for the Cultivation of Science, for carrying out the BET surface area and micro-pore analysis, and the Post Graduate Department of Chemistry, Rashbehari Siksha Prangan, University of Calcutta, for carrying out the Fourier transform infrared spectroscopy (FTIR) analysis.

NOMENCLATURES

a_{RP}	Redlich–Peterson isotherm
ARE	Average relative error, $\dfrac{100}{n}\sum\limits_{i=1}^{n}\left|\dfrac{(q_{e,calc}-q_{e,meas})}{q_{e,meas}}\right|_i$
b_T	Temkin constant related to the heat of sorption indicating $1/b_T$ as the adsorption potential of the adsorbent, J/mol
C_e	Equilibrium concentration of MB, mg/L
C_f	Final concentration, mg/L
C_O	Initial concentration of MB, mg/L
C_t	Concentration of MB at any time t, mg/L
E	Activation energy, kJ/mol
$F_{(t)}$	Fractional attainment of adsorption equilibrium $\left[=\dfrac{qt}{qe}\right]$, dimensionless
g	Redlich—Peterson isotherm exponent
HYBRID	Hybrid fractional error function, $\dfrac{100}{(n-p)}\sum\limits_{i=1}^{n}\left|\dfrac{(q_{e,calc}-q_{e,meas})^2}{q_{e,meas}}\right|_i$
k_1	Pseudo-first order adsorption kinetic rate constant, min^{-1}
k_2	Pseudo-second order adsorption kinetic rate constant, g/mg L min
K_C	Thermodynamic equilibrium constant, dimensionless
K_F	Adsorption capacity [Freundlich isotherm], dimensionless
k_{fd}	Film diffusion constant, sec^{-1}
k_H	Halsey constant, dimensionless
k_L	Langmuir constant, L/mg
k_{RP}	Redlich—Peterson isotherm constant, L/g
k_S	Sips isotherm model constant, L/g
k_T	Temkin isotherm constant, L/gm

m	Mass of the adsorbent per unit volume, W/V, g/mL
MPSD	Marquardt's percent standard deviation,

$$100\sqrt{\frac{1}{(n-p)}\sum_{i=1}^{n}\left\{\frac{(q_{e,calc}-q_{e,meas})}{q_{e,meas}}\right\}_i^2}$$

n	Number of data points
n_F	Freundlich constant, dimensionless
n_H	Halsey constant, dimensionless
n_S	Sips isotherm model exponent
$1/n_F$	Adsorption intensity [Freundlich isotherm], dimensionless
P	Number of parameters
Q_a	Air flow rate, m^3/s
q_{mDR}	Theoretical monolayer sorption capacity, mg/g
q_e or $q_{e,meas}$	Amount of MB adsorbed at equilibrium, mg/g
$q_{e,cal}$	Amount of MB adsorbed calculated at equilibrium, mg/g
q_m	Maximum amount of MB adsorbed, mg/g
q_t	Amount of MB adsorbed at time t, mg/g
r	Correlation coefficient, dimensionless
r^2	Coefficient of determination, dimensionless
R	Universal gas constant, 8.314 J/mol K
%R	Percentage removal in terms of equilibrium concentration, dimensionless
SAE	Sum of the absolute errors, $\sum_{i=1}^{n}\left|q_{e,calc}-q_{e,meas}\right|_i$
SSE	Sum of the squares of the errors, $\sum_{i=1}^{n}\left(q_{e,calc}-q_{e,meas}\right)_i^2$
t	Contact time, min
T	Absolute temperature, K
V	Volume of dye solution, L
W	Amount of adsorbent, g

GREEK LETTERS

ΔG^0	Gibbs free energy change, kJ/mol
ΔH^0	Enthalpy change, kJ/mol
ΔS^0	Entropy change, J/mol K
χ^2	Chi-square, dimensionless
ε	Dubinin—Radushkevich isotherm constant

REFERENCES

Alberghina, G., Bianchini, R., Fichera, M., & Fisichelia, S. (2000). Deminerization of Cibacron Blue F3GA and other dyes: Influence of salts and temperature. *Dyes and Pigments, 46,* 129–137.

Attia, A. A., Girgis, B. S., & Fathy, N. A. (2008). Removal of methylene blue by carbons derived from peach stones by H_3PO_4 activation: Batch and column studies. *Dyes and Pigments, 76,* 282–289.

Banat, F., Al-Asheh, S., Al-Anbar, S., & Al-Rafaie, S. (2007). Microwave- and acid-treated bentonite as adsorbents of methylene blue from a simulated dye wastewater. *Bulletin of Engineering Geology and the Environment, 66*(1), 53–58.

Batzias, F. A., & Sidiras, D. K. (2007). Simulation of dye adsorption by beech sawdust as affected by pH. *Journal of Hazardous Materials, 141*(3), 668–679.

Bhattacharyya, K. G., & Sharma, A. (2005). Kinetics and thermodynamics of methylene blue adsorption on Neem (*Azadirachta indica*) leaf powder. *Dyes and Pigments, 65*(1), 51–59.

Boyd, G. E., Adamson, A. W., & Myers, L. S. (1947). The exchange adsorption of ions from aqueous solutions by organic zeolites. II. Kinetics. *Journal of American Chemical Society, 69*, 2836–2848.

Chabani, M., Amrane, A., & Bensmaili, A. (2006). Kinetic modelling of the adsorption of nitrates by ion exchange resin. *Chemical Engineering Journal, 125*, 111–117.

Chowdhury, A., Sarkar, A., & Bandyopadhyay, A. (2009). Rice husk ash as a low cost adsorbent for the removal of methylene blue and congo red in aqueous phases. *Clean—Soil, Air, Water, 37*(7), 581–591.

Dubinin, M. M., & Radushkevich, L. V. (1947). The equation of the characteristic curve of the activated charcoal. *Proceedings of the Academy of Sciences of the USSR. Physical Chemistry Section, 55*, 331–337.

El Qada, E. N., Allen, S. J., & Walker, G. M. (2006). Adsorption of methylene blue onto activated carbon produced from steam activated bituminous coal: A study of equilibrium adsorption isotherm. *Chemical Engineering Journal, 124*(1–3), 103–110.

Freundlich, H. M. F. (1906). Uber die adsorption in losungen. *Zeitschrift fur Physikalische Chemie-Leipzig, 57*, 385–470.

Ghosh, S. K., & Bandyopadhyay, A. (2016). Characterizing acidic fly ash with and without biomass combustion residue for adsorptive removal of Crystal Violet with optimization of mixed adsorbent by response surface modeling. *Environmental Earth Sciences, 75*(9), Article 820. https://doi.org/10.1007/s12665-016-5571-z

Ghosh, S. K., & Bandyopadhyay, A. (2017a). Novel air agitated tapered adsorber for crystal violet removal on biomass combustion residue with process optimization using response surface modeling. *Journal of Environmental Chemical Engineering, 5*(3), 2415–2430.

Ghosh, S. K., & Bandyopadhyay, A. (2017b). Adsorption of methylene blue onto citric acid treated carbonized bamboo leaves powder: Equilibrium, kinetics, thermodynamics analyses. *Journal of Molecular Liquids, 248*, 413–424.

Ghosh, S. K., Hajra, A. K., & Bandyopadhyay, A. (2017). Air agitated tapered bubble column adsorber for hazardous dye (crystal violet) removal onto activated ($ZnCl_2$) carbon prepared from bamboo leaves. *Journal of Molecular Liquids, 240*, 313–321.

Halsey, G. (1948). Physical adsorption on non-uniform surface. *The Journal of Chemical Physics, 16*, 931–937.

Hamdaoui, O. (2006). Batch study of liquid-phase adsorption of methylene blue using cedar sawdust and crushed brick. *Journal of Hazardous Materials, 135*(1–3), 264–273.

Hameed, B. H., Din, A. T. M., & Ahmad, A. L. (2007). Adsorption of methylene blue onto bamboo-based activated carbon: Kinetics and equilibrium studies. *Journal of Hazardous Materials, 141*, 819–825.

Han, R., Wang, Y., Yu, W., Zou, W., Shi, J., & Liu, H. (2007). Biosorption of methylene blue from aqueous solution by rice husk in a fixed-bed column. *Journal of Hazardous Materials, 141*, 713–718.

Ho, S. Y., & McKay, G. (1999). Pseudo second order model for sorption process. *Process Biochemistry, 34*, 455–465.

Kavitha, D., & Namasivayam, C. (2007). Experimental and kinetic studies on methylene blue adsorption by coir pith carbon. *Bioresource Technology, 98*(1), 14–21.

Lagergren, S. (1898). About the theory of so-called adsorption of soluble substances. Kungliga Svenska Vetenskapsakademiens. *Handlingar, 24*(4), 1–39.

Langmuir, I. (1918). The adsorption of gases on plane surfaces of glass, mica and platinum. *Journal of American Chemical Society, 40*(9), 1361–1403.

Liu, Y. (2009). Is the free energy change of adsorption correctly calculated? *Journal of Chemical Engineering Data, 54*, 1981–1985.

Machado, F. M., Bergmann, C. P., Fernandes, T. H. M., Lima, E. C., Royer, B., Calvete, T., & Fagan, S. B. (2011). Adsorption of reactive red M-2BE dye from water solutions by multiwalled carbon nanotubes and activated carbon. *Journal of Hazardous Materials, 192*(3), 1122–1131.

Ncibi, M. C., Mahjoub, B., & Seffen, M. (2007). Kinetic and equilibrium studies of methylene blue biosorption by *Posidonia oceanic* (L.) fibres. *Journal of Hazardous Materials, B139*, 280–285.

Ozcan, A., Oncu, E. M., & Ozcan, S. (2006). Adsorption of acid blue 193 from aqueous solutions 640 onto DEDMA-sepiolite. *Journal of Hazardous Materials, 129*, 244–252.

Özer, D., Dursun, G., & Özer, A. (2007). Methylene blue adsorption from aqueous solution by dehydrated peanut hull. *Journal of Hazardous Materials, 144*, 171–179.

Redlich, O., & Peterson, D. L. (1959). A useful adsorption isotherm. *Journal for Physical Chemistry, 63*, 1024–1026.

Roy, S., Ghosh, S. K., & Bandyopadhyay, A. (Fall 2015). Adsorptive removal of crystal violet on biomass combustion residue: Equilibrium, isotherm, kinetics and mass transfer analyses. *Environmental Quality Management, 25*(1), 55–79.

Sips, R. (1948). Combined form of Langmuir and Freundlich equations. *The Journal of Chemical Physics, 16*, 490–495.

Temkin, M. I., & Pyzhev, V. (1940). Kinetic of Ammonia synthesis on promoted Iron catalyst. *Acta Physicochimica U.R.S.S., 12*, 327–356.

Tsai, W.-T., Hsu, H.-C., Su, T.-Y., Lin, K.-Y., & Lin, C.-M. (2008). Removal of basic dye (methylene blue) from wastewaters utilizing beer brewery waste. *Journal of Hazardous Materials, 154*(1–3), 73–78.

Vilar, V. J., Botelho, C. M., & Boaventura, R. A. (2007). Methylene blue adsorption by algal biomass based materials: Biosorbents Characterization and process behavior. *Journal of Hazardous Materials, 147*(1–2), 120–132.

Zhao, M., Tang, Z., & Liu, P. (2008). Removal of methylene blue from aqueous solution with silica nano-sheets derived from vermiculite. *Journal of Hazardous Materials, 158*(1), 43–51.

Index

Note: Page numbers in *italics* indicate a figure, and page numbers in **bold** indicate a table on the corresponding page.